Lecture Notes of the Institute for Computer Sciences, Social Informatics and Telecommunications Engineering 79

Luigi Atzori   Jaime Delgado
Daniele Giusto (Eds.)

Mobile Multimedia Communications

7th International ICST Conference
MOBIMEDIA 2011
Cagliari, Italy, September 5–7, 2011
Revised Selected Papers

 Springer

Volume Editors

Luigi Atzori
Daniele Giusto
University of Cagliari
Department of Electrical and Electronic Engineering
Piazza d'Armi, 09123 Cagliari, Italy
E-mail: l.atzori@diee.unica.it; ddgiusto@unica.it

Jaime Delgado
Universitat Politècnica de Catalunya
Department of Computer Architecture
Tànger, 122-140, 08018 Barcelona, Spain
E-mail: jaime.delgado@ac.upc.edu

ISSN 1867-8211   e-ISSN 1867-822X
ISBN 978-3-642-30418-7   ISBN 978-3-642-30419-4 (eBook)
DOI 10.1007/978-3-642-30419-4

Springer Heidelberg Dordrecht London New York

Library of Congress Control Number: 2012937596

CR Subject Classification (1998): H.4, C.2, H.3, D.2, I.2, H.5

Typesetting: Camera-ready by author, data conversion by Scientific Publishing Services, Chennai, India

Printed on acid-free paper

Springer is part of Springer Science+Business Media (www.springer.com)

Preface

Successfully deploying multimedia services and applications in mobile environments requires adopting an interdisciplinary approach where multimedia, networking and physical layer issues are addressed jointly. Content features analysis and coding, media access control, multimedia flow and error control, cross-layer optimization as well as mobility management and security protocols are research challenges that need to be carefully examined when designing new architectures. We also need to put a great effort in designing applications that take into account the way the user perceives the overall quality of the provided service.

Within this scope, MobiMedia intends to provide a unique international forum for researchers from industry and academia, working on multimedia coding, mobile communications and networking fields, to study new technologies, applications and standards.

In this MobiMedia edition, particular emphasis was put on the issue of quality of experience (QoE) in pervasive media networks and applications. Indeed, the quality of the user experience, the perceived simplicity of accessing and interacting with systems and services, and the effective and acceptable hiding of the complexity of underlying technologies are certainly determining factors for success or failure of the multimedia services, as well as graceful degradation. With this intent, the conference featured a special session on QoE as well as a panel for discussing the importance of this subject for different types of multimedia applications, related standards and issues in management. Other than a session on QoE, the conference included sessions on dynamic-spectrum-access wireless networks in the TV white spaces, media streaming, mobile visual search, image processing and transmission, multimedia in human–machine interaction, and mobile applications. This book presents a selection of the revised papers.

September 2011   Luigi Atzori

Organization

MobiMedia 2011 was organized by the Multimedia Communications Lab at the Department of Electrical and Electronic Engineering, University of Cagliari.

Organizing Committee

Conference General Co-chairs

Michele Battelli   Google, USA
Daniele D. Giusto   University of Cagliari, Italy

Technical Program Co-chairs

Luigi Atzori   University of Cagliari, Italy
Zhibo Chen   Technicolor, China
Jaime Delgado   Universitat Politecnica de Catalunya, Spain

Local Chair

Cristian Perra   University of Cagliari, Italy

Special Session Co-chairs

Pablo Angueira   University of the Basque Country, Spain
Maurizio Murroni   University of Cagliari, Italy

Publicity Chair

Giaime Ginesu   University of Cagliari, Italy

Web Chair

Michele Nitti   University of Cagliari, Italy

Conference Management

Aza Swedin   EAI, Europe

Steering Committee

Luigi Atzori   DIEE – University of Cagliari, Italy
Imrich Chlamtac   Create-Net, Italy

Advisory Board

Tasos Dagiuklas	TEI of Mesolonghi, Greece
Jyrki Huusko	VTT, Finland
George Kormentzas	University of the Aegean, Greece
Maria G. Martini	Kingston University, UK
Christos Politis	Kingston University, UK
Jonathan Rodriguez	Instituto de Telecomunicações, Portugal

Technical Program Committee

Michele Albano	Instituto de Telecomunicações, Portugal
Nancy Alonistioti	University of Athens, Greece
Luis Alonso	UPC, Spain
Faouzi Bader	CTTC, Spain
Marco Cagnazzo	ENST, France
Pietro Camarda	Politecnico di Bari, Italy
Periklis Chatzimisios	TEI Thessaloniki, Greece
Tao Chen	VTT, Finland
Ioannis Chochliouros	OTE, Greece
Tasos Dagiuklas	TEI of Mesolonghi, Greece
Filip De Turck	Ghent University IBBT, Belgium
Marco Di Renzo	CNRS/Supelec, France
Tapio Frantti	VTT, Finland
Carlo Giannelli	University of Bologna, Italy
Apostolos Gkamas	Computer Technology Institute, Greece
Atanas Gotchev	Tampere University of Technology, Finland
Oliver Hoffmann	TU-Dortmund, Germany
Georgios Kormentzas	University of the Aegean, Greece
George Koudouridis	Huawei, Sweden
Maria Martini	Kingston University, UK
Alberto Nascimento	Universidade da Madeira, Portugal
Petros Nicopolitidis	Aristotle University, Greece
Eleni Patouni	University of Athens, Greece
Fernando Pereira	IST Lisbon, Portugal
Nikopolitidis Petros	AUTH, Greece
Nitendra Rajput	IBM Research,USA
Gianluca Reali	University of Perugia, Italy
Claudio Sacchi	University of Trento, Italy
Lambros Sarakis	NCSR 'Demokritos', Greece
Wan-Chi Siu	Hong Kong Polytechnic University, SAR China

Thomas Stockhammer   Nomor Research, Germany
Christian Timmerer   Klagenfurt University, Austria
Popescu Vlad   University of Brasov, Romania
Qin Xin   Simula Research Laboratories, Norway
George Xylomenos   Athens University of Economics and Business,
                              Greece
Nizar Zorba   University of Jordan, Jordan

Table of Contents

Media Streaming

Mobile Visual Search

Image Processing and Transmission

Mobile Applications

Tackling the Sheer Scale of Subjective QoE

Vlado Menkovski, Georgios Exarchakos, and Antonio Liotta

Eindhoven University of Technology,
P.O. Box 513, 5600MB Eindhoven, The Netherlands
{v.menkovski,g.exarchakos,a.liotta}@tue.nl

Abstract. Maximum Likelihood Difference Scaling (MLDS) used as a method for subjective assessment of video quality alleviates the inconveniencies associated with high variation and biases common in rating methods. However, the number of tests in a MLDS study rises fairly quickly with the number of samples that we want to test. This makes the MLDS studies not scalable for the diverse video delivery environments commonly met in pervasive media networks. To tackle this issue we have developed an active learning approach that decreases the number of MLDS tests and improves the scalability of this method.

Keywords: Maximum Likelihood Difference Scaling, adaptive MLDS, Video Quality Assessment, Quality of Experience, QoE.

1   Introduction

As video is becoming highly pervasive, pervasive media networks are being developed as an underlying delivery technology to handle the newly arisen technical requirements. Pervasive media networks deliver and adapt video and other multimedia content to the context, environment and purpose for which the content is being used. Efficient adaptation of the different video parameters necessitates understanding of the effect of these parameters on the delivered Quality of Experience (QoE). For example, depending on the context, type of content and screen characteristics a person might not perceive any more improvement if the video bit-rate is larger than 512kbps. On the other hand, for a low cost service a 256kbps video could offer only slightly lower quality than 512kbps (again in the specific context) and be the optimal setting. Calculating these utilities requires understanding of the costs, but more importantly it requires understanding of the perceived quality for these resources. To determine the utility of these resources an accurate estimation of quality is necessary. This needs to be achieved through subjective testing, because of the subjective nature of perceived quality of video.

Our focus is on Maximum Likelihood Difference Scaling (MLDS) because of its superior performance as a subjective testing methodology. MLDS is based on two-alternative-forced choice (2AFC) tests that suffer significantly less from bias and variability [1]. However, MLDS studies require all the combinations of four for a given set of samples. As the number of parameter or characteristic of interest

L. Atzori, J. Delgado, and D. Giusto (Eds.): MOBIMEDIA 2011, LNICST 79, pp. 1–15, 2012.

increases, so does the number of samples and in turn the number of MLDS tests. Even though each of these 2AFC tests is simple and straight forward the overall subjective study is not scalable. To tackle the scale of this type of subjective studies we have developed an adaptive test selection procedure for MLDS that improves the learning rate. The adaptive approach iteratively inputs new data by asking the participant to do specific tests, instead of randomly going through all of the combinations of samples. Because of the built in redundancy and high correlation in MLDS, some tests become more informative than others over the course of the experiment. The adaptive MLDS estimates the responses of the unknown tests from the information collected by the answered ones. The tests estimated with less confidence are more informative and are selected as next. Additionally, the confidence for the remaining unknown tests is an indication of how much more tests are necessary, and provides for early stopping capability.

The adaptive MLDS algorithm implemented in a software test bed and executed over subjective test data showed significant improvement in the learning rate and substantial decrease in the number of tests that are necessary.

2   Video Quality Assessment

Estimation of video quality is highly diverse area with many methods, which fall within the two main categories of objective or subjective. Objective methods estimate the quality by focusing on the signal fidelity or measuring the distortions of the video compared to the original. These objective methods are referred to as full-reference (FR) methods. Some effective FR methods include MultiScale-Structural SIMilarity index (MS-SSIM) [2], Perceptual Video Quality Metric (PVQM) [3], and the perceptual spatio-temporal frequency-domain based MOtion-based Video Integrity Evaluation (MOVIE) [4]. These vary in accuracy compared to the subjective reference estimation and complexity. In these methods there is always a trade-off between accuracy and complexity and memory requirements. In addition to the FR methods there are the reduced-reference (RR) and no-reference (NR) objective methods. The RR methods have only partial information on the original signal. Although less accurate this makes them more practical than FR and applicable to continuous assessment of video quality while FR are mostly used in offline estimation. One such method is [5], which examines local harmonic strength features. These features are correspond to artifacts such as blockiness and blurriness. By examining the loss of these features the method estimate the video quality.   The NR methods are the most practical because the hold no information on the source of the signal, but also most challenging to implement.

The subjective methods include some type of tests with actual human participants. Evidently objective tests are more practical and therefore with significantly more widespread use. However, objective tests commonly are not designed to consider all the factors that affect the perceived quality of the video or the QoE [6]. In this manner the subjective methods are regarded as more accurate and are usually used as a benchmark for the objective methods. One such study by Seshadrinathan et al. [7] analyzes the different objective video quality assessment algorithms by correlating their output with the differential mean opinion score (DMOS) of a subjective study they executed. This type of undertaking is costly, time consuming and necessitates considerable amount of tests to achieve statistical significance. The bias and the variability of subjective testing arise from the fact that subjective tests rely in rating as the estimation procedure. Rating is inheritably biased due to the variance in the internal representation of the rating scale by the subjects [8][9][10].

In [8] we describe the use of a two-alternative-forced-choice (2AFC) method to estimate the relative differences in quality. The method Maximum Likelihood Difference Scaling (MLDS) delivers the ratio of subjective quality between a video with different levels of resource provided. Because the method is a 2AFC method, meaning the participant has to answer a single binary question of the 'which is bigger' type, the amount of bias and variability is significantly lower than in rating [1].

In the case of video quality estimation the 2AFC test is discriminating between different levels of quality. More particularly, four videos or two pairs of video are presented and the participant needs to select which pair has the bigger difference in quality. This might sounds as a particularly difficult and time-consuming effort, but in most cases the difference in video quality is quite evident. The video is typically short (less than 10 seconds) and uniformly impaired, so very often the participant is confident enough to vote after only watching a part of each of the video. Many of the tests are quite obvious and derivative, i.e. based on previous responses the following are apparent. Nevertheless, the number of tests is combination of all the samples over four, so the number of tests is a function that is forth order polynomial of the number of samples. For one or two parameters that affect the video the number of samples is not very big, but as number of samples grows the tests become unfeasible.

Motivated by the effectiveness of MLDS in estimating difference in quality or the utility of the resources and the possibilities for improving the efficiency of the method we have developed an adaptive test selection procedure for MLDS that improves the learning rate and provides for possibilities for executing a subset of the subjective tests while estimating the rest with a given confidence.

2.1   MLDS

To better understand the mechanics of the adaptive MLDS we need to start with a discussion on MLDS itself. The goal of this method is to map the objectively measurable scale of video quality to the internal psychological scale of the viewers. The output is a quantitative model for this relationship based on a psychometric function [11] as depicted in Figure 1.

The horizontal axis of the Figure 2 represents the physical intensity of the stimuli – in our study the bit-rate of the video. The vertical axis represents the psychological scale of the perceived difference in quality. The perceptual difference of quality ψ_1 of the first (or reference) sample x_1 is fixed to 0 and difference of quality ψ_{10} of the last sample x_{10} is fixed to 1 without any loss in generality [12]. In other words there is 0% difference in quality between x_1 and x_1 (itself), while there is 100% difference in quality between x_1 and x_{10}. The MLDS method estimates the relative distances of the rest of the videos ψ_2 through ψ_9 and therefore models the viewers' internal quality scale.

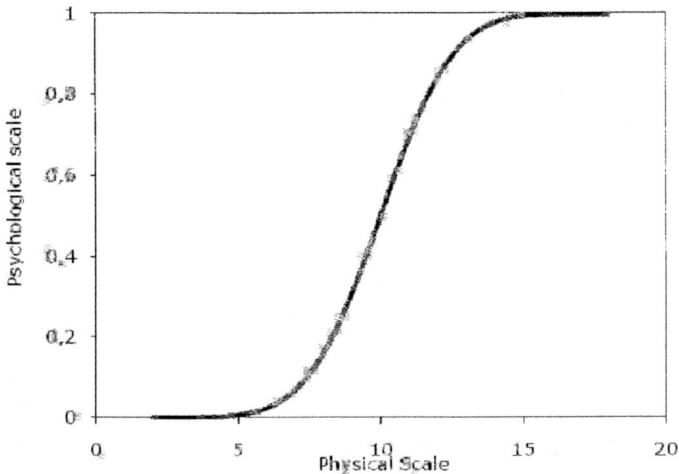

Fig. 1. Psychometric function

This 2AFC test is designed in the following manner; two pairs of videos are presented to the viewers $\{x_i, x_j\}$ and $\{x_k, x_l\}$ where the indexes of the samples are selected as $1 \leq i < j < k < l \leq 10$, so that the ranges of quality does overlap. The video with smaller index has higher quality. The viewer then selects the pair of videos that have bigger difference in quality. For a given test T_n the viewer selects the first pair (sets $R_n=1$) if she perceived the qualities of videos in the quadruple as $|\psi_j - \psi_i| - |\psi_l - \psi_k| > 0$, otherwise she chooses the second pair ($R_n=0$). These comparisons between the quality distances of video pairs allow for design of a quality distance model between all of the presented videos. The method calculates the quality differences ψ_2 through ψ_9 as parameters in maximum likelihood estimation (MLE).

The MLE is a method for estimating the parameters of a statistical model. Using signal detection theory (SDT) [13] MLDS models each response as sampled from a Gaussian distribution with unknown parameters. The difference of differences of quality between the four videos is the signal contaminated by Gaussian noise or the mean of a Gaussian distribution (1). When executing a test the participant calculates the value.

$$\delta(i,j,k,l)_n + \varepsilon = \psi_{j_n} - \psi_{i_n} - \psi_{l_n} + \psi_{k_n} + \varepsilon \tag{1}$$

Where ε is value sampled from a Gaussian distribution with zero mean and standard deviation of 1.

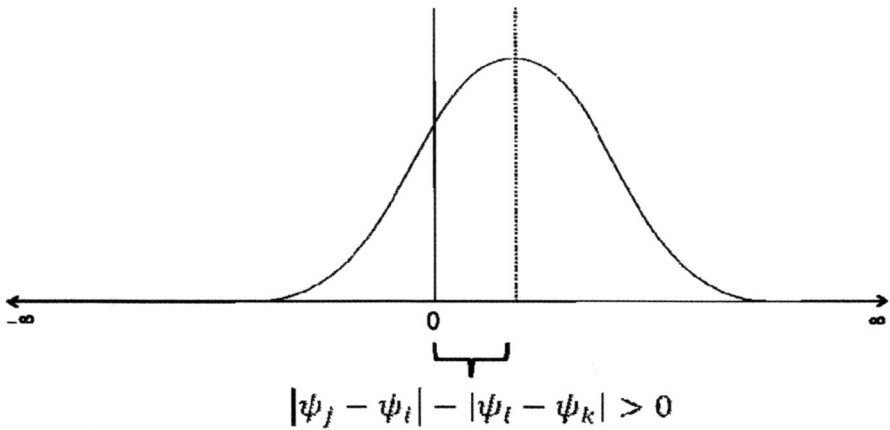

$$|\psi_j - \psi_i| - |\psi_l - \psi_k| > 0$$

Fig. 2. Stimuli intensity contaminated with Gaussian noise

Using this assumption, the probability of each response is given in (2).

$$P(R_n = 1; \delta_n, \sigma^2) = 1 - \Phi\left(\frac{0 - \delta_n}{\sigma}\right) = \Phi\left(\delta_n\right) \tag{2}$$

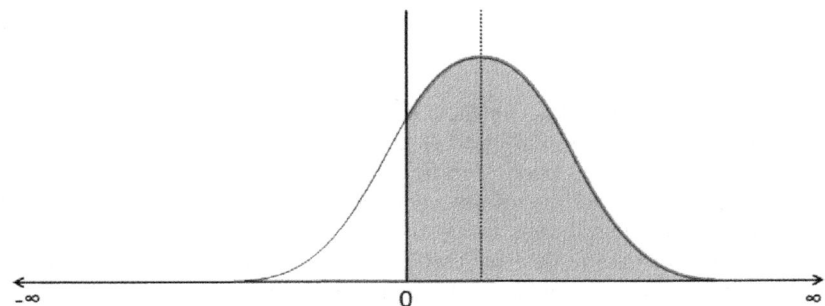

Fig. 3. Probability of first pair being selected

For a test where the first pair is selected the probability is given in (2).

$$P(R_n = 0; \delta_n) = 1 - P(R_n = 1; \delta_n) = 1 - \Phi\left(\delta_n\right) \tag{3}$$

For a test where the second pair is selected the probability is given in (3). The likelihood of all the responses is accordingly as equation (4).

$$L(\Psi \mid \vec{R}) = \prod_{n=1}^{N} \Phi\left(\delta_n\right)^{R_n} \left(1 - \Phi\left(\delta_n\right)\right)^{1-R_n} \tag{4}$$

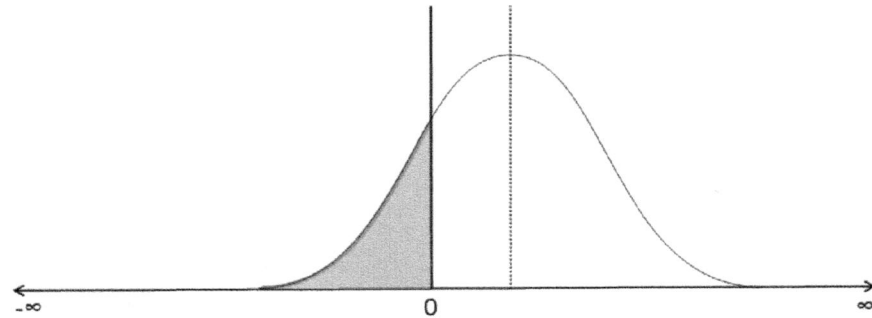

Fig. 4. Probability of second pair being selected

There is no closed form for such a solution, so a direct numerical maximization method needs to be used to compute the estimates (5).

$$\widehat{\Psi} = \arg\max_{\overline{\Psi}} L(\overline{\Psi} \,|\, \overline{R}) \tag{5}$$

More details on MLDS for video quality can be found in [14] and on MLDS for image quality in [15].

A fitter curve through the $\hat{\Psi}$ also represents the utility of the bit-rate as a resource or how much we can improve the quality by increasing the bit-rate over the tested range assuming that the cost of increasing the bit-rate is constant over the same range.

3   Adaptive MLDS

The MLDS method is appealing for their simplicity and efficiency, however one full round of tests for ten levels of stimuli (i.e. video qualities) requires 210 individual tests. The full range of tests carry significant redundancy and removing some of it should not necessarily make the results significantly less reliable; even more so it can have only negligible effects on the end result.

In this adaptive procedure we have two aims, to improve the rate of learning and to decrease the number of required tests. The approach is based on the idea that with the knowledge acquired by executing a small number of tests we can estimate the answers of the remaining tests with some confidence. Then using these estimates together with the known responses we execute the MLDS method. Executing the MLDS with more responses helps the argument maximization procedure. The estimates rely on the characteristics of the psychometric curve (such as its increasing monotonicity), so that the overall performance of MLDS is improved.

The idea comes from the notion that some of the tests are covering the range of others. In fact, all of the tests are being covered by others in one way or the other. The approach makes use of the characteristics of the psychometric curve. The psychometric curve is a monotonously increasing function $\overline{\Psi} = f\left(\overline{X}\right)$. Consequently, for $k < l < m$, $x_k > x_l > x_m$ if $x_k - x_l > x_k - x_m$ in the physical domain then $\psi_k - \psi_l \geq \psi_k - \psi_m$ in the psychological domain Figure 5.

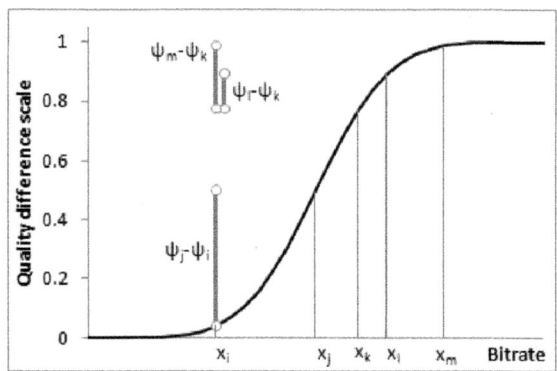

Fig. 5. Monotonicity of the psychometric curve

If we now observe five samples x_i, x_j, x_k, x_l, x_m such that $i < j < k < l < m$ and we observe two tests $T_1(x_i, x_j; x_k, x_l)$ and $T_2(x_i, x_j; x_k, x_m)$, the perceived qualities in the psychological domain are $\psi_i \leq \psi_j \leq \psi_k \leq \psi_l \leq \psi_m$. If in T_2 the first pair is bigger or $\psi_j - \psi_i > \psi_m - \psi_k$ that would mean that $\psi_j - \psi_i > \psi_m - \psi_k \geq \psi_l - \psi_k$. In other words, if in T_2 the first pair is selected with a bigger difference, then in T_1 the first pair has a bigger difference as well (Figure 5).

There are many different combinations of tests that have this dependency for the first pair or the second pair. We can generate a list of dependencies for each pair based on two simple rules:

- Let us assume test $T_1(a, b, c, d)$ such that $a < b < c < d$, $\psi_b - \psi_a > \psi_d - \psi_c$ and test $T_2(e, f, g, h)$ with $e < f < g < h$. If $e \leq a < b \leq f$ and $c \leq g < h \leq d$ then $\psi_f - \psi_e > \psi_h - \psi_g$ (Figure 6).

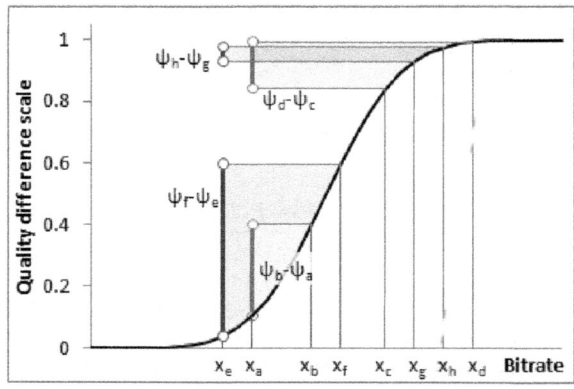

Fig. 6. If first pair in T_1 is bigger than first pair of T_2 is bigger as well

- Let us assume that for test $T_1(a, b, c, d)$ with $a < b < c < d$, $\psi_b - \psi_a < \psi_d - \psi_c$.
  If for test $T_2(e, f, g, h)$ with $e < f < g < h$ the following hold: $a \le e < f \le b$ and $g \le c < d \le h$ then $\psi_f - \psi_e < \psi_h - \psi_g$.

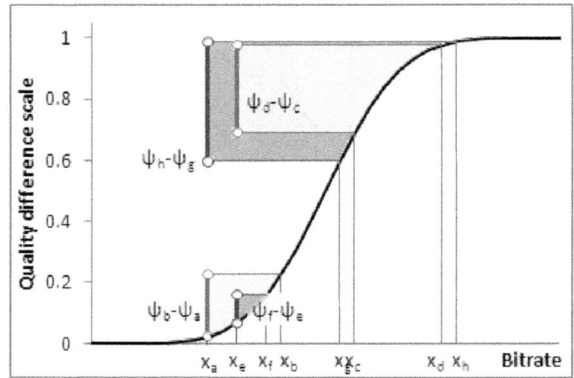

Fig. 7. If second pair in T_1 is bigger than second pair of T_2 is bigger as well

After introducing an initial set of responses we can estimate the probabilities of the rest, however first we need to learn the probabilities of each of the known responses to be actually valid. MLDS estimates the values of the psychological parameters $\Psi = (\psi_1, ..., \psi_{10})$ such that the combined probabilities of each response or the overall likelihood of the dataset is maximized. Nevertheless, after the argument maximization is finished the different responses have different probabilities of being true.

Having a set of initial quality Ψ values as the prior knowledge about the underlying process coming from the data, we generate the estimations for the rest of the tests. The interdependencies from the tests are far more complex, of course.

Let us assume, for example, a test T_1 that depends on tests T_2 and T_3. If the answer from T_2 indicates that the first pair has a larger difference in T_1 and the answer from T_3 indicates the opposite then we need to calculate the combined probability of T_2 and T_3 to estimate the answer of T_1.

Assuming that the responses of T_2 and T_3 are independent and that the probability of giving the first and second answer is the same, the combined probability of T_2 and T_3 is given in (6).

$$P(T_1) = \frac{P(T_2)(1 - P(T_3))}{P(T_2)(1 - P(T_3)) + (1 - P(T_2))P(T_3)} \tag{6}$$

Of the remaining tests that have no responses, some will have higher estimates than others. In other words we have better estimations for some of tests than others. To improve the speed of learning, the adaptive MLDS method, focuses on tests that have smaller confidence in the estimations. This way when we receive the next batch of responses the overall uncertainty in the estimates should be minimized.

The goal of the adaptive MLDS is to develop a metric that will indicate how sufficient the amount of tests is for determining the psychometric curve. We can obtain this indication from the probabilities of the estimations. As we get more responses by asking the right questions the estimation for the rest of the tests improves. At some point adaptive MLDS will have very high probabilities of estimating correctly all of the remaining tests. This is a good indication that no more tests are necessary.

4   Experimental Setup

To show the performance of the adaptive MLDS we have developed a software simulation. The software simulates the learning process of the adaptive MLDS algorithm by sequentially introducing data from a previously [14] executed subjective study. The simulation test-bed is a Java application that loads the subjective data from a file, and then sequentially introduces new datapoints. The datapoints are selected by the adaptive MLDS algorithm and the estimated values are used to calculate the psychometric curve in each iteration. The output is compared to the output of running MLDS on the full dataset and the root mean square error (RMSE) is computed on the differences. In parallel a random introduction of data is also executed as a baseline for comparison. The adaptive MLDS algorithm is implemented in Java, while the MLDS software from [12] is used directly from R using a Java to R interface. To account for the variation in the results due to the random start and random data introduction in the comparison process, the simulation is repeated 100 times and results averaged. Finally the simulation process was computationally very demanding. Each numerical optimization was bootstrapped 1000 times. This was repeated for each step in the introduction of new batch of data and for each video. All this for a single simulation. To handle the computational demand the simulation was executed on a high performance computing grid.

5   Results

Adaptive MLDS as an active learning algorithm explores the space of all possible 2AFC tests with the goal of optimizing the learning process. It also provides indication of confidence in the model built on the subset of the data, which provides for early stopping of the experiment. The performance of the adaptive MLDS is presented in Figure 8, 9, 10 and 11. In Figure 8 we present the accuracy of the estimations for ten types of videos against the number of introduced datapoints. In Figure 9 we observe the leaning rate of adaptive MLDS against the classical MLDS. The horizontal axis represents the number of points introduced at the time the calculation was executed and the vertical axis the RMSE between the estimated curve and the curve built on the whole dataset. We can clearly observe that for this datapoints adaptive MLDS brings significant improvement in the learning rate.

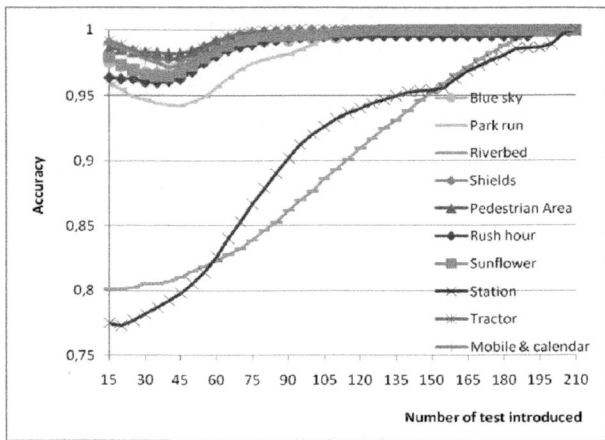

Fig. 8. Accuracy of the estimations

In Figure 10 we present the standard deviation of the different value for the RMSE at each point. Figure 11 presents the distribution of the confidence or the probabilities of those estimations. Starting from the initial 15 data points most of the unknown 195 test are estimated with 0.5 accuracy, but soon after introducing more data the estimations rapidly improve. Between 40 and 60 collected answers the confidence in the estimations was close to 1, suggesting that the rest of the tests are not necessary and that we can correctly estimate the psychometric curve without them. This also evident in Figure 9. The accuracy of the predicted psychometric curves is high for all

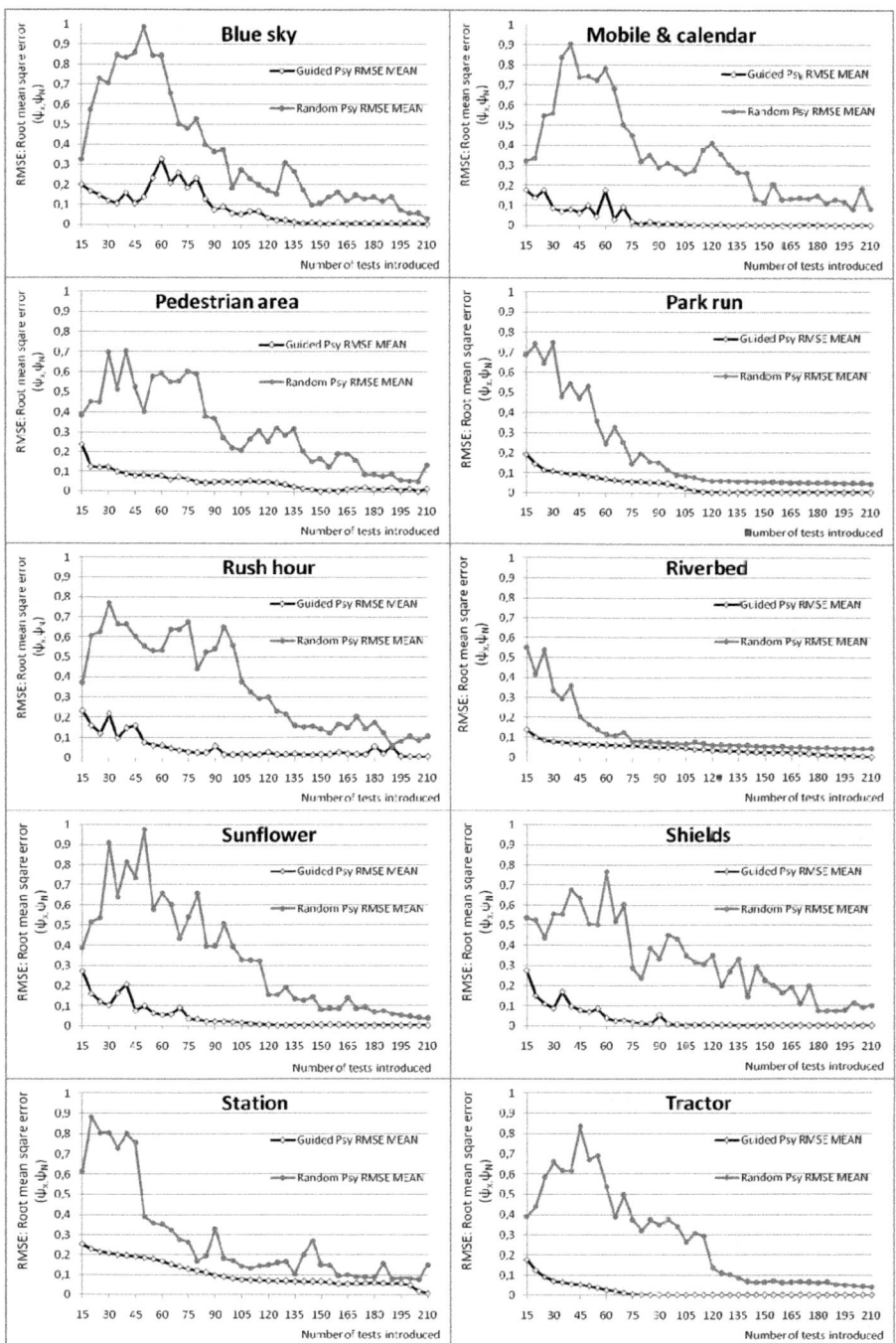

Fig. 9. Mean RMSE for the ten types of video

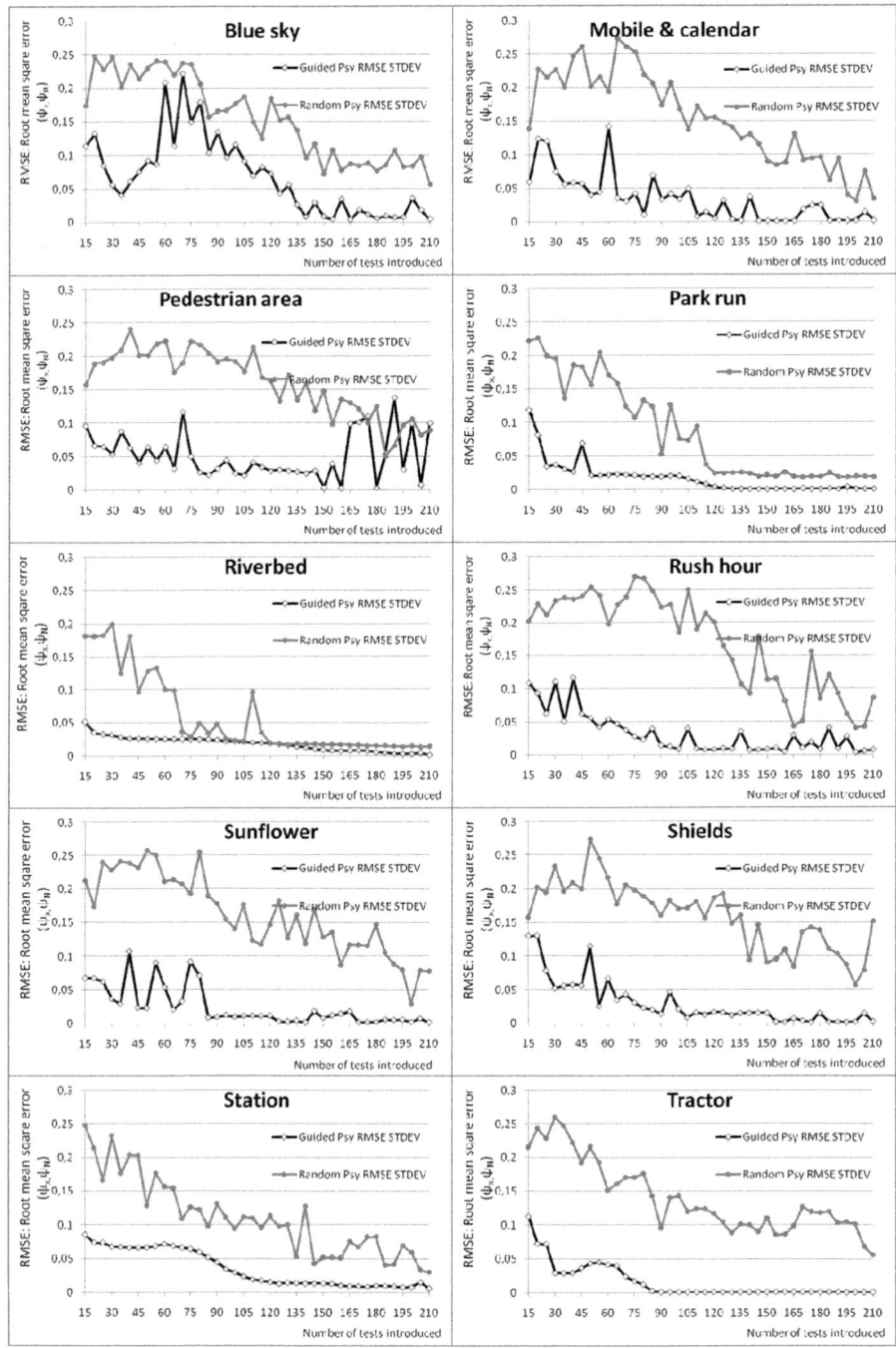

Fig. 10. Standard deviation of the RMSE for the three types of video

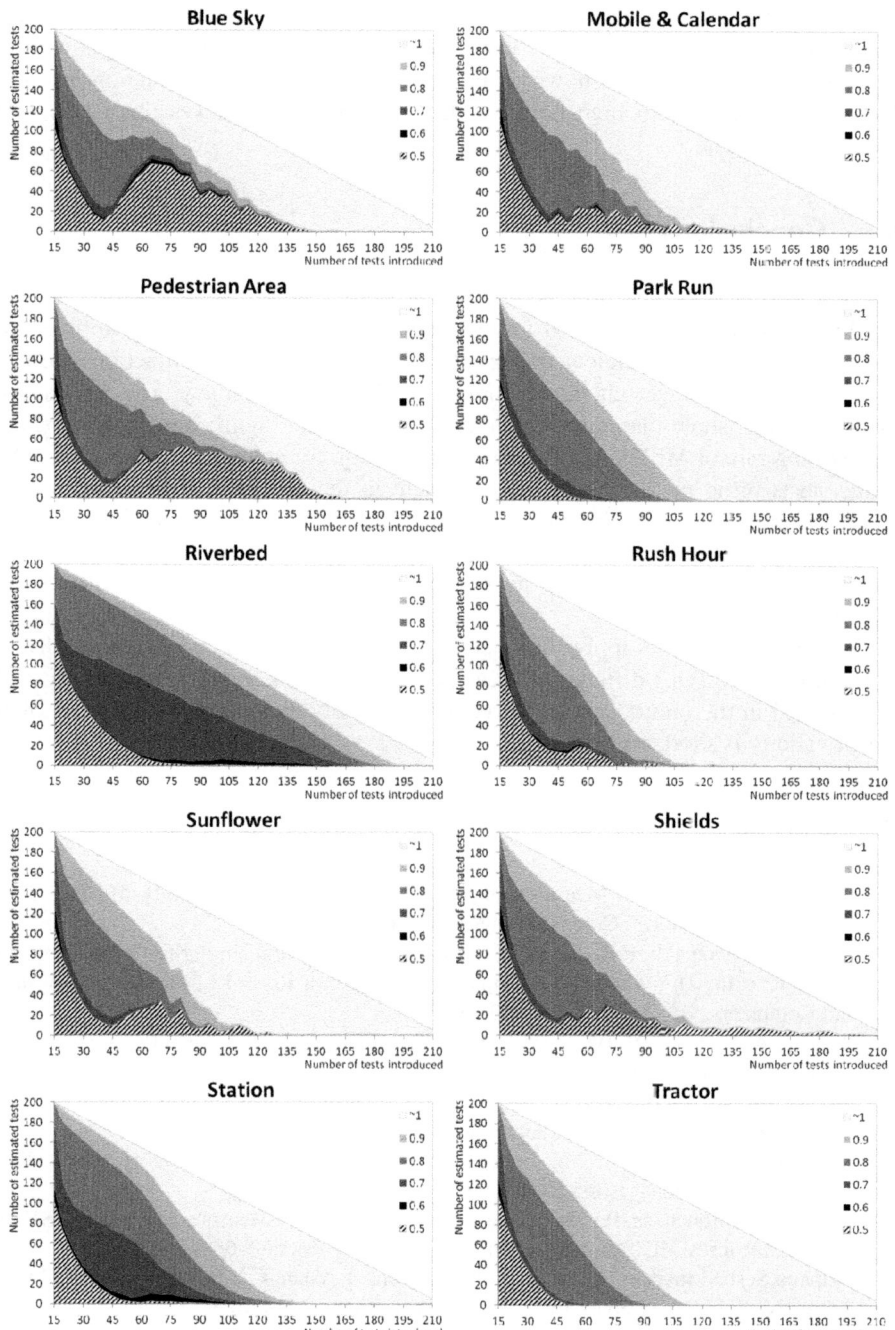

Fig. 11. Estimation confidences for the three types of videos over the number of introduced datapoints

datasets in this range. The RMSE is bellow 0.3 between the 10 given and predicted Ψ values. The accuracy in the prediction is generally very high and improves with the introduction of more data, shown in Figure 8. The Riverbed and Station are more difficult to learn due to high noise in the answers, which makes them also more difficult to estimate.

6   Conclusions

The adaptive MLDS algorithm is an active learning algorithm specifically designed for the MLDS method, a method for estimating a psychometric curve. Motivated by the fact that MLDS is efficient in estimating video quality utility functions we have developed this adaptive scheme to improve its learning efficiency. The results from the simulations show that adaptive learning provides for significant improvement in the learning rate of MLDS and gives solid indication for stopping the test early when further tests bring no significant improvement in the accuracy of the psychometric curve. Overall this approach adds to the efficiency of MLDS into tackling the issues that arise with subjective estimations of video quality. This further makes this method an excellent candidate for use in management of video delivery services and optimizing the QoE.

Further we intend to apply this method in a subjective study of a more diverse environment, involving different devices and modes of use. Finally the method could be extended in the online learning direction for highly dynamic environments where model validity is short termed.

References

1. Watson, A.B.: Proposal: Measurement of a JND scale for video quality. IEEE G-2.1. 6 Subcommittee on Video Compression Measurements (2000)
2. Wang, Z., Simoncelli, E.P., Bovik, A.C.: Multiscale structural similarity for image quality assessment. In: Thirty-Seventh Asilomar Conference on Record of the Signals, Systems and Computers, vol. 2, pp. 1398–1402 (2003)
3. Hekstra, A.P., et al.: PVQM-A perceptual video quality measure. Signal Processing: Image Communication 17(10), 781–798 (2002)
4. Seshadrinathan, K., Bovik, A.C.: Motion-based perceptual quality assessment of video. In: Proc. SPIE-Human Vision and Electronic Imaging (2009)
5. Gunawan, I.P., Ghanbari, M.: Efficient reduced-reference video quality meter. IEEE Transactions on Broadcasting 54(3), 669–679 (2008)
6. Winkler, S., Mohandas, P.: The Evolution of Video Quality Measurement: From PSNR to Hybrid Metrics. IEEE Transactions on Broadcasting 54(3), 660–668 (2008)
7. Kalpana Seshadrinathan, A., Rajiv Soundararajan, B., Alan, C.B.B., Lawrence, K.C.B.: A Subjective Study to Evaluate Video Quality Assessment Algorithms
8. Krantz, D.H., Luce, R.D., Suppes, P., Tversky, A.: Foundations of measurement, vol. 1: Additive and polynomial representations. Academic, New York (1971)
9. Shepard, R.N.: On the status of direct'psychophysical measurement. Minnesota Studies in the Philosophy of Science 9, 441–490

10. Shepard, R.N.: Psychological relations and psychophysical scales: on the status of direct. Journal of Mathematical Psychology 24(1), 21–57 (1981)
11. Ehrenstein, W.H., Ehrenstein, A.: Psychophysical methods. Modern Techniques in Neuroscience Research, 1211–1241 (1999)
12. Knoblauch, K., Maloney, L.T.: MLDS: Maximum likelihood difference scaling in R. Journal of Statistical Software 25(2), 1–26 (2008)
13. Green, D.M., Swets, J.A.: Signal detection theory and psychophysics (1966)
14. Menkovski, V., Exarchakos, G., Liotta, A.: The value of relative quality in video delivery. Eindhoven University of Technology, Eindhoven (2011)
15. Charrier, C., Maloney, L.T., Cherifi, H., Knoblauch, K.: Maximum likelihood difference scaling of image quality in compression-degraded images. Journal of the Optical Society of America A 24(11), 3418–3426 (2007)

On the Quality Assessment of H.264/AVC Video under Seamless Handoffs

Ilias Politis, Tasos Dagiuklas, and Lampros Dounis

Dept. of Telecommunication Systems and Networks, TEI of Mesolonghi,
National Road Nafpaktos-Antirrion, 30300, Greece
{Ilpolitis,dounis}@gmail.com, ntan@teimes.gr

Abstract. This paper examines the quality assessment of video streaming applications in a heterogeneous wireless environment, where the user hands off across inter-technology radio access networks. Three different scenarios have been considered: scenario with seamless handover using the media handover framework to initiate handover, seamless handoff combined with rate adaptation that is based on Rate-Distortion and seamless handoff with rate adaptation that is optimized using network bandwidth and packet loss parameters. The results from two video sequences have shown that both objective quality evaluation and the subjective evaluation (double stimulus-SDSCE, DSCQS) are optimized under the combined seamless handover and rate adaptation functionalities.

Keywords: Mobile Video Delivery, Subjective and Objective Video Quality Evaluation.

1 Introduction

Due to rapid growth of wireless communications, multimedia applications are becoming increasingly popular. In the recent years, this progress has also been aided by the proliferation of technologies such as 3G/3G+/LTE/WiFi, and the trend has been to allocate these services more and more also on mobile users [1]. Video transmission over heterogeneous wireless networks poses many challenges, including the issue of coping with losses due to physical impairments and network congestion, as well as maintenance of Quality of Service (QoS) and session continuity [2]. A key change is to support and maintain video quality while the user moves across heterogeneous networks. It has been reported that in order to support and maintain video quality, handover functions may be triggered not only from the physical layer but also from the network (rapid increase to packet loss) and the application layer (PSNR drops substantially).

The aim of this paper is to evaluate the quality of video streaming applications in different mobility scenarios using both objective and subjective methodologies. Three different scenarios have been considered: The first one considers an on-going video session that is seamless transferred in a vertical handoff function under the Media Independent Handover (MIH) functionality. In the second scenario, handover is

L. Atzori, J. Delgado, and D. Giusto (Eds.): MOBIMEDIA 2011, LNICST 79, pp. 16–30, 2012.

combined with rate adaptation using the Rate-Distortion functionality of the encoder. In the third scenario handover is combined with a novel rate adaptation that is based on the optimizing technique by taking into account the available bandwidth of the new access network and the packet loss.

The paper is structured as follows: Seamless Handoff and Video Rate Adaptation are presented in section 2; section 3 presents the testbed platform in order to carry out the experimentation results. Subjective evaluation using double stimulus methodologies is presented in section 4. Conclusions are presented in section 5.

2   Seamless Handoff

Handover is the process of network association of a Mobile Terminal while it moves across different access point [3]. Handover aims to accomplish the following goals:

- Nomadicity: It is the ability of the user to change his network point of attachment while he/she is on the move.
- Session Continuity: It is the ability that the mobile terminal can switch to a new network point of attachment while maintaining the ongoing session towards the new point of attachment.
- Seamless handoff: It aims to minimize the packet loss while the session is associated with the new point of attachment. It is sometimes referred to as smooth handoff.

One of the standards that have specified a framework for seamless mobility is the IEEE 802.21 standard. This section describes a seamless mobility framework that combines the Media Independent Handover framework (MIH), which is responsible for seamless handover management across heterogeneous access networks and the available BW based rate control scheme, described above.

2.1   Media Independent Handover Framework Overview

IEEE created the 802.21 standard in order to challenge one of the main issues in wireless mobility, seamless handovers across inter-technology RATs (Radio Access Technology) networks [4]. In particularly, mobility protocols such as Mobile IP are suffering from sensible latency and they have not knowledge about the application layer parameters and candidate network conditions. IEEE 802.21 proposes the MIH framework where mobile nodes and the network exchange information and commands for an optimal handover. Moreover, for hiding the heterogeneity of the MAC and physical layers, MIH inserts an intermediate layer between layer 3 (and above) and the divert Layer 2 technology specifics, the Media Independent Handover Function (MIHF). The MIH framework describes three different types of communication that act as services: Event Service, Command Service and Information Service, as illustrated in Fig. 1.

The Media Independent Event Service (MIES) is a communication procedure where indications for handoff (events) are passed to the MIH users for further handling. The Media Independent Command Service (MICS) provides a set of handover commands in order for the MIH users to be able to implement their handover decisions. The Media Independent Information Service (MIIS) is a database that contains all the available information about the network ranging from channel parameters to presence of application layer services. It is used by mobility protocols in order to find appropriate networks that can facilitate a handover.

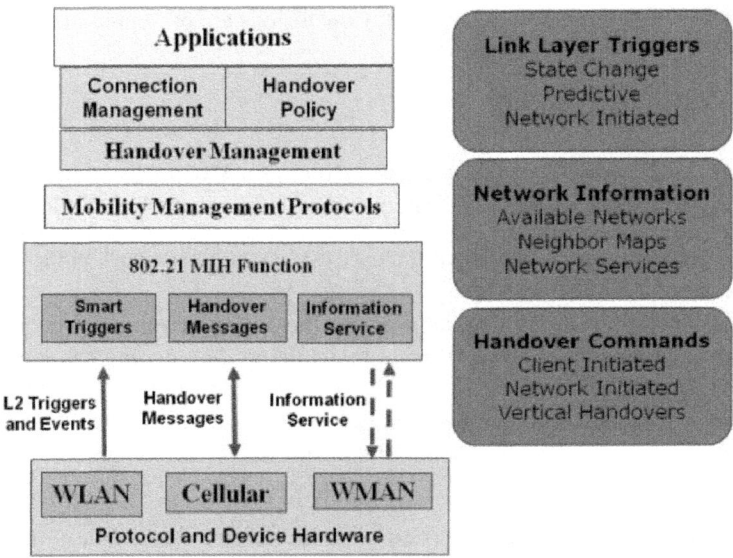

Fig. 1. IEEE 802.21 Media Independent Handover framework

MIH can be considered as a co-operative decision making scheme for QoE-aware handoff policy. This means that triggers from both MT and the radio access networks for the handover decisions. There are scenarios for handover from QoE based triggering [5], [6].

- Network Load increases at the current wireless network leading to congestion and packet loss.
- Application Deteriorates. This can be verified through video QoS monitoring (MOS, PSNR) generating alarms when these parameters are sharply altered.

2.2   Application QoE Triggering

In contrast to previous studies that consider only physical and network layer statistics, this paper proposes a handover functionality that can be triggered also by estimating PSNR. Monitoring of QoE is based on the RTP Control Protocol Extended Report (RTCP-XR) as defined in [7], which provides a useful set of fields providing information for video performance analysis. Important information that is related to, video is the following:

- The packet loss within I/B/P frames.
- Knowledge of GoP structure and key coding parameters to estimate PSNR

The process of estimating the PSNR of real time video streaming requires the video client to send an RTCP-XR report to the application server with the ID numbers of the lost RTP packets, per video frame. The application server can then estimate the current value of PSNR by a video distortion model that calculates the distortion of the received video due to packet losses in real time. Specifically, the proposed distortion prediction model is a recursive formula that takes into account the correlation among video frames during the intra-frame period. The distortion model incorporates the random behavior of losses in the wireless medium (isolated losses, burst of losses, losses separated by a lag). More information about the predicted distortion model and Video QoS Monitoring can be found in [8] and [9].

2.3   Vertical Handoff Policy

A handover scheme, as part of the MIH framework is responsible for deciding whether a handover is needed based on physical, network and application layer statistics, collected from both the MT and the access networks. The handover policy includes three phases [4]:

- Decision phase – all the handover related information (e.g Signal-to-Noise-Ratio, delay, jitter, Peak Signal-to-Noise-Ratio, packet loss, etc), from the mobile terminal, the currently selected access network and the already discovered neighboring networks are retrieved from the MIIS through the MIH entity.
- Initiation phase – the collected parameters are evaluated and compared against a set of predetermined threshold values. These thresholds are either determined by the network provider or are specified in the user profile. The MIES service is responsible for comparing the collected statistics with the threshold values and for informing the command service when one or more thresholds are violated and the handover criteria are matched.
- Execution phases – the MIH triggers the Mobile IP module which is responsible for performing the actual handover and bidding with the new point of attachment, ensuring seamless service continuity.

2.4   Network Selection

In the context of MIH, it is necessary to incorporate a mechanism that selects the access technology that is the most suitable according to the needs of the user at each moment. This decision is based on QoS parameters/criteria that must be optimized depending on the available access networks. In this paper, the network selection scheme combines two Multi Attribute Decision Making (MADM) algorithms methods, the Analytic Hierarchy Process (AHP) method [10] and the Total Order Preference by Similarity to the Ideal Solution (TOPSIS) method [11]. The first one determines weights of the criteria and the second one calculates the final access network ranking.

AHP Method

In the case of AHP, the following parameters are considered: Throughput, Packet Loss and SNR. All the handover parameters are compared pairwise according to their levels of influence with respect and the comparison results are inserted in a square matrix using the following rule: A=[aij]nxn where n are the number of factors. Each element represents a handover parameter with a value that implies the extent, at which an element is more important to another from 1 to 9. The value 1 defines equal importance between the two elements and value 9 defines extreme importance.

The weights vector w is calculated through the following repetitive process:

- The elements of each line of the matrix are added up: $s_i = \sum a_{ij}$ for each i.

- In each line of matrix, the weight of each element is estimated by calculating the quotient of the value s_i via the sum of all elements of the matrix:

  $w_i = s_i / \sum_i \sum a_{ij}$. The elements of the received vector w are normalized, so that their sum equals to 1.

- The square of the matrix is calculated and all the procedure steps are repeated until two successive approaches do not differ considerably in the frame of the desirable precision.

TOPSIS Method

In TOPSIS method, the best radio access must have the shortest distance from the positive ideal solution and the longest distance from the negative ideal solution. It comprise the following steps:

- Based on the scores achieved for each one of the selected criteria (attributes), the Network Matrix is expressed as:

$$NW_{ij} = \begin{array}{c} A_1 \\ A_2 \\ M \\ A_m \end{array} \begin{bmatrix} d_{11} & d_{12} & \cdots & d_{1m} \\ d_{21} & d_{22} & \cdots & d_{2m} \\ M & M & \cdots & M \\ A_{m1} & A_{m2} & \cdots & A_{mm} \end{bmatrix}$$

$$C_1 \quad C_2 \quad \cdots \quad C_m$$

where A_1, A_2, ... , A_n are the possible network alternatives and C_1, C_2, ..., C_m are the criteria, which measure the performance of the alternatives. Each element dij of the Network Matrix NW_{ij} is the rating of the alternative Ai with respect to the criterion C_j.

- The normalized value of r_{ij} is computed as: $r_{ij} = \dfrac{d_{ij}}{\sqrt{\sum_{i=1}^{n} d_{ij}^2}}$ where i={1,2, ...,n},

j={1,2, ..., m}

- The normalized weights are determined according to the following: $NW_{normij} = NW_{ij} * r_{ij} * w_j$

- Determination of the positive and negative ideal solutions using the following formula: $A^+ = \left\{ \max_j \left(NWnorm_{ij} \right) \mid j \in I_b, \min_j \left(NWnorm_{ij} \right) \mid j \in I_c \right\}$

  and $A^- = \left\{ \min_j \left(NWnorm_{ij} \right) \mid j \in I_b, \max_j \left(NWnorm_{ij} \right) \mid j \in I_c \right\}$ where I_b denotes the set with the benefit criteria, and I_c denotes the set with the cost criteria

- The distance of each alternative from the ideal and the negative ideal solution is given by the following formulas: $S_{+i} = \sqrt{\left(NWnorm_{ij} - NWnorm_j^+ \right)^2}$ and

  $$S_{-i} = \sqrt{\left(NWnorm_{ij} - NWnorm_j^- \right)^2}$$

- The relative closeness determines the relative closeness of each alternative Ai from the ideal solution using the following formula: $C_i = \dfrac{S_i^+}{S_i^+ + S_i^-}$, i={1,…,n}

- The best candidate networks are ranked according to the Ci values in descending order.

2.5   Seamless Handoff and Rate Adaptation

Seamless handover can be combined with rate adaptation in order to optimize QoS/QoE of an on-going video session. In case, where the mobile user moves to a network with less available bandwidth than the current one, seamless handoff can be combined with rate control so that QoE/QoS is optimized. This can be transformed to an optimization problem (determine the best Quantization Parameter that optimizes PSNR under certain networking conditions) [12].

In the context of this paper, a rate control scheme has been used that maximises perceived video quality based on the currently available bandwidth (BW). The bandwidth availability is estimated based on RTCP feedback from the mobile terminal.

The proposed real time video rate control framework requires that pre-encoded video sequences have been tested over different network condition. This is required in order to extract important statistical information. Such statistical information regards the relationship between encoding distortion and QP, sending bit rate and QP. The above relationships are obtained through experimentation.

A Rate Control Module (RCM) is defined as an entity within the video encoder that stores the aforementioned statistical information, collects real-time information from the network and sends feedback to the encoder in order to control the sending rate. Without loss of generality, we assume that this control is applied to each video frame. RCM receives periodic feedback with information regarding current available

BW and packet losses, and decides upon the optimum QP value that maximizes the perceived video QoS (PSNR at the decoder). RCM's decision is forwarded to the video encoder, which selects the optimum QP parameter, as shown in Fig. 2.

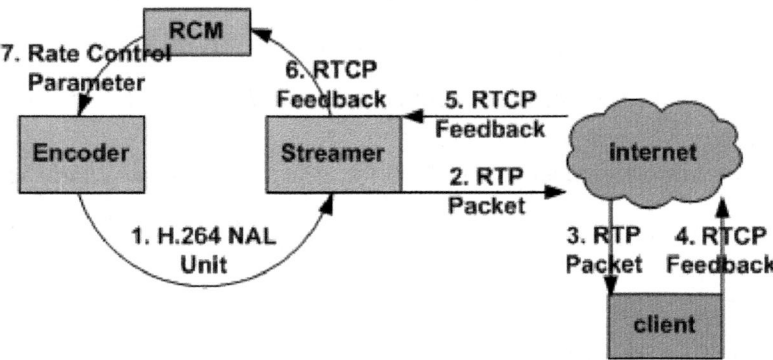

Fig. 2. Video Rate Control Scheme

The perceived video QoS in terms of PSNR is maximized by selecting the optimum value of the QP parameter, according to the current available bandwidth (BW). Without loss of generality, the term available bandwidth refers to the capacity of the access network that becomes available to the user. That is the bottleneck between the video encoder and the end-user. This network capacity is periodically monitored by probing the network with a predefined stream of dummy RTP packets. The user is informed for the available network capacity (or available BW) by the periodic RTCP messages that carry this information to the message header. In order to determine the optimum QP, the RCM is based on pre-stored PSNR versus QP data for different BW conditions [12].

This relationship is illustrated in Fig. 3, where a video sequence has been encoded at different QP values and is transmitted multiple times over a network with varying load conditions. It is evident that for low available bandwidth, perceived PSNR increases with QP, due to the fact that larger QP results in lower video rate. The PSNR reaches a certain peak value, which is the maximum perceived quality that a video user can have for specific network conditions. This is the point where both coding distortion and packet loss have the least impact on the perceived video quality for a given available bandwidth. Any further increase to the QP value will result in higher coding distortion and smaller video transmission rates (packet loss increases) that deteriorates the perceived video QoS. As the available BW becomes higher, PSNR reaches its peak value earlier (i.e. at smaller QP), shifting this point towards the left. Furthermore, Fig. 4 illustrates the surface fitted model of Fig. 3. This curve is true for one particular video sequence (NTIA gold fish pond) [13], however, the algorithm can be extended to other video sequences as well.

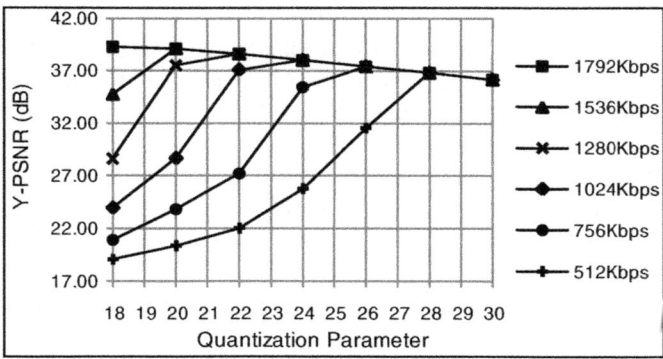

Fig. 3. Perceived PSNR vs. QP under different network conditions (available BW for video transmission)

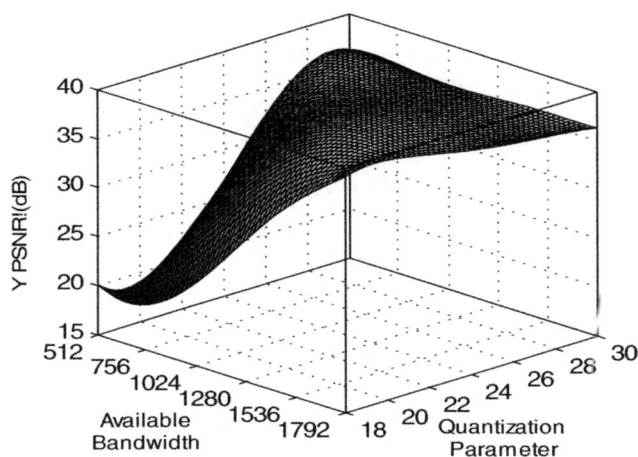

Fig. 4. Surface fitted model of perceived PSNR vs. QP under different network conditions (available BW for video transmission)

For the surface-fitting model a number of different polynomials and rational equations have been used, which resulted in the following polynomial equation with a reasonable fitting goodness.

$$PSNR = 605 - 1.2*BW - 57.6*QP + 4*10^{-3}*BW^2 + 0.12*BW*QP + 1.26*QP^2 - 3.7*10^{-3}*BW*QP^2 + 0.02*QP^3 - 6*10^{-4}*QP^4 \qquad (1)$$

According to the BW based rate control algorithm, both the currently available BW and the PSNR versus QP relationship over different available BWs are regarded as input to the algorithm. BW conditions are collected from RCM via RTCP reports [14]. The algorithm optimizes the perceived PSNR by selecting the optimum QP value according to the network conditions. Depending on network conditions optimum QP can either be higher or lower than the current QP. As the network load increases (decreases), QP should increase (decrease) so that the sending rate is adapted according to the available bandwidth. The solution of the partial derivative of the three-dimensional function of Fig. 4 (Eq. 1), with respect to the QP given that the available BW information is collected by RTCP, returns the critical points (QP values) that maximize the perceived PSNR. Fig. 5 outlines the proposed BW based rate control algorithm.

Input
BW → *Available bandwidth*
#*F(BW,QP)*→ *Function describes the surface of the Bandwidth based rate control algorithm.*

Output
#*bestQP* →*Best Quantization Parameter for available BW*
#*maxPSNR* → *Estimated average PSNR for Best QP*

$$[solutions] = \frac{\partial}{\partial QP} F(BW, QP) = 0$$
$$\forall iQP \in solutions$$
$$if\ iQP > 0\ \&\ \frac{\partial}{\partial iQP^2} F(BW, iQP) < 0$$
$$bestQP = iQP$$

Fig. 5. Proposed available BW based rate control scheme

Another approach of rate control is to use a Rate Distortion Model that is inherently implemented within the video encoder [15]. In this model, the targeted bit rate is provided to the encoder in order to select on the fly the appropriate QP parameter using the Rate Distortion Model. In this approach, when the user is handed over to a wireless network with less bandwidth than the current one, a signal is sent towards the encoder to adapt QP parameter accordingly.

3   Testbed Platform

In order to study the perceived video quality, an experimental test-bed was implemented.

- Two fully configurable 802.11e access points and one 3G access that allow the monitoring and collection of physical and network layer statistics.
- A MIH capable mobile terminal that can connect to any access network through two corresponding adapters. The MT monitors the status of the current connection and the availability of any other candidate RATs in each vicinity.

This information is reported to the MIH (MIIS) through a client-server application, and based on the decision from the vertical handover functions it will be instructed by the MIP core

- A MIH server that hosts all MIH services MIIS, MIES, MICS as well as, the handover decision functionality
- A video server that consists of a fully configurable H.264/Advance Video Coding encoder and streamer [15], capable of exchanging RTCP-XR messages through a clients server application
- During the experiments, the network is stressed with background traffic based on a statistical video traffic model, which regards a number of multiplexed homogeneous and mutually independent video sources that transmit simultaneously. This model can accurately simulate the effect of aggregate video traffic from multiple video sources. Moreover, Dummynet [16] is used in order to emulate packet losses and network load.

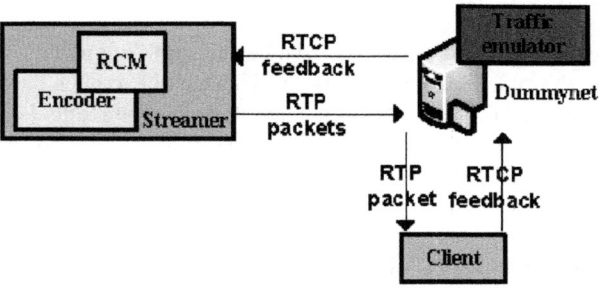

Fig. 6. Test-bed Platform

Fig. 7 depicts the objective quality (PSNR) of the 'Fish' video sequence temporal basis [13]. We consider that both WLANs and 3G Network are stressed with background traffic ranging from 50% to 75% of the total capacity.

The paper focuses on measuring the effect of seamless handoff on the perceived video quality. To this end three test-bed experiments have been carried out in a control laboratory environment. The handover is initiated by both application layer and network layer triggers (including PSNR drop and Packet loss increase due to mobility). The seamless handoff is executed by the MIH platform, described above.

In the first scenario, the impact of seamless handoff functionality on the perceived video quality is presented. Two other scenarios are considered where seamless handoff is combined with rate control. The first one considers the optimization functionality presented in section 2. The second one considers a rate distortion functionality that is inherent in the video encoder. It is obvious that the seamless handoff functionality and optimized rate control gives the optimum perceived video quality.

.

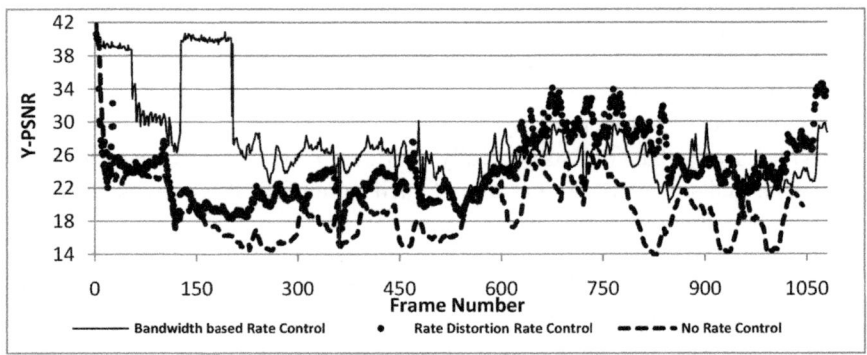

Fig. 7. Objective Video QoS for three video mobility scenarios

4   Subjective Evaluation

The subjective quality evaluation tests have been carried out using two different video sequences with the same spatial resolutions. The tests have been carried out using the recommendations by ITU-T BT.500 for laboratory environments [17]. The following parameters have been considered: daylight conditions, mid gray background using appropriate curtains. High quality LCD displays have been used for the subjective evaluation [18], [19].

We have used two high-quality raw videos that have been recorded with professional equipment using YUV 4:2:0 format. All videos are freely available for downloading from the Internet.

All subjects that participated at the evaluation are undergraduate students and faculty members of the TEI of Mesolonghi, Department of Telecommunication Systems and Networks, Greece[1]. A total of 80 subjects have been used to evaluate the videos. During the test setup phase, each subject gets familiar with scoring procedure and video artifacts. This will ensure that subjects will get familiar with the testing procedure and score video artifacts accordingly. Training videos have been used for this purpose. Both training and test videos have been impaired.

The subjective evaluation study uses simultaneous double stimulus continuous evaluation that is described below. Under Double Stimulus framework, two video sequences are shown simultaneously to each subject. The first one is the original video sequence and the second one is the transmitted in a heterogeneous wireless environment. The subject is informed about the presence of the reference video (Stimulus A) and the distorted video (stimulus B), so that he/she continuously evaluates the test material.

4.1   SDSCE

The votes are sampled every 0.5, as described from the Simultaneous Double Stimulus for Continuous Evaluation   (SDSCE) in ITU BT-500. Two different

---
[1] http://www.tesyd.teimes.gr/cones

impaired video segments haven been evaluated. Each impaired video segment corresponds to a different wireless video transmission policy. The aim to evaluate the video sequences under the aforementioned framework: wireless video transmission in a heterogeneous wireless environment, wireless video transmission + rate adaptation (using the R-D of the encoder and the proposed bandwidth adaptation algorithm).

In this voting procedure, the next step regards the removal of vote cutliers. These votes refer to the cases where the difference between mean subject vote and the mean vote for this test case from all other subjects exceeds 15%. This is a general rule that has been also used in other research works [18], [19].

The following figure illustrates the max, min and average MOS for both "Fish" and "Tea" video sequences. The best quality is experienced under the Bandwidth Based Rate Control, which is the seamless handover using the optimised rate adaptation functionality and the worst scenario is the one where the seamless handover is not combined with rate adaptation.

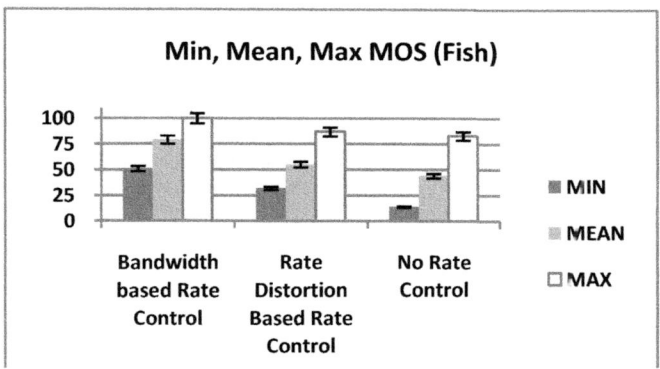

(a) 'Fish' Video Sequence Min, Mean, Max MOS (SDSCE)

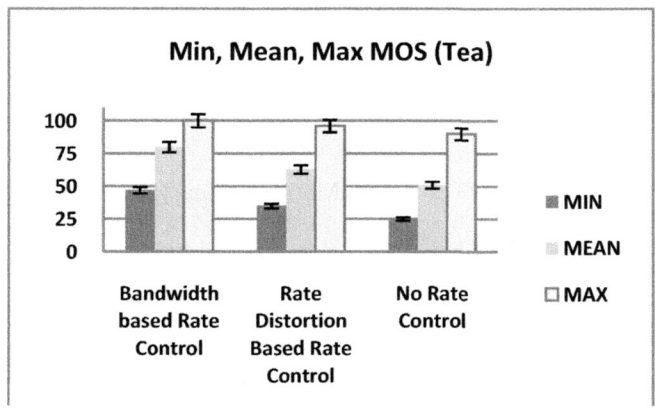

(b) 'Tea' Video Sequence Min, Mean, Max MOS (SDSCQE)

Fig. 8. Max, Mean and Min MOS under the three handover schemes

The above observations are verified by illustrating average MOS over frame for the three scenarios. Under the seamless handover scenario, the MT may switch to a network with much less available bandwidth. In the scenario, where rate adaptation is carried out using Rate Distortion, there is an improvement in the MOS due to the fact that video encoding is adapted to bandwidth constraints. This improvement is optimized when using novel rate control functionality.

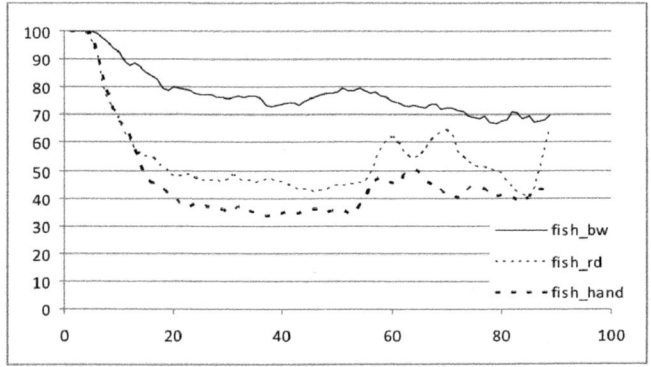

(a) 'Fish' Video MOS over time (SDSCE)

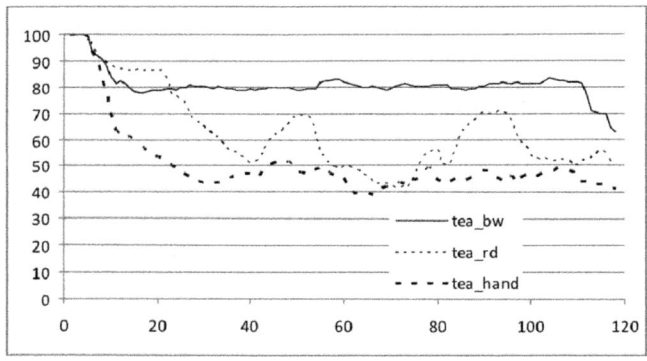

(b) 'Tea' Video MOS over time (SDSCE)

Fig. 9. MOS versus time under the three different schemes

4.2   DSCQS

Within the Double Stimulus Continuous Quality Scale (DSCQS) methodology, two consecutive presentations of two stimuli take places [17]. There is a 10s duration of the reference and the test (distorted) video, separated by 1s grey frame. The above procedure is repeated three times, and at the last round the subject must vote for both the reference and the test video in the scale 0 to 100.

Similar to the above procedure (SDSCE), the outliers are removed in order to remove subjects whose scores deviate considerably from the votes of the other subjects. Under this methodology, the Differential MOS computed using the above

formula: $DMOS_i = \dfrac{\sum\limits_{j=1}^{N}\left(s_{ji}^A - s_{ji}^B\right)}{N}$,   where N is the number of valid subjects and

s_{ji}^A, s_{ji}^B,   are the scores of the test and the reference video respectively. Using the DMOS, MOS can be computed for the i test using the following formula:

$$(MOS)_i = \dfrac{100-(DMOS)_i}{10}$$

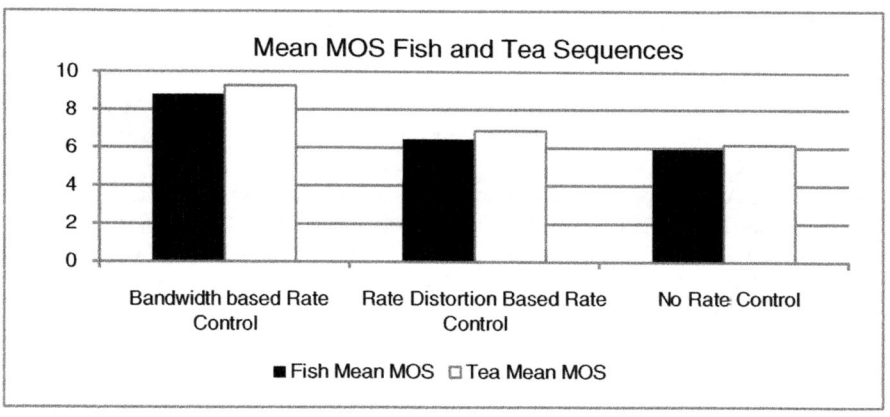

Fig. 10. Average MOS using DSCQS Methodology

5   Conclusions

This paper examines subjective and objective evaluation of video sequences under seamless handoff schemes. Handover decision is based on the Media Independent Handover Framework by collecting information from physical, network layer and application layer from both MT and network entities. Video Quality has been evaluated using three different cases: in the first cases video is received by considering only mobility function, in the second case mobility is combined with rate adaptation that uses Rate-Distortion function of the video encoder and in the third case mobility is combined with rate adaptation that is optimized by taking into account available bandwidth and packet loss of the network that the MT will handoff. Through experimentation from a testbed, both subjective and objective quality is optimized when handover is combined with rate control.

References

[1] Etoh, M., Yoshimura, T.: Advances in Mobile Video Delivery. Proceedings of the IEEE 93, 111–122 (2005)

[2] Zhang, Q., Zhu, W., Zhang, Y.: End-to-End QoS for Video Delivery over Wireless Internet. Proceedings of the IEEE 93, 123–133 (2005)

[3] Nasser, N., Hasswa, A., Hassanein, H.: Handoffs in fourth generation heterogeneous networks. IEEE Communications Magazine 44, 96–103 (2006)

[4] IEEE 802.21/D10.0. Draft Standard for Local and Metropolitan Area Networks: Media Independent Handover Services, IEEE Draft (2008)

[5] Rodriguez, J., Tsagaropoulos, M., Politis, I., Kotsopoulos, S., Dagiuklas, T.: A Middleware Architecture Supporting Seamless and Secure Multimedia Services across Inter-Technology Radio Access Networks. IEEE Wireless Communications Magazine 16, 24–31 (2009)

[6] Dounis, L., Tsagkaropoulos, M., Politis, I., Dagiuklas, T.: On the impact of MIH triggering techniques on the performance of video streaming across heterogeneous RATs. In: 6th International Mobile Multimedia Communications Conference (Mobimedia), Lisbon, Portugal (2010)

[7] IETF draft-ietf-avt-rtcpxr-video-02.txt, RTCP XR Video Metrics (2007), http://www.ietf.org/internet-drafts/draft-ietf-avt-rtcpxr-video-02.txt

[8] Tao, S., Apostolopoulos, J., Guerin, R.: Real-Time Monitoring of video quality in IP networks. IEEE/ACM Transactions on Networking 16, 1052–1065 (2008)

[9] Politis, I., Tsagkaropoulos, M., Pliakas, T., Dagiuklas, T.: Distortion Optimized Packet Scheduling and Prioritization of Multiple Video Streams over 802.11e Networks. In: Advances in Multimedia (2007)

[10] Saaty, T.L.: The Analytic Hierarchy Process. RWS Publications, Pittsburgh (1990)

[11] Hwang, C.L., Yoon, K.: Multiple attribute decision making:Methods and applications, A State of the Art Survey. Springer, New York (1981)

[12] Dounis, L., Dagiuklas, T., Politis, I.: On the Comparison of Real-Time Rate Control Schemes for H.264/AVC Video Streams over IP-based networks using network feedbacks. In: IEEE ICC, Kyoto, Japan (2011)

[13] The consumer digital video library, http://www.cdvl.org

[14] Huitema, C.: Real Time Control Protocol (RTCP) attribute in Session Description Protocol (SDP), RFC 3605 (October 2003)

[15] Vanguard Software Solutions Inc. (2010), http://www.vsofts.com

[16] Rizzo, L.: Dummynet: a simple approach to the evaluation of network protocols. ACMSIGCOMM Computer Communication Review 27 (1997)

[17] ITU-R BT.500-11, Methodology for the Subjective Assessment (2002)

[18] Seshadrunathan, K., Soundararajan, R., Bovik, A., Cormack, L.: Study of Subjective and Objective Quality of Assessement of Video. IEEE Transactions On Image Processssing 19, 1427–1441 (2010)

[19] Oelbaum, T., Schwarz, H., Wien, M., Wiegand, T.: Subjective performance evaluation of the SVC Extension of H.264/AVC. In: 15th IEEE International Conference on Image Processing ICIP 2008, pp. 2772–2775 (2008)

Reduced-Reference Image Quality Assessment Based on Edge Preservation

Maria G. Martini[1], Barbara Villarini[2,*], and Federico Fiorucci[2]

[1] WMN Research Group, Kingston University London
Penrhyn Road, KT12EE, London, UK
[2] DIEI, University of Perugia
via Duranti, Perugia, Italy

Abstract. Assessing the subjective quality of processed images through an objective quality metric is a key issue in multimedia processing and transmission. In some scenarios, it is also important to evaluate the quality of the received images with minimal reference to the transmitted ones. For instance, for closed-loop optimisation of image and video transmission, the quality measure can be evaluated at the receiver and provided as feedback information to the system controller. The original images - prior to compression and transmission - are not usually available at the receiver side, and it is important to rely at the receiver side on an objective quality metric that does not need reference or needs minimal reference to the original images.

The observation that the human eye is very sensitive to edge and contour information of an image underpins the proposal of our reduced reference (RR) quality metric, which compares edge information between the distorted and the original image.

Results highlight that the metric correlates well with subjective observations, also in comparison with commonly used full-reference metrics and with a state-of-the-art reduced reference metric.

Keywords: Edge detection, image quality assessment, reduced-reference, quality index, mean opinion score (MOS), SSIM.

1   Introduction

The quality of images can be assessed based on a variety of methodologies. Subjective tests, such as the mean opinion score (MOS) [1], are difficult to reproduce. In addition, they are expensive, time consuming and typically cannot be implemented algorithmically. On the other side, pure objective measurements such as mean square error (MSE) and peak signal-to-noise ratio (PSNR) do not adequately represent the subjective quality [2].

For both performance assessment and on-the-fly system adaptation, it is crucial in image/video compression and transmission to develop an objective quality

---

* B. Villarini was with Kingston University London, UK.

L. Atzori, J. Delgado, and D. Giusto (Eds.): MOBIMEDIA 2011, LNICST 79, pp. 31–45, 2012.

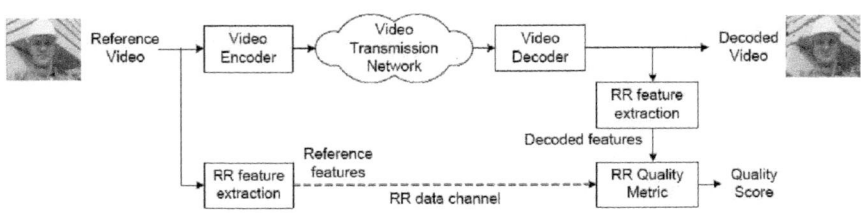

Fig. 1. Reduced reference scheme

assessment metric which accurately represents the subjective quality of compressed and corrupted images. Objective methods based on subjective measurements are based either on a perceptual model of the Human Visual System [3], or on a combination of relevant parameters tuned with subjective tests [4] [5].

It is also important to evaluate the quality of the received images with minimal reference to the transmitted ones [6]. For closed loop optimisation of image/video transmission, the image quality measure can be provided as feedback information to a system controller [7]. The original images - prior to compression and transmission - are not usually available at the receiver side, hence an objective quality metric that does not need reference or needs minimal reference to the original images is required.

The human eye is very sensitive to edge and contour information of an image, *i.e.*, the edge and contour information give a good indication of the structure of an image and it is critical for a human to capture the scene [8]. Some works in the literature proposed full-reference (FR) image quality metrics considering edge structure information. For instance in [9] the structural information error between the reference and the distorted image is computed based on the statistics of the spatial position error of the local modulus maxima in the wavelet domain.

We propose here a reduced reference quality metric which compares the edge information between the distorted image and the original one. We consider the Sobel operator [10] for edge detection, due to its simplicity and efficiency. Further details on this choice are reported in the following Section.

2   Edge Detection

There are many methods to perform edge detection. The most used among these can be grouped into two categories: gradient and Laplacian. The gradient method detects the edges by finding the maximum and minimum in the first derivative of the image. The Sobel method is an example of these. A pixel location is declared an edge location if the value of the gradient exceeds a threshold. Edges will have higher pixel intensity values than those surrounding it. Once a threshold is set, the gradient value can be compared to the threshold value and an edge is detected when the threshold is exceeded. When the first derivative is at a maximum, the

-1	-2	-1
0	0	0
1	2	1

-1	0	1
-2	0	2
-1	0	1

Fig. 2. Sobel masks

second derivative is zero. As a result, an alternative to finding the location of an edge is to locate the zeros in the second derivative. This method is known as the Laplacian, since it makes use of the Laplacian operator.

The aforementioned methods can be extended to the two-dimensions case. The Sobel operator performs a 2-D spatial gradient measurement on an image. The Sobel edge detector uses a pair of 3×3 convolution masks, one estimating the gradient in the x-direction (columns) and the other estimating the gradient in the y-direction (rows). The mask is then slid over the image, manipulating a square block of pixels at a time. The Sobel operator can hence detect edges by calculating the partial derivatives in 3×3 neighborhood. The main reason for using the Sobel operator is that it is relatively insensitive to noise and it has relatively smaller masks with respect to other operators such as the Roberts operator and the two-order Laplacian operator.

The partial derivatives in x and y directions are given as:

$$S_x = f(x+1, y-1) + 2f(x+1, y) + f(x+1, y+1) + \tag{1}$$
$$-[f(x-1, y-1) + 2f(x-1, y) + f(x-1, y+1)]$$

and

$$S_y = f(x-1, y+1) + 2f(x, y+1) + f(x+1, y+1) + \tag{2}$$
$$-[f(x-1, y-1) + 2f(x, y-1) + f(x+1, y-1)]$$

The gradient of each pixel is calculated according to $g(x, y) = \sqrt{S_x^2 + S_y^2}$ and a threshold value t is selected. If $g(x, y) > t$, the point is regarded as an edge point.

The Sobel operator can also be expressed in the form of two masks as shown in Figure 2: the two masks are used to calculate S_y and S_x, respectively.

3   Proposed Metric

Since structural distortion is tightly linked with edge degradation, our reduced reference (RR) quality metric compares edge information between the distorted image and the original one. We propose to apply Sobel filtering locally, only for some blocks of the entire image, after subsampling the images. The different steps for the calculation of the metric are reported in detail below, together with the relevant motivation.

Subdivision of the Image in Sub-windows. The first step consists of the subdivision of images in sub-windows. For instance, if images have size 512 x 768 we could subsample of a factor of 2 and consider 16×16 macroblocks of size 16×24 each, or we can subsample of a factor 1.5 and consider 18×16 macroblocks with size 19×32 each, as in the example reported in Figure 3.

Selection of the Visual Attention Area. In order to reduce the overhead associated with the transmission of side information, only 12 blocks are selected to represent the different areas of the images. The block pattern utilized for our tests is chosen after several investigations based on visual attention (VA). Various experiments have been proposed in the literature for VA modeling and salient region identification, aiming at the detection of salient regions in an image. Models on visual attention are often developed and validated by visual fixation patterns through eye tracking experiments [11] [12]. In [13] a framework is proposed in order to extend existing image quality metrics with a simple VA model. In this work a subjective Region Of Interest (ROI) experiment was performed, with 7 images, in which the viewers' task was to select within each image the region mostly attracting their attention. For simplicity, in this experiment, only rectangular-shaped ROIs were allowed. Considering the obtained ROI as a random value, it is possible to calculate the mean value and the standard deviation. It was observed that the ROI's center coordinates are around the image center for most of the images, and the mean values of the ROI dimensions are very similar in both x and y directions. This confirms that the salient region, which include the most important informative content of the image, is often placed in the center of the picture.

Following these observations we selected the block pattern as a subset of the ROI with a central symmetry, by minimizing the number of blocks in order to reduce the overhead associated to the transmission of side information. Figure 3 shows the block pattern considered below.

Edge Comparison. We propose to compare the edge structure of the blocks of the corrupted image to the structure of the corresponding blocks in the original image. For this purpose, we apply Sobel filtering locally in the selected blocks.

For each pixel in each block we obtain a bit value, where one represents an edge and zero means that there are no edges. If m and n are the block dimensions, we denote the corresponding blocks l in the original and in the possibly corrupted image as the $m \times n$ matrices O_l and C_l respectively, and the Sobel-filtered version of blocks l as the $m \times n$ binary matrices $SO_l = \mathcal{S}(O_l)$, with elements $so_{i,j}$, with $i = 1, ..., m, j = 1, ..., n$, and $SC_l = \mathcal{S}(C_l)$, with elements $sc_{i,j}$, with $i = 1, ..., m, j = 1, ..., n$. We denoted above with $\mathcal{S}()$ the Sobel operator. The similarity of two images can be assessed based on the similarity of the edge structures, $i.e.$, by comparing the matrices SO_l, associated to the filtered version of the block in the original image, and SC_l, associated to the filtered version of the block in the possibly corrupted image.

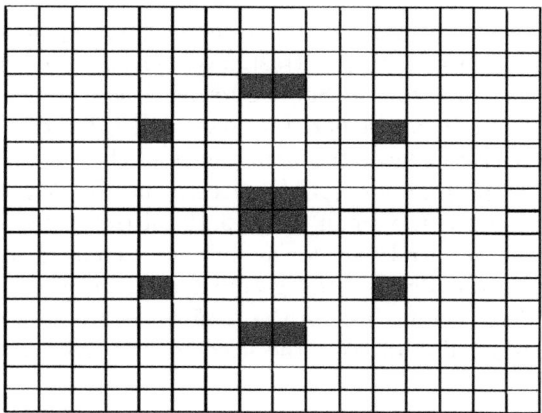

Fig. 3. Example of block pattern selected based on visual attention models

The following step is thus the quantification of the edge similarity between the reference and the processed images, done by summing the zeros and ones which are unchanged after compression or lossy transmission of the image.

Hence, for each block the similarity index can be computed as

$$I_{s,l} = n_l / p_l \tag{3}$$

where

$$n_l = p_l - \sum_{i=1}^{m} \sum_{j=1}^{n} |sc_{l,ij} - so_{l,ij}| \tag{4}$$

is the number of zeros and ones unchanged in the $l - th$ block and $p_l = m \times n$ is the total number of pixels in the $l - th$ block.

If N_b is the number of blocks in the selected block pattern, the similarity index I_s for an image is defined here as

$$I_s = \frac{1}{N_b} \sum_{l=1}^{N_b} I_{s,l} \tag{5}$$

For images decomposed in blocks of equal size, as considered here, the proposed quality index is thus:

$$I_s = \frac{1}{N_b} \sum_{l=1}^{N_b} \left(1 - \frac{\sum_{i=1}^{m} \sum_{j=1}^{n} |sc_{l,ij} - so_{l,ij}|}{mn} \right) \tag{6}$$

3.1 Threshold Selection

The threshold value is an important parameter that depends on a number of factors, such as image brightness, contrast, level of noise, and even edge direction. The selection of the threshold in Sobel filtering is associated to the sensitivity of the filter to edges. In particular, the lower the value of the threshold, the higher the sensitivity to edges. Too high values of the threshold do not detect edges which are important for quality assessment. On the other side, if the value of the threshold is too small, large parts of the image are considered as edges, whereas these are irrelevant for quality assessment. The threshold can be selected following an analysis of the gradient image histogram. Based on this consideration and on the analysis of the performance of Sobel filtering for the images of the considered databases, the selected threshold value is $t = 0.001$.

3.2 Complexity

The selection of Sobel filtering results in a low complexity metric. The Sobel algorithm is characterized, in fact, by a low computational complexity and consequently high calculation speed. In [14] some edge detection techniques are compared for an application which uses a DSP implementation: the Sobel filter exhibits the best performance in terms of edge detection time in comparison with the other wavelet-based edge detectors. Sobel filtering has been implemented in hardware and used in different areas, often when real-time performance is required, such as for real-time volume rendering systems, and video assisted transportation systems [15] [16]. This makes the proposed metric suitable for real-time implementation, an important aspect when an image/video metric is used for the purpose of "on the fly" system adaptation as in the scenario considered here.

3.3 Overhead

In order to perform the proposed edge comparison, we should transmit the matrices composed of one's and zeros's in the reference blocks. By considering the pattern in Figure 3, this would result in the transmission of $19 \times 32 \times 12 = 7.29$ kbits per image. Note that the size of the original image (not compressed) is $3 \times 512 \times 768 \times 8 = 9.4$ Mbits.

Since side information is in our case composed of a large number of zeros appearing in long runs, it is possible to reduce the overhead by compressing the relevant data, e.g., through run-length encoding, or to transmit only the positions of ones in the matrix.

In order to reduce both overhead and complexity, we propose to calculate the metric on the luminance frame only.

4   Simulation Set-Up and Results

In order to test the performance of our quality assessment algorithm, we considered publicly available databases.

Fig. 4. Images in the LIVE [18] database

The first one [17] is provided by the Laboratory for Image & Video Engineering (LIVE) of the University of Texas Austin (in collaboration with the Department of Psychology at the same University). An extensive experiment was conducted to obtain scores from human subjects for a number of images distorted with different distortion types. The database contains 29 high-resolution (typically 768×512) original images (see Figure 4), altered with five types of distortions at different distortion levels: besides the original images, images corrupted with JPEG2000 and JPEG compression, white-noise, Gaussian blur and JPEG2000 compression and subsequent transmission over Rayleigh fading are considered. Our interest is in particular on the data representing transmission errors in the JPEG2000 bit stream using a fast-fading (FF) Rayleigh channel model, since our goal is to assess the quality of images impaired by both compression and transmission errors. Our quality metric is tested versus the subjective quality values provided in the database. Subjective results reported in the database were obtained with observers providing their quality score on a continuous linear scale that was divided into five equal regions marked with adjectives Bad, Poor, Fair, Good and Excellent. Two test sessions, with about half of the images in each session, were performed. Each image was rated by 20-25 subjects. No viewing distance restrictions were imposed, and normal indoor illumination conditions were provided. The observers received a short training before the session. The raw scores were converted into difference scores (between the test and the reference) and then converted to Z-scores [20], scaled back to 1 - 100 range, and finally a difference mean opinion score (DMOS) for each distorted image was obtained.

The second database, IRCCyN/IVC [19], was developed by the *Institut de Recherche en Communications et Cyberntique de Nantes*. It is a 512×512

Fig. 5. Images in the IRCCyN/IVC [19] database

pixels color images database. This database is composed by 10 original images and 235 distorted images generated by 4 different processing methods / impairments: JPEG, JPEG2000, LAR (Locally Adaptive Resolution) coding and blurring. Subjective evaluations were made at a viewing distance of 6 times the screen height, by using a Double Stimulus Impairment Scale (DSIS) method with 5 categories and 15 observers. The images in the database are reported in Figure 5.

With the aid of the databases above, we compare the performance versus subjective tests for the following metrics:

– Structure SIMilarity Index (SSIM) [4] (full reference);
– PSNR (full reference);
– Reduced-reference metric in [6];
– Proposed Sobel-based metric (reduced reference).

To apply the SSIM metric, the images have been modified according to [21].

We report our results in terms of scatter plots, where each symbol in the plot refers to a different image: Figures 6, 7, 8, and 9 report scatter plots for the metrics above in the case of compression according to the JPEG2000 standard and subsequent transmission over a fast fading channel.

The figures report, besides scatter plots, the linear approximation best fitting the data using the least-squares method, the residuals and the norm of residuals L for the linear model, i.e., $L = \sqrt{\sum_{i=1}^{N}(d_i)^2}$, where the residual d_i is the difference between the predicted quality value and the experimental subjective quality value for image i, and N is the number of the considered images. The values of the norms of residuals enable a simple numerical comparison among the different metrics. Note that in the case of the SSIM metric we have provided a non-linear approximation, better fitting the data.

A summary of the results for the LIVE database [18] in terms of norms of residuals is reported in Table 1.

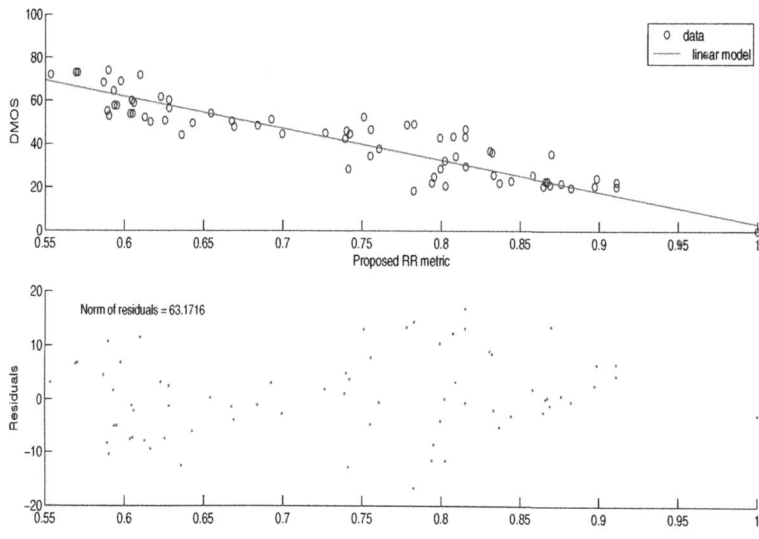

Fig. 6. Fast fading, LIVE image database [18] - Proposed metric. Above: scatter plot between difference mean opinion score and proposed metric. Below: residuals for the linear approximation and norm of residuals.

We can observe that our metric well correlates with subjective tests, with results comparable to those achieved by full reference metrics. Our metric outperforms the considered state-of-the-art reduced reference metric in all the considered scenarios, except JPEG2000 compression for images in the LIVE database. In the latter case, the benchmark reduced reference metric, based on the wavelet transform, provides a better performance.

However, for the same type of impairment (JPEG2000 compression) our metric performs slightly better than the benchmark one when the images in the IRCCyN/IVC database [19] are considered. The relevant results are reported in Table 2, where results for JPEG compression are also reported. Figures 10, 11, 12, and 13 present in detail the relevant results for the case of JPEG compression.

Table 1. Norm of residuals versus DMOS, LIVE image database [18]

	PSNR	RR [6]	Proposed RR	SSIM
JPEG 2000 + Fast fading	69.80	85.03	63.17	53.64
White noise	25.25	63.68	60.82	31.44
Gaussian blur	90.08	70.63	55.96	34.06
JPEG compression	82.83	115.06	83.04	65.58
JPEG2000 compression	70.23	63.94	84.33	74.06

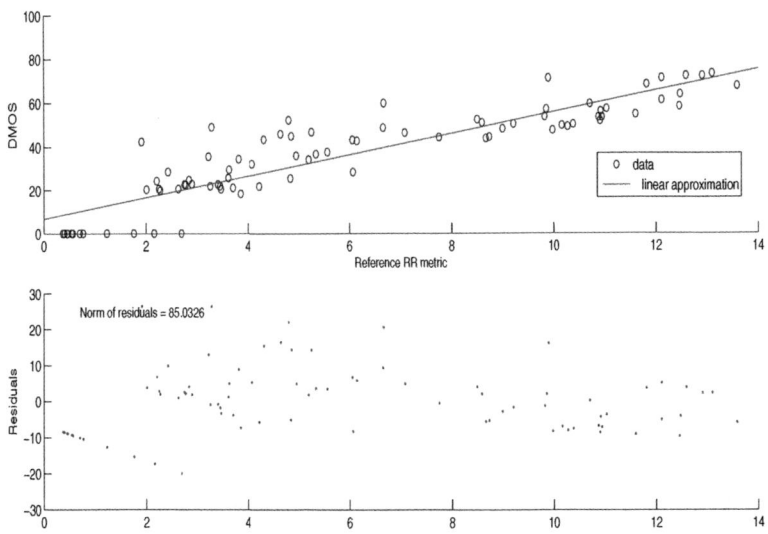

Fig. 7. Fast fading, LIVE image database [18] - Reduced reference metric in [6]. Above: scatter plot between difference mean opinion score and metric in [6]. Below: residuals for the linear approximation and norm of residuals.

We can observe that, in this case, our metric performs better not only with respect to the benchmark reduced-reference metric, but also with respect to PSNR, regardless of the need in the latter for full reference information.

Table 2. Norm of residuals versus MOS, IRCCyN/IVC image database [19]

	PSNR	RR [6]	Proposed RR	SSIM
JPEG compression	6.60	7.29	5.11	3.75
JPEG2000 compression	5.30	5.42	5.27	5.28

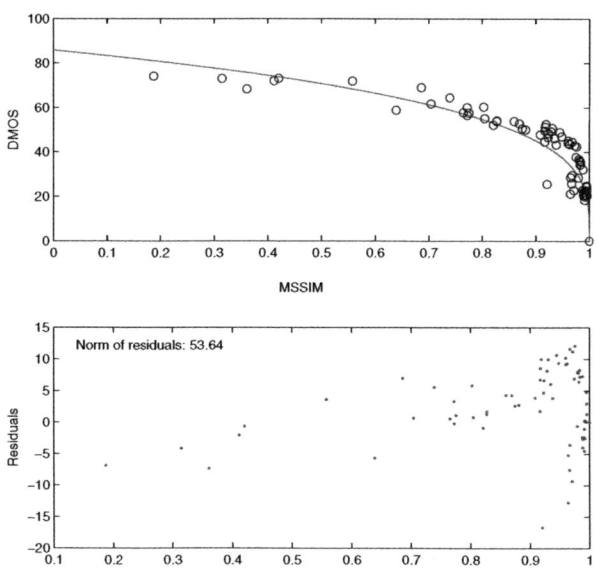

Fig. 8. Fast fading, LIVE image database [18] - MSSIM. Above: scatter plot between difference mean opinion score and MSSIM. Below: residuals for the considered approximation and norm of residuals.

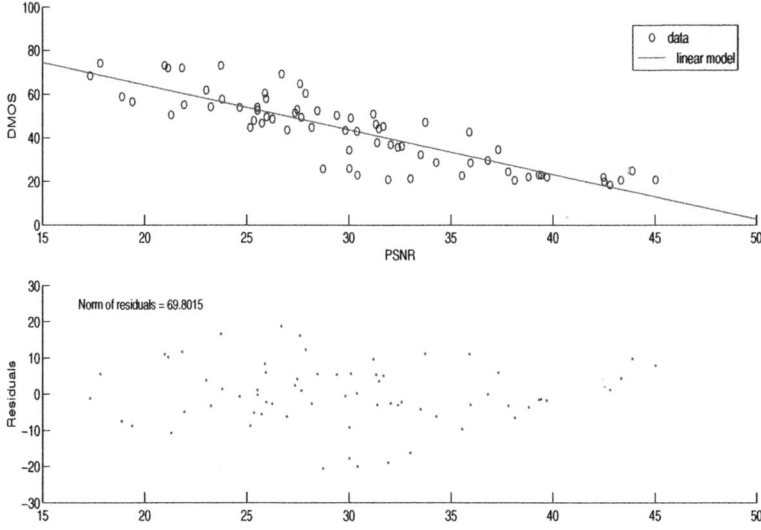

Fig. 9. Fast fading, LIVE image database [18] - PSNR. Above: scatter plot between difference mean opinion score and PSNR. Below: residuals for the linear approximation and norm of residuals.

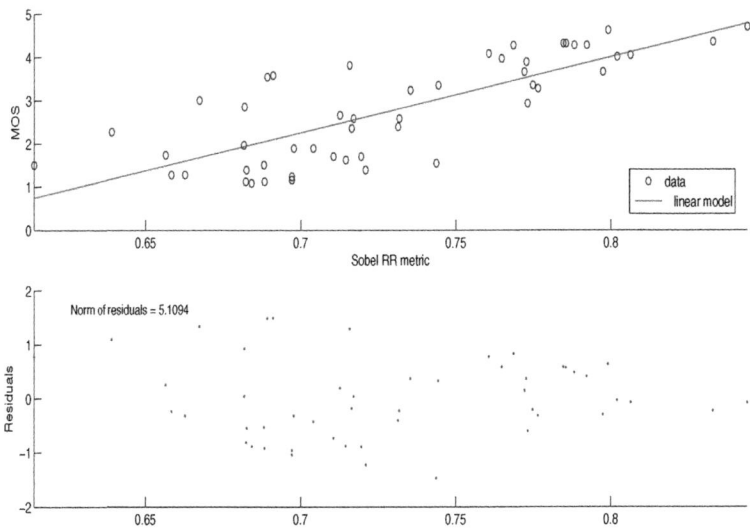

Fig. 10. JPEG compression, IRCCyN/IVC image database [19] - Proposed metric. Above: scatter plot between mean opinion score and proposed metric. Below: residuals for the linear approximation and norm of residuals.

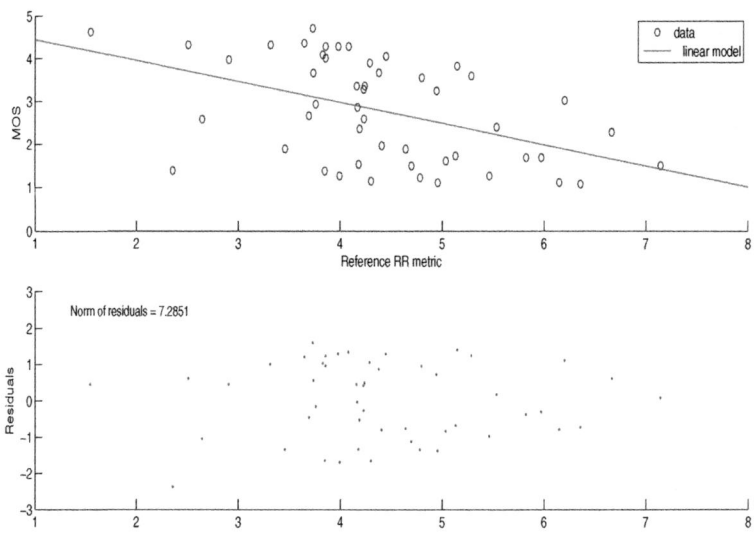

Fig. 11. JPEG compression, IRCCyN/IVC image database [19] - Reduced reference metric in [6]. Above: scatter plot between mean opinion score and metric in [6]. Below: residuals for the linear approximation and norm of residuals.

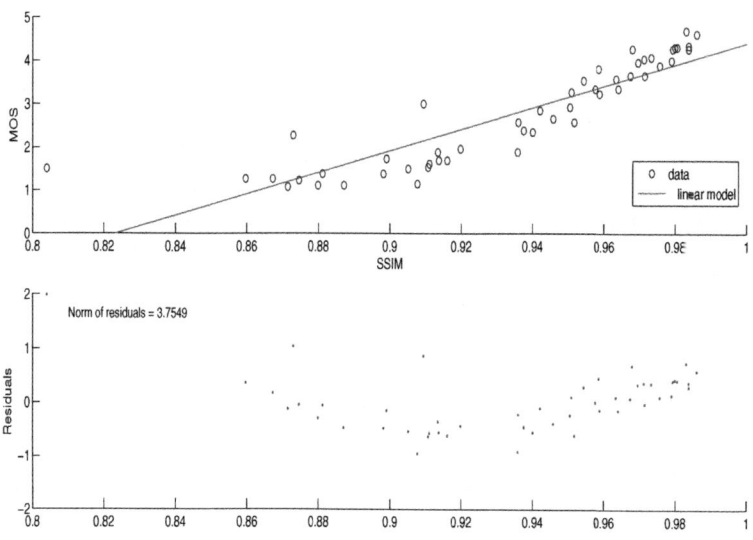

Fig. 12. JPEG compression - IRCCyN/IVC image database [19], MSSIM. Above: scatter plot between mean opinion score and MSSIM. Below: residuals for the linear approximation and norm of residuals.

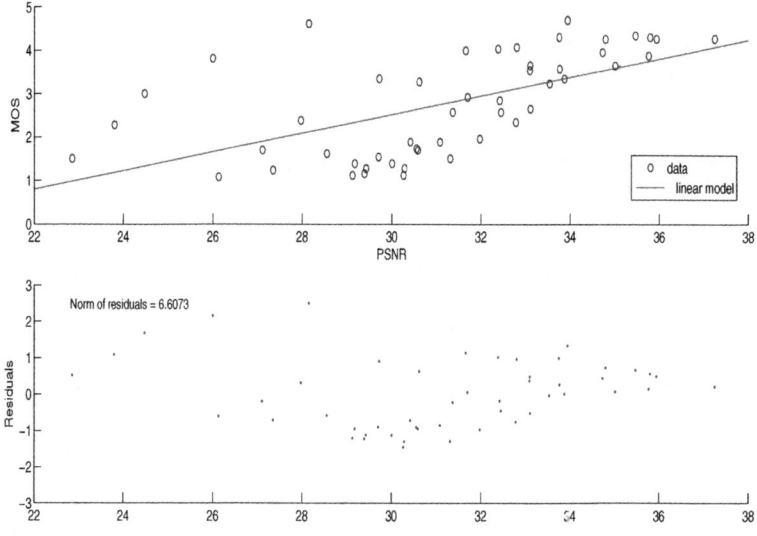

Fig. 13. JPEG compression - IRCCyN/IVC image database [19] - PSNR. Above: scatter plot between mean opinion score and PSNR. Below: residuals for the linear approximation and norm of residuals.

5   Conclusion

We proposed in this paper a perceptual reduced reference image quality metric which compares edge information between portions of the distorted image and of the original one by using Sobel filtering. The algorithm is simple and has a low computational complexity. Results highlight that the proposed metric well correlates with subjective observations, also in comparison with commonly used full-reference metrics and with state-of-the-art reduced-reference metrics. Hence, it appears suitable for the assessment of the quality experienced by the user without the need for full reference information.

References

1. Pinson, M.H., Wolf, S.: Comparing subjective video quality technologies. In: Proc. of SPIE Video Communication and Image Processing, Lugano, Switzerland (September 2003)
2. Eskicioglu, A.M., Fisher, P.S.: Image quality measures and their performance. IEEE Transactions on Comms. 43, 2959–2965 (1995)
3. Pinson, M.H., Wolf, S.: A new standardized method for objectively measuring video quality. IEEE Transactions on Broadcasting 50(3), 312–322 (2004)
4. Wang, Z., Bovik, A., Sheikh, H., Simoncelli, E.: Image quality assessment: from error measurement to structural similarity. IEEE Trans. Image Processing 13(4), 600–612 (2004)
5. Sheikh, H.R., Sabir, M., Bovik, A.C.: A statistical evaluation of recent full reference image quality assessment algorithms. IEEE Trans. Image Processing 15(11), 3440–3451 (2006)
6. Wang, Z., Simoncelli, E.P.: Reduced-reference image quality assessment using a wavelet-domain natural image statistic model. In: Human Vision and Electronic Imaging, pp. 149–159 (March 2005)
7. Martini, M.G., Mazzotti, M., Lamy-Bergot, C., Huusko, J., Amon, P.: Content adaptive network aware joint optimization of wireless video transmission. IEEE Communications Magazine 45(1), 84–90 (2007)
8. Marr, D., Hildreth, E.: Theory of edge detection. Proceedings of the Royal Society of London. Series B (1980)
9. Zhang, M., Mou, X.: A psychovisual image quality metric based on multi-scale structure similarity. In: Proc. IEEE International Conference on Image Processing (ICIP), San Diego, CA, pp. 381–384 (October 2008)
10. Woods, J.: Multidimensional Signal, Image and Video Processing and Coding. Elsevier (2006)
11. Yarbus, A.L.: Eye Movements and Vision. Plenum Press, New York (1967)
12. Privitera, C.M., Stark, L.W.: Algorithms for defining visual regions-of-interest: comparison with eye fixations. IEEE Trans. Pattern Anal. Mach. Intell. 22(9), 970–982 (2000)
13. Engelke, U., Zepernick, H.: Framework for optimal region of interest-based quality assessment in wireless imaging. Journal of Electronic Imaging 19(1), 011 005-1 – 011 005-13 (2010)
14. Musoromy, Z., Bensaali, F., Ramalingam, S., Pissanidis, G.: Comparison of real-time DSP-based edge detection techniques for license plate detection. In: Sixth International Conference on Information Assurance and Security, Atlanta, GA, pp. 323–328 (August 2010)

15. Zhou, W., Xie, Z., Hua, C., Sun, C., Zhang, J.: Research on edge detection for image based on wavelet transform. In: Proceedings of the 2009 Second International Conference on Intelligent Computation Technology and Automation. Washington, DC, USA, pp. 686–689 (2009)
16. Kazakova, N., Margala, M., Durdle, N.G.: Sobel edge detection processor for a real-time volume rendering system. In: Proc. of the 2004 International Symposium on Circuits and Systems (ISCAS 2004), pp. 913–916 (May 2004)
17. Seshadrinathan, K., Soundararajan, R., Bovik, A.C., Cormack, L.K.: LIVE video quality assessment database (2010),
    http://live.ece.utexas.edu/research/quality/live_video.html
18. Sheikh, H.R., Wang, Z., Cormack, L., Bovik, A.C.: Live image quality assessment database (2008), http://live.ece.utexas.edu/research/quality
19. Callet, P.L., Autrusseau, F.: Subjective quality assessment IRCCyN/IVC database (2005), http://www.irccyn.ec-nantes.fr/ivcdb/
20. van Dijk, A.M., Martens, J.B., Watson, A.B.: Quality assessment of coded images using numerical category scaling. In: Proc. SPIE, vol. 2451, pp. 99–101 (1995)
21. Wang, Z., Bovik, A.C., Sheikh, H.R., Simoncelli, E.P.: The SSIM index for image quality assessment (2008),
    http://www.ece.uwaterloo.ca/~z70wang/research/ssim/#usage

On Measuring the Perceptual Quality of Video Streams over Lossy Wireless Networks

Ilias Politis, Michail Tsagkaropoulos, Tasos Dagiuklas, and Lampros Dounis

Department of Telecommunications Systems and Networks,
Technological Educational Institute of Messolonghi, Nafpaktos 30300, Greece
{ipolitis,mtsagaro}@ece.upatras.gr, ntan@teimes.gr,
dounis@gmail.com

Abstract. This paper studies the perceptual quality of video streams over lossy wireless networks. The focus is on investigating the impact on the perceived video quality of both physical error impairments and packet losses due to network congestion, by using objective and subjective evaluation methods. Extensive video quality assessments have shown that packet losses due to congestion are more severe than packet losses due to the physical error on the objective video quality. Furthermore, the comparison of MOS among different spatial resolution video sequences of the same bit rate indicates that a better perceived video quality can be achieved for lower resolution when the network is characterized by both high BER and network load.

Keywords: Simultaneous Double Stimulus for Continuous Evaluation method (SDSCE), Mean Opinion Score (MOS), Peak Signal to Noise Ratio (PSNR), video quality estimation (VQE), video streaming, wireless networks.

1 Introduction

Due to rapid growth of wireless communications, multimedia applications such as video conferencing, digital video broadcasting (DVB), streaming video, and audio over networks are becoming increasingly popular. In the recent years, this progress has also been aided by the proliferation of technologies such as IEEE 802.11X, 3G, LTE and WiMAX, and the trend has been to allocate these services more and more on mobile users. Mobile video delivery across heterogeneous wireless networks poses many challenges, including the issue of coping with losses due to both physical impairments and network congestion, as well as, maintenance of QoS and session continuity.

Wireless channels are prone to errors due to fading and interference effects resulting in error bursts. The bursty error characteristics of the wireless channel can be modeled by Gilbert-Elliot model using a discrete two state Markov Chain. The original two-state Markov model considers one "Good" and one "Bad" state where no errors occur in the "Good" State. The above model has been enhanced where errors occur in "Good" State. There are many authors that have studied the impact of physical error characteristics on the video quality [1] and [2], however most of these

L. Atzori, J. Delgado, and D. Giusto (Eds.): MOBIMEDIA 2011, LNICST 79, pp. 46–54, 2012.

research works focused on the monitoring of the distortion due to network impairments based on objective quality evaluation methods.

The deployment of real-time multimedia applications over wireless networks proved that the simple network parameters like bandwidth, loss, delay, jitter, etc. are inadequate to assess accurately the perceived by the human viewer, quality of service. As mobile users expect high perceptual quality that depends not only on technical parameters, but also on user experience. The operators need to control resources and at the same time maintain user satisfaction. To this end, several objective and subjective video quality assessment (VQA) methodologies have been used to evaluate video quality that can be introduced at any stage in the end-to-end video delivery system, including coding distortion, network impairments (congestion/packet loss, physical impairments) and decoding process (i.e. error concealment). Objective VQA methods to calculate video distortion in terms of parameters such as MSE, PSNR and SSIM have been studied extensively by [3]. On the other hand, subjective QoS measurements evaluate video as perceived by users, i.e., what is their opinion on the quality of particular audio/ video sequences, have been extensively studied by [4], [5] and described in ITU-T recommendation BT 500-11 [6].

As opposed to studies that consider the impact of physical layer and network layer impairments on the video quality, separately, this paper focus is on examining through extensive VQA tests (40 subjects-evaluators both experts in video QoS and non-experts, three different video sequences) how packet losses due to network congestion and physical channel errors, affect the perceived video quality using both objective and subjective evaluation methodologies. This methodology has been applied for different spatial resolution video sequences.

The rest of the paper is organized as follows. Section 2 outlines the wireless model that is used during the simulations. In Section 3 the simulation setup is described, while Section 4 includes an analysis of the selected objective and subjective methods used during the VQA procedure. The quality assessment scores are presented and discussed in Section 5 and Section 6 concludes the paper.

2   Modeling Wireless Error

Apart from losses that are due to network congestion, we are interested in studying the impact of physical impairments on the perceived video quality. The classical two-state Gilbert–Elliott model for bursty noisy channels [7],[8] has been extensively studied by many researchers [9]-[12]. In [9], a finite-state Markov channel is presented for packet transmission where the received instantaneous Signal-to-Noise Ratio (SNR) is partitioned into K disjoint intervals. The channel is in state k when the SNR takes a value within the kth interval. Clearly, each state is characterized by a different BER in PHY layer. In this paper, the Rayleigh fading channel is reflected by a two-state Markov model. A low mobility scenario is assumed (5 Km/h) [13] and the SNR threshold has been used in order to determine the steady state probabilities, the average BERs within each state and the transitional probabilities between the two states as it is shown in the following Table 1. In the rest of the paper bad, medium and good channel qualities will be referred to as bad, medium and low physical channel BER, accordingly.

Table 1. Physical channel simulation parameters

	Bad Channel Quality	Medium Channel Quality	Good Channel Quality
SNR (Threshold)	25	30	35
BER^B	$1.29\ 10^{-2}$	$1.29 10^{-2}$	$1.25\ 10^{-2}$
BER^G	$4.1\ 10^{-12}$	$1.3\ 10^{-13}$	$4.13\ 10^{-14}$
$P_{G->B}$	0.013	0.007	0.004
$P_{B->G}$	0.198	0.360	0.664
$P_{B \to B}$	0.802	0.64	0.336
$P_{G \to G}$	0.987	0.993	0.996

In Table 1, BER^B and BER^G refer to the bit error rate at the BAD and the GOOD states of the two-state GE model, accordingly. Obviously, BER^B is always larger than BER^G. In order to simulate three different wireless channel qualities (BAD, MEDIUM, GOOD), the values of BER^B and BER^G have been selected in such a way that when the channel is at a GOOD state the BER^B and BER^G are assigned their lowest values. Moreover, Table 1 includes the transition probabilities between the two states of the GE model.

3   Simulation Setup

In the process of video quality evaluation (VQE), three high quality uncompressed video sequences have been used named "highway", "deadline" and "paris", which are freely available by PictureTel at [15]. Both YUV 4:2:0 color CIF and QCIF spatial resolution of the three video sequences were used at a frame rate of 30fps. All video sequences have been compressed by the H.264/AVC reference encoder (JM12) available at [16]. The encoding configuration parameters include a GOP size of 12 frames (GOP has the form of IPP...I) a number of 5 reference frames and different QP values for every encoded video sequence in order to achieve the same (or almost the same) average encoded video bit rate among the CIF and QCIF resolutions of each video sequence.Table 2 summarizes the video sequences characteristics.

Table 2. Video sequence characteristics

Video Sequence	CIF (352×288)	QCIF(176×144)
Highway	700kbps@30Hz, QP=12	700kbps@30Hz, QP=6
Deadline	800kbps@30Hz, QP=12	800kbps@30Hz, QP=4
Paris	1.1Mbps@30Hz, QP=12	900kbps@30Hz, QP=2

A unicast H.264 video transmission (one video server and one video client) is simulated and a single NAL unit packetization scheme (one RTP packet – one NAL unit) is adapted with an RTP packet size of 1024 bytes (payload). The generated video packets are delivered through the simulated wireless network.

A NS-2 based simulation environment with the appropriate extensions for simulating 802.11b WLANs is adopted [17]. Additionally to the video server, a second server generates background traffic at Constant Bit Rate (CBR) over UDP in order to overload the simulated 802.11b network. The background CBR traffic is transmitted at three different transmission rates 2.5Mbps, 3.75Mbps and 4.5Mbps that correspond to 50%, 75% and 90% network load respectively.

4   Video Quality Evaluation

The aim of this study is to measure the perceived video quality of video streams over error prone wireless channels, using both objective and subjective video quality assessment (VQA) methods. In particularly, emphasis has been given on estimating the impact of both physical errors due to physical impairments and packet losses due to congestion, on the perceived video quality. Moreover, focus is given also on comparing the quality assessments of CIF and QCIF spatial resolution, under the same network conditions. The aim is to provide evidence that although the perceived video QoS under a lossless transmission environment is significantly better when the video sequence resolution is CIF instead of QCIF, in case of severe network conditions with high BER and network loads, the perceived QoS of a QCIF video sequence can be significantly better compared with the perceived QoS of the same video sequence at CIF resolution. To this end all the test video sequence have been encoded at the same (or similar) bit rate, thus the impact of background traffic will be almost the same in every case.

4.1   Objective Evaluation

As objective video quality evaluation method the Peak Signal to Noise Ratio (PSNR) is selected, since it is the most common, widely used by the research community and simple objective VQE scheme. In short PSNR is the ratio of the maximum (peak) power of the signal over the power of the signal's noise. In order to calculate the PSNR of a video sequence at the receiver, first Mean Square Error (MSE) between the original frame $F(i, j)$ and the distorted frame $F'(i, j)$ needs to be defined as:

$$MSE = \frac{1}{MN} \sum_{i=1}^{M} \sum_{j=1}^{N} |F(i,j) - F'(i,j)|^2$$

where, every video frame consists of $M \times N$ pixels. PSNR is then defined as the logarithm of the ratio of the peak signal value over the MSE due to noise. This also implies that PSNR calculation requires full video reference (i.e. reconstructed frame at the receiver), hence its use is limited in real time applications. Without loss of generality, in this study the distortion introduced to the video at the receiver is considered to be due to physical errors occurred at the wireless channel and packet losses due to congestion at the transport layer.

4.2   Subjective Evaluation

In the case of the subjective assessment of the perceived video quality, the tests have been carried out according to the ITU-T BT.500 recommendations for laboratory environments [6]. The simultaneous double stimulus for continuous evaluation method (SDSCE) has been preferred over the single stimulus schemes, since it is more appropriate for evaluating time varying degradations on the fidelity of visual information. According to SDSCE method, the original reference sequence and the test sequence are displayed simultaneously side by side. The subject is informed about the reference video (Stimulus A) and the distorted video (Stimulus B) and is allowed to evaluate continuously the test material in a scale from 0 (Bad) to 100 (Excellent) during the testing session. The votes from the voting bar are sampled every 0.5 sec, as described from the SDSCE in ITU BT-500.

In order to produce reliable and repeatable results, the tests have been conducted in a controlled testing environment provided by the Converged Networks and Services Group (CONES) of the TEI of Mesolonghi[1], Department of Telecommunication Systems and Networks, Greece. The facilities include high quality LCD displays, controlled light conditions and mid-gray background using appropriate curtains. During the video quality evaluation test 40 subjects were asked to evaluate the test videos. These subjects-evaluators include academic staff and students of the department. According to the subjective video assessment recommendations, the pool of the evaluators consists of a small number of experts on video quality, and the rest of them have no expertise in video evaluation. In accordance to the SDSCE specifications the test included three phases:

- a training and a demonstration phase – where subjects get familiar to the testing procedure and understand how to recognize artifacts.
- a pseudo-test phase – where selected represented conditions are shown with a different video sequence than the ones used for the test.

Moreover, to avoid subject's fatigue and decreasing level of attention, the test sessions lasted less than 30 minutes, including the training, demonstration and pseudo test phases. Since the entire set of test material presented as a single test session exceeds 30 minutes, multiple sessions were scheduled so that each subject could perform all sessions and rate all the test material.

Finally, the resulting scores need to be statistically processed before presented as final results. There is the need to remove subjects whose scores deviate from the scores of the other subjects, thus the technique of outliers detection was performed. The outlier detection refers to the detection and removal of scores in cases where the difference between mean subject vote and the mean vote for this test case from all other subjects exceeds 15%. This is a general rule that has been also used in other research works [14].

5   Processing of Results

In this section both objective and subjective evaluation results are illustrated and discussed. Since the aim is to identify the impact of physical errors and packet losses,

---

[1] http://www.teimes.tesyd.gr/cones

all video perceived QoS measurements for all six testing video sequences are compared against different bit error rates and network load conditions.

5.1   Objective Scores

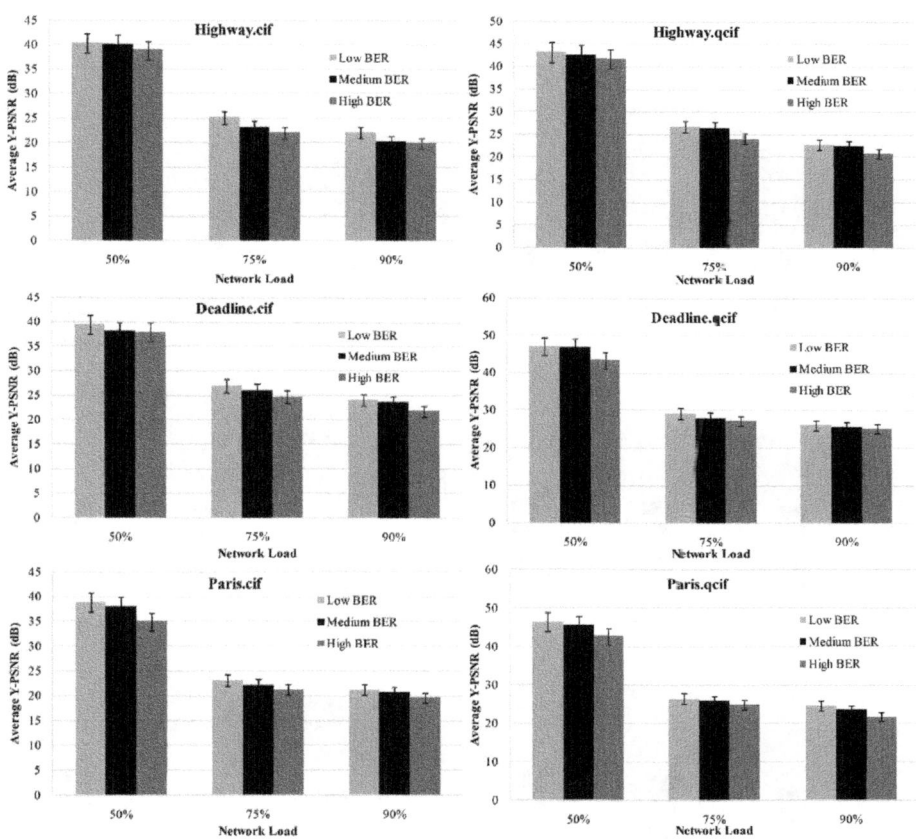

Fig. 1. PSNR measurements for CIF and QCIF spatial resolutions of the three test video sequences

The PSNR at the receiver for different network and channel conditions and a 5% error bar for each measurement, are illustrated in Figure 1. From the PSNR measurements, it can be seen that the increase in network load from 50% to 75% of the network capacity due to the background traffic, causes significant drop to the perceived video quality (e.g. from almost 40dB at load 50% to almost 25 dB at 75% load, in the case of "highway.cif"). A further increase of the network load from 75% to 90% of the network capacity results in a marginal drop of the average PSNR at the receiver (e.g. from 25dB at 75% load to 20dB at 90% in the case of "highway.cif"). Moreover, the impact of BER to the video quality at the receiver is limited mainly due to the fact the physical errors occur randomly and last only for short periods in time compared to packet losses from network congestion. In addition, the errors in the physical channel can be recovered using FEC mechanisms that are inherent at the physical and MAC layers. Similar conclusions can be derived from the QCIF measurements, as well. It must be mentioned that the average PSNR of the QCIF sequences measured at any physical and network conditions is higher than the corresponding average PSNR of the CIF sequences.

5.2   Subjective Scores

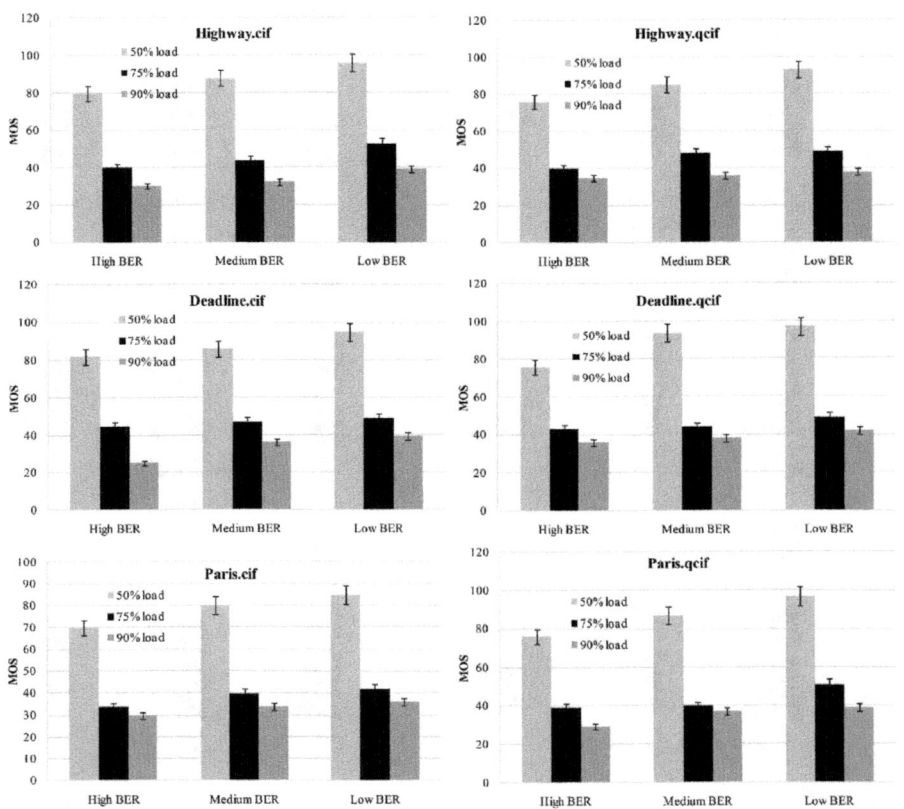

Fig. 2. MOS obtained with the SDSCE method for CIF and QCIF spatial resolutions of the three test video sequences

In Figure 2, the MOS obtained with the SDSCE video quality estimation method are shown. In particular, it is evident that the perceived video quality deteriorates fast as the network load increases, for the same channel conditions (BER). This MOS behavior is similar to the PSNR measurements, which means that the human viewer is more sensitive to the distortion that is introduced to the decoded video due to packet loss, rather than the distortion due to errors in the physical channel. Moreover, the comparison between MOS for CIF and QCIF, as shown in Figure 3, indicates that video streaming with lower spatial resolution may result in better-perceived QoS under high BER and increased network loads. However, this conclusion is not final as it also depends on the context of the video sequence, thus further investigation is required and more experiments to deeper understand the effect of visual context on perceived QoS are planned.

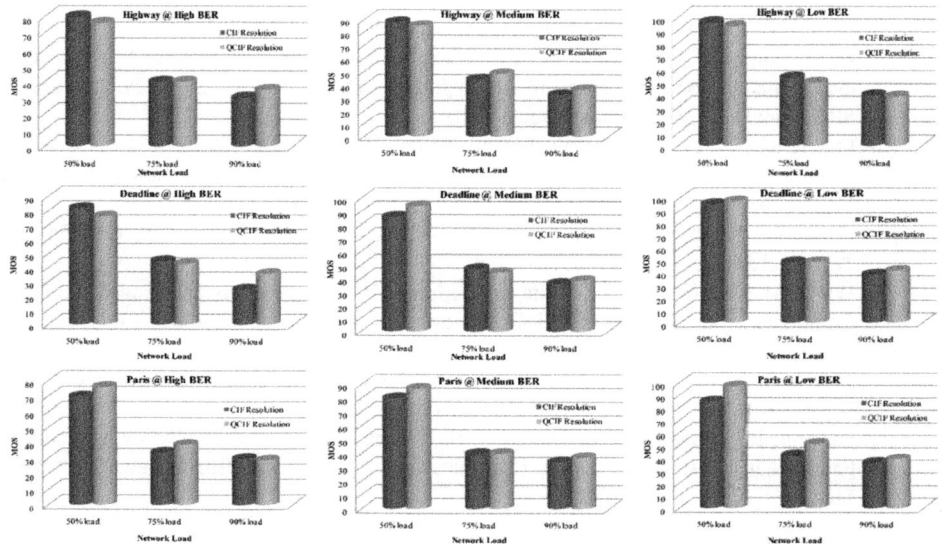

Fig. 3. Comparison of average MOS scores for CIF and QCIF sequences

6   Conclusions

This paper studies the impact of wireless physical channel impairments and packet losses due to network congestion on the perceived video quality of video streams with different spatial resolution. Extensive video quality assessments with objective measurements based on PSNR, as well as, subjective test according to the SDSCE method indicate that an increase of the BER has limited impact on the perceived video quality, as opposed to the impact on the QoS resulted by an increase of the network congestion. Moreover, the MOS comparisons among different network conditions indicate that better perceived video QoS can be achieved if lower spatial resolution video is transmitted over a network characterized by high BER and network load. Further experiments and real test-bed experiments are already undergoing, which will help to better understand the effect that specific visual context of the video sequence has to the video evaluator.

References

[1] Zhang, Q., Zhu, W., Zhang, Y.-Q.: End-to-End QoS for Video Delivery Over Wireless Internet. Proceedings of the IEEE 93(1), 123–134 (2005)
[2] Li, F., Liu, G.: Transmission Distortion Analysis for H.264 Video over Wireless Networks. In: 4th IEEE International Conference on Circuits and Systems for Communications, Shanghai, China, May 26-28, pp. 477–481 (2008)

[3] Chikkerur, S., Sundaram, V., Reisslein, M., Karam, L.J.: Objective Video Quality Assessment Methods:A Classification, Review, and Performance Comparison. IEEE Transactions onBroadcasting 57(2), 165–182 (2011)

[4] De Simone, F., Naccari, M., Tagliasacchi, M., Dufaux, F., Tubaro, S., Ebrahimi, T.: Subjective Quality Assessment of H.264/AVC Video Streaming with Packet Losses. EURASIP Journal on Image and Video Processing 2011, 1–12 (2011)

[5] De Simone, F., Goldmann, L., Lee, J.-S., Ebrahimi, T., Baroncini, V.: Subjective evaluation of next-generation video compression algorithms: a case study. In: Proc. of SPIE, San Diego, California (August 2010)

[6] ITU-R BT.500-11, Methodology for the Subjective Assessment (2002)

[7] Gilbert, E.N.: Capacity of a burst-noise channel. Bell Systems Technical Journal 39, 1253–1265 (1960)

[8] Elliot, E.: Estimates of error rates for codes on burst-noise channels. Bell Systems Technical Journal 42, 1977–1997 (1963)

[9] Wang, H.S., Moayeri, N.: Finite-state Markov Channel—A useful Model for Radio Communication Channels. IEEE Transactions on Vehicular Technology 44(1), 163–171 (1995)

[10] Zhang, Q., Kassam, S.: Finite-State Markov Model for Rayleigh Fading Channels. IEEE Transactions on Communications 47(11), 1688–1692 (1999)

[11] Sadeghi, P., Kennedy, R.A., Rapajic, P.B., Shams, R.: Finite-state Markov Modeling of Fading Channels: A Survey of Principles and Applications. IEEE Signal Processing Magazine 25(5), 57–80 (2008)

[12] Zhu, H., Karachontzitis, S., Toumpakaris, D.: Low Complexity Resource Allocation in Downlink Distributed Antenna Systems. IEEE Wireless Communications Magazine 17(3), 44–50 (2010)

[13] Aspelin, K.: (2005-05-25). Establishing Pedestrian Walking Speeds. Portland State University (retrieved August 24, 2009)

[14] Oelbaum, T., Schwarz, H., Wien, M., Wiegand, T.: Subjective performance evaluation of the SVC Extension of H.264/AVC. In: IEEE ICIP Conference, San Diego (October 2008)

[15] http://media.xiph.org/video/derf/

[16] H.264/AVC Software Coordination (2007), http://iphome.hhi.de/suehring/tml/

[17] The Network Simulator–NS-2, http://www.isi.edu/nsnam/ns/

The Correlation Dimension:
A Video Quality Measure

Bogdan Budescu, Alexandru Căliman, and Mihai Ivanovici

MIV Imaging Venture Laboratory, Department of Electronics and Computers,
Faculty of Electrical Engineering and Computer Science, Transilvania University, Str.
Politehnicii nr. 1, 500019 Brasov, Romania
{bbudescu,alexandru.caliman,mihai.ivanovici}@gmail.com
http://www.miv.ro

Abstract. Correlation dimension is a measure of the multidimensional complexity of an object. Stemming from the area of chaos theory and having several applications involving the study of the convergence and the recurring patterns of random signals, it has been proven to be a possible way to assess video quality. Based on its meaning in the multidimensional space of color fractals, it can be used, in the context of a fractal's intrinsic similarity to natural shapes and colours, to quantify the aesthetic and harmonic properties of an image. Our approach in the assessment of the perceived quality of a video stream is based on the analysis of the fractal dimension of video signals expressed in the CIE L*a*b* color space. This colour space has a strong resemblance to the human visual perception system, thus making its ΔE_{2000} norm relevant for the measurement of the perceptual difference between colours, and hence useful for image quality assessment. The fractal dimension is computed through the correlation dimension definition. In this paper we expose the experimental results obtained in a simulation of a real-life scenario: the streaming of a video of a football game over a busy network.

Keywords: Correlation dimension, colour fractal dimension, video quality.

1 Introduction

Human perceived quality of visual data is a complex metric influenced by subjective parameters aside from many objective aspects, like spatial frequency, topology and colour. In order to self-adjust the transmission parameters of digital video streams, an automatic assessment mechanism of the perceptual quality of images is required. In order to quantify this metric, there have been several approaches involving statistical observation of the perceived quality, but for the definition of a valid scale it was necessary to minimize the influence of the subjective factors in the study. The techniques that emerged were burdened, for instance, with the need to provide the same environment for every sampling session (see the ITU-T recommendations [1, 2, 3, 4] on video quality measurements regarding viewing distance and room lighting).

L. Atzori, J. Delgado, and D. Giusto (Eds.): MOBIMEDIA 2011, LNICST 79, pp. 55–64, 2012.
© Institute for Computer Sciences, Social Informatics and Telecommunications Engineering 2012

However, objective metrics have also been developed. They may be grouped into three categories, depending on the set of data available at the time of assessment. The full reference results of traditional signal processing methods, like RMSE [5] and PSNR, along with the classical Quality of Service metrics have the disadvantage that they are not highly correlated with the magnitude of distortion perceived by the human visual system. Full reference techniques that also take the specific characteristic of the human perception into consideration have also been defined [6] [7] leading to Quality of Experience.

The second category from the reference data availability point of view is represented by reduced reference metrics. These are based on the comparison of a set of synthesized characteristics or features of both the reference and distorted images, thus reducing the information necessary for degradation quantification. Carnec [8] employs a model of the human visual system (HVS) in order to detect points of interest from which features are extracted. A correspondence coefficient based on structural and color information (expressed in the Krauskopf perceptual color space) is then computed for the feature from both the distorted and the reference images. Cheng [9] proposes a reduced-reference metric based on natural image statistical prior computed on the gradient.

There are significantly fewer representatives of the third class of objective metrics, the ones that do not use any reference signal in the assessment process. Lu et al. [10] introduce a no-reference technique based on image structural information gathered after multiscale and multidirectional spectral decomposition.

We shall focus in the following on reduced-reference metrics and propose the usage of fractal measures for the assessment of video quality. The study of fractals have proven that many of the shapes and colours encountered in nature respect fractal-like patterns, providing this way a formal model for the perception of natural, aesthetic and harmonic visual data. We propose to take advantage of this model and to use its formalized metrics to quantify the perceptual quality of images.

The usage of the fractal dimension as a metric for video quality was proved in [11], the argumentation relies on the relationship between the complexity and the power spectrum of a signal. The degradation that affects the video signal is a mixture of several impairments, like blockiness and the sudden occurrence of new colours. The modifications of the image content reflect both in the colour histograms—a larger spread of the histogram due to the presence of new colours—and the spectral representation of the luminance and chrominance (new higher frequencies due to blockiness). The complexity of a continuous random signal can be defined based on its power density spectrum: for a random fractal signal $v(t)$, the power density function varies upon a power law in $\frac{1}{f^\beta}$, thus the Fourier transform $V(f, T)$ computed on a time interval T of $v(t)$ allows to express the spectral density function $S_V(f)$ as:

$$S_V(f) \propto T|V(f,T)|^2 \text{ as } T \to \infty \tag{1}$$

Therefore the intimate relationship between the power law and the fractal dimension [12] allows us to use the fractal dimension as a measure of signal complexity and, eventually of signal quality - based on the hypothesis that the degraded signal exhibits an increase of complexity.

2 Correlation Dimension Estimation

There are two important fractal metrics that are relevant in image quality assessment: fractal dimension and fractal lacunarity [11], and this article focuses on a new way to estimate the fractal dimension of color images. The fractal dimension indicates the complexity of a fractal set, by computing the fraction of its bounding area that it covers. Given the computational complexity of the reference theoretical Hausdorff dimension computation algorithm, implementations usually use more simple algorithms, like the box-counting algorithm, that provide the expected result (with notable exceptions, which are to be carefully taken into consideration) [13] [14] [15] [16].

First formalized in [17] and then applied in fractal theory in [18], correlation dimension is a measure of how much space is occupied by a set of random points. Although this measure has been defined in the context of chaos theory, its intimate connection to fractal dimension has provided an effective computation approach. The correlation integral $C(r)$ for a set of points $X_1, X_2, ..., X_N$ is defined in [18] as:

$$C(r) = \lim_{N \to \infty} \frac{2q}{N(N-1)} \tag{2}$$

where q is the number of pairs (i, j) whose distance $d(X_i, X_j)$ is less than r. The correlation integral is related to the standard correlation function, being the definite integral of it. In [18], the relationship between the Hausdorff dimension and the correlation dimension (ν) is proven, which for a continuous random vector process X, in certain conditions, is smaller than the Hausdorff dimension $dim_H X : 0 \leq \nu \leq dim_H X \leq d$. The theoretical Hausdorff dimension is not used in practice and it is approximated by various equivalent definitions in the discrete domain. For the definition of the Hausdorff dimension the reader is invited to consult [13].

In order to use the correlation dimension for the assessment of the complexity of a colour image, the definition 2 has to be extended to a 5-dimensional space and a colour representation should be chosen. In our approach, the extension to the colour domain is based on the ΔE_{2000} colour distance between the pixels of a colour video frame, which is the latest CIE standard: the CIE recommendations for colour distance evolved from the initial ΔE, to the ΔE_{94} and finally ΔE_{2000}. The equation of colour distance ΔE_{2000} is the following:

$$\Delta E_{2000} = \sqrt{\left(\frac{\Delta L'}{K_L S_L}\right)^2 + \left(\frac{\Delta C'}{K_C S_C}\right)^2 + \left(\frac{\Delta H'}{K_H S_H}\right)^2 + R_T \left(\frac{\Delta C'}{K_C S_C}\right)\left(\frac{\Delta H'}{K_H S_H}\right)} \tag{3}$$

The parameters K_L K_C and K_H weight the formula depending on the conditions of observation. The following terms were added in ΔE_{2000} in order to bring several corrections, and ultimately the ΔE_{2000} has a better behavior that suits the human vision than ΔE and ΔE_{94} for small colour differences:

- S_L: Compensation for lightness, corrects for the fact that Delta gives the predictions larger than the visual sensation for light or dark colours;
- S_C: Compensation for chroma, mitigates for the significant elongation of the ellipses with the chromaticity;
- S_H: Compensation for hue, that corrects for the magnification of ellipses with the chromaticity and hue;
- R_T: To take into account the rotation of the ellipses in the blue.

The steps of the algorithm for the estimation of colour correlation dimension involve the computation of colour distances histogram and the calculus of their cumulative density function, which is then represented in a log-log space. The resulted regression line's slope represents an estimate of the correlation dimension. The regression line is computed through Matlab `robustfit()` function which generates 9 regression lines (based on the flavor of the least square classical technique) from which we chose the mean slope. In [19] the authors concluded that the Andrews approach for the computation of the regression line led to the largest fractal dimension and implicitly to the largest range of values. However, the fractal dimension in that case was estimated using a box-counting definition.

3   Experiments

In order to study the relevance of fractal dimension computed by means of correlation dimension in the perceptual quality of images, we compared the measurements obtained from a video sequence transported over a network with zero loss to the measurements obtained from the received video sequence that has been transported over the same network, but in heavy load conditions in which a certain amount of packet loss also occurred.

3.1   Experimental Setup

We chose an MPEG-4 video streaming application - very sensitive to packet loss, due to the fact that neither UDP itself nor the video streaming application implement a retransmission mechanism: any lost packet in the network will cause missing bits of information in the MPEG video stream. In addition, the generated traffic is inelastic, because the transmission rate is not adapted to network conditions in any way. Therefore, the packet loss is the major issue for an MPEG-4 video streaming application.

In our experiments the induced packet loss percentage varied from 0% to 1.3% - the maximum amount of loss for which the application still functions and tests cannot be performed. We used the Helix streaming server from Real Networks (http://www.realnetworks.com) as MPEG-4 streaming server and the MPEG-4 client was mpeg4ip (http://mpeg4ip.sourceforge.net). The source code of the client was modified in order to record the received video sequence as individual frames in bitmap format. The test sequence we used was the widely-used 10-second long video test sequence football (250 frames, of 320×240 pixel each). The average transmission rate was approximately 1 Mb/s, which was a constraint imposed by the use of a trial version of the MPEG-4 video streaming server; however it represents a realistic scenario. In the next section the results for a video stream with an induced 1% normally distributed packet loss are presented.

3.2   Experimental Results

In order to quantify the perceived colour diversity of the frames in the original and altered video streams we computed the histograms of ΔE_{2000} distance in the CIELAB colour space between all pairs of different pixels within each frame. We have observed that due to the high amount of irregularities and colours generated artificially by erroneous estimations determined by the lack of data (lost in transmission), the histograms of affected frames expose a much wider distribution than the ones of unaltered frames. Fig. 2, for instance, illustrates the histograms of one original video frame of the unaltered stream (Fig. 1(a)) and the corresponding degraded video frame of the altered stream (Fig. 1(b)).

(a) Original video frame   (b) Degraded video frame

Fig. 1. Corresponding Original and Degraded Frames in Unaltered/Altered Video Streams

The cumulative distribution functions of the ΔE_{2000} distance in L*a*b* space encountered within the selected frames expressed in a logarithmic reference system are shown in Figure 3. The robust multilinear regressions of the CDFs are computed through the implementation provided by Mathworks' MATLAB. By default, the algorithm uses iteratively reweighted least squares with a bisquare weighting function, but for precision purposes we experimented with the 9 available weighting functions: Andrews, bisquare(default), Cauchy, fair, Huber, logistic, ordinary least squares (no weighting function), Talwar and Welsch. We also computed the average slopes of the regression lines obtained by using these weighting functions, which we used in further statistical analysis. Table 1 shows the regression slopes for the above mentioned frames. One can notice that for the degraded video frame, clearly exhibiting an increase in complexity, the average estimated correlation dimension is larger than the one of the corresponding original video frame.

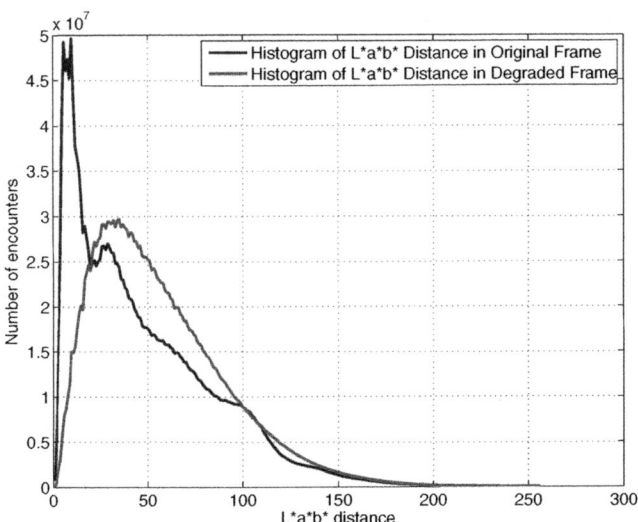

Fig. 2. Histograms of inter-pixel ΔE_{2000} Distance in L*a*b* Space of Corresponding Frames in Unaltered/Altered Video Streams

Table 1. Regression slopes of CDF of ΔE_{2000} distance in L*a*b* space computed in logarithmic reference system (i.e. Correlation Dimension) using several weighting functions

	Original Video Frame	Degraded Video Frame
Andrews	0.42511	0.39097
bisquare	0.42615	0.39672
Cauchy	0.47903	0.76571
Fair	0.54682	0.9248
Huber	0.52227	0.87212
Logistic	0.52217	0.87241
OLS	0.73887	1.1498
Talwar	0.45088	0.46192
Welsch	0.43661	0.47543
Average	**0.50532**	**0.7011**
σ	**0.098689705**	**0.2763793**

The average slope of the regression line (i.e. the correlation dimension), along with its variation by computation method is illustrated in Figure 4, for both the original (Fig. 4(a)) and the altered (Fig. 4(b)) video streams. The average is represented with a red line and the variation of the correlation dimension with green. One can notice the intervals for which both the larger average correlation dimension and its larger variance indicate the degradation of the video stream, e.g. around frames #50, #130 and #150.

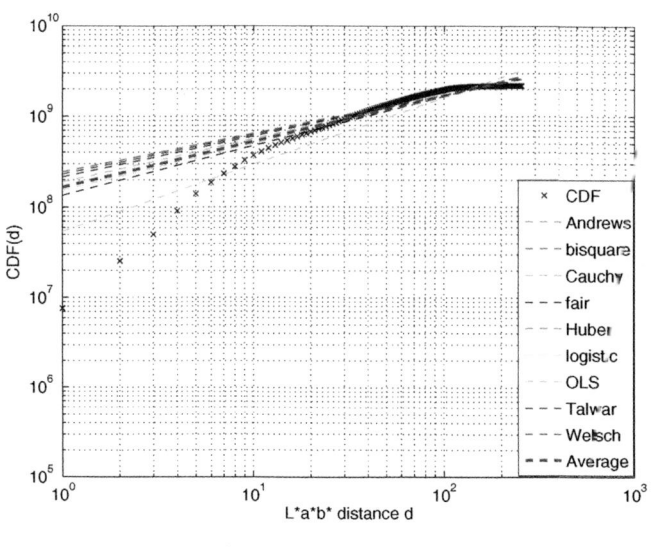

(a) Original video frame

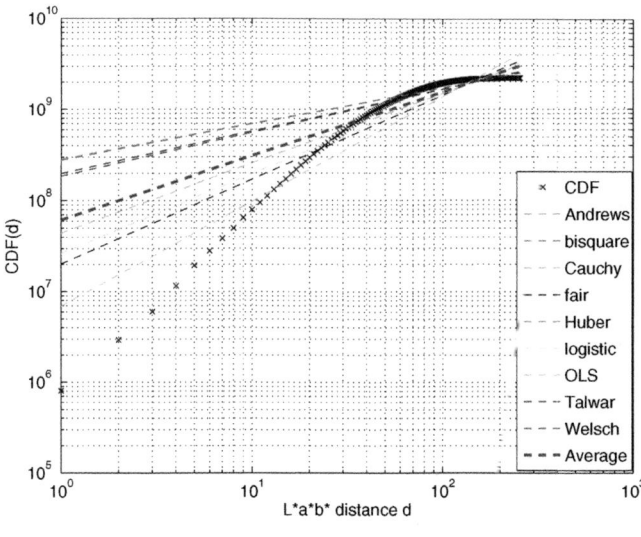

(b) Degraded video frame

Fig. 3. CDFs of inter-pixel ΔE_{2000} Distance in L*a*b* Space of Corresponding Frames in Unaltered/Altered Video Streams

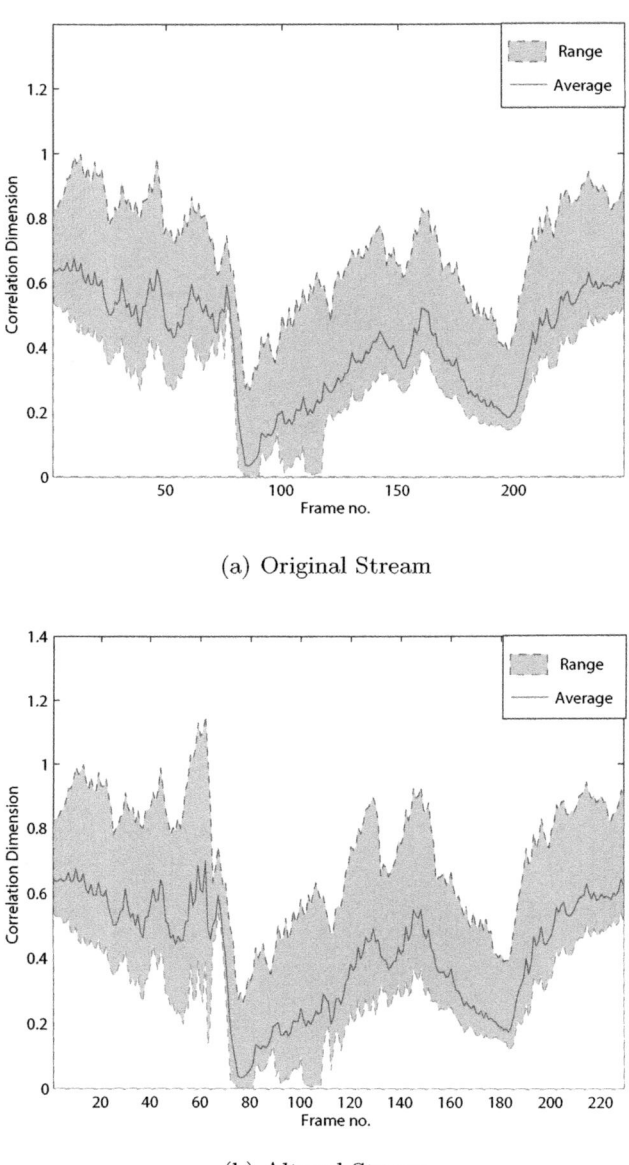

(a) Original Stream

(b) Altered Stream

Fig. 4. Regression Slopes of CDFs of inter-pixel ΔE_{2000} Distance in L*a*b* Space between pixels within correspondent frames of Original/Degraded Video Streams

4   Conclusions

We presented in this article an extension of the correlation dimension to the colour domain, thus we are able to estimate the fractal complexity of a colour image. As a possible application, in the assessment of the video quality, we present the results we obtained for a video sequence, in the case of an MPEG-4 video streaming application. We conclude that the colour correlation dimension can be used for the objective video quality assessment in a reduced-reference scenario. However, the real-time assessment of video quality through fractal metrics remains an open question, given the not-neglectable algorithm complexity. A possible solution we envisage is the use of GPU for the acceleration of the approach.

Acknowledgement. This work and the paper were partially supported by the Sectoral Operational Programme Human Resources Development (SOP HRD), ID76945 financed from the European Social Fund and by the Romanian Government.

References

[1] ITU-R Recommendation BT.500. Subjective quality assessment methods of television pictures. International Telecommunications Union (1998)

[2] ITU-T Recommendation P.910. Subjective video quality assessment methods for multimedia applications. International Telecommunications Union (1996)

[3] ITU-R Recommendation J.140. Subjective assessment of picture quality in digital cable television systems. International Telecommunications Union (1998)

[4] ITU-T Recommendation J.143. User requirements for objective perceptual video quality measurements in digital cable television. International Telecommunications Union (2000)

[5] Wang, Z., Bovik, A.C.: Mean squared errorl: Love it or leave it? IEEE Signal Processing Magazine, 98–117 (January 2009)

[6] Wang, Z., Bovik, A.C., Sheikh, H.R., Simoncelli, E.P.: Image quality assessment: From error visibility to structural similarity. IEEE Transactions on Image Processing 13(4), 600–612 (2004)

[7] Sampat, M.H., et al.: Complex wavelet structural similariy: A new image similarity index. IEEE Transactions on Image Processing 18(11), 2385–2401 (2009)

[8] Carnec, M., Le Callet, P., Barba, D.: Objective quality assessment of color images based on a generic perceptual reduced reference. Image Commun. 23, 239–256 (2008) ISSN: 0923-5965,
http://portal.acm.org/citation.cfm?id=1371260.1371548

[9] Cheng, G., et al.: Image quality assessment using natural image statistics in gradient domain. AEU - International Journal of Electronics and Communications 65(5), 392 (2011) ISSN: 1434-8411,
http://www.sciencedirect.com/science/article/pii/S143484111000172X

[10] Lu, W., et al.: No-reference image quality assessment in contourlet domain. Neurocomput. 73, 784–794 (2010) ISSN: 0925-2312,
http://dx.doi.org/10.1016/j.neucom.2009.10.012

[11] Ivanovici, M., Richard, N., Fernandez-Maloigne, C.: Towards video quality metrics based on colour fractal geometry. EURASIP Journal on Image and Video Processing (2010) (2010)

[12] Voss, R.: Random fractals: characterization and measurement. Scaling phenomena in disordered systems 10(1), 51–61 (1986)

[13] Falconer, K.: Fractal Geometry, mathematical foundations and applications. John Wiley and Sons (1990); ISBN: 0-471-9228-70

[14] Allain, C., Cloitre, M.: Characterizing the lacunarity of random and deterministic fractal sets. Physical Review A 44(6), 3552–3558 (1991)

[15] Plotnick, R.E., Garnder, R.H., Hargrove, W.H., Prestgaard, K., Perlmutter, M.: Lacunarity analysis: a general technique for the analysis of spatial patterns. Physical Review E 53(3), 5461–5468 (1996)

[16] Maragos, P., Sun, F.K.: Measuring the fractal dimension of signals: morphological covers and iterative optimization. IEEE Transactions on Signal Processing 41(1), 108–121 (1993)

[17] Grassberger, P., Procaccia, I.: Measuring the strangeness of strange attractors. Physica D: Nonlinear Phenomena 9(1-2), 189–209 (1983)

[18] Bardet, J.-M.: Dimension de corrélation locale et dimension de haussdorf des processus vectoriels continus / local correlation dimension and hausdorff dimension of continuous random fields. Comptes Rendus de l Acadmie des Sciences - Series I - Mathematics 326(5), 589–594 (1998)

[19] Ivanovici, M., Richard, N.: Fractal dimension of colour fractal images. IEEE Transactions on Image Processing 20(1), 227–235 (2011)

TV White Spaces Exploitation for Signal Distribution

Mauro Fadda[1], Maurizio Murroni[1], Vlad Popescu[1], and Vlad Cristian Stoianovici[2]

[1] Department of Electrical and Electronic Engineering,
University of Cagliari, Piazza d'Armi, 09123 Cagliari, Italy
{mauro.fadda,murroni,vlad.popescu}@diee.unica.it
[2] Department of Electronics and Computers,
University of Transilvania, Str. Politehnicii 1-3 500019 Brasov, Romania
stoianovici.vlad@unitbv.ro

Abstract. The new spectrum regulation policies for dynamic spectrum access, especially those concerning the use of the white spaces in the Digital Terrestrial Television (DTT) bands, arise the need for fast and reliable signal identification and classification methods. In this paper we present a two-stage identification method for signals in the white spaces, using combined energy detection and feature detection. The band of interest is divided by means of the Discrete Wavelet Packet Transformation (DWPT) in sub-bands where the signal power is calculated. Modulation classifiers taking into account the statistical parameters of the signal in the wavelet domain are used as features for identifying the modulation schemes, in this case specifically for the DVB-T broadcast standard. As a possible application we are considering an indoor short-range distribution system for video signals.

Keywords: Signal processing for transmission, Dynamic spectrum access, TV white spaces, spectrum sensing.

1   Introduction

Within the cognitive radio paradigm, as a highly praised alternative for overcoming the inherent limitations of the RF spectrum, the current worldwide situation of the VHF and UHF TV channels is an excellent application scenario. In the US, the complete switchover to digital television in 2009 opened an entire new topic of the usability of the so-called TV white spaces (TVWS) for short-range wireless consumer devices. Moreover, the gradual global passage to digital television poses new specific challenges to the white spaces detection.

Within this topic, spectrum sensing for DTT broadcasting signals plays a crucial role, along with geolocation databases [1] (GL-DBs). In the US, the FCC has already commissioned the creation of geo-location databases, free to access for any CR device. The database entries provide, for a certain location (geographical coordinates), the list of available channels and the allowable maximum effective isotropic radiated power (EIRP) useful to transmit without providing harmful interference [2]. Even if the GL-DBs are up-to date, the values provided for a specific geographical point are

L. Atzori, J. Delgado, and D. Giusto (Eds.): MOBIMEDIA 2011, LNICST 79, pp. 65–72, 2012.
© Institute for Computer Sciences, Social Informatics and Telecommunications Engineering 2012

still the results of applying a signal propagation models and estimated power levels. Due to this static approach, the provided data might be inaccurate for different reasons such as variable atmospheric conditions or multipath and fading phenomena [3, 4]. Therefore, there is still the need of a validation in terms of frequency occupancy and maximum EIRP of the free frequency channels provided by the GL-DBs, using specific spectrum sensing methods.

As known, spectrum sensing techniques mainly focus on primary transmitter detection and can be classified in three categories: matched filter, energy detection and signal feature detection [2]. Combinations of these methods are used for achieving good results in terms of sensitivity, computational time and signal classification, in so-called two-stage spectrum sensing schemes proposed initially in [3] and then refined in [4] and especially in [5]. The mentioned two-stage schemes perform coarse sensing based on energy detection, followed by a feature detection performed on the signals in the sub-bands declared free by the previous stage.

This work presents a different spectrum sensing approach in a two-stage scheme using the Discrete Wavelet Packet Transformation (DWPT) for dividing the analyzed frequency band and calculating the signal power in the resulting sub-bands (channels). The subbands identified as free can be used directly for transmission. The remaining subbands with a signal power higher than a pre-defined threshold are subsequently analyzed by the feature detector, for distinguishing between primary users (PU) and possible secondary users (SU). The feature detection method used in the second stage of the spectrum sensing exploits the statistical properties of the DWPT's coefficients.

The remainder of the paper is organized as follows: in Section 2, we first present the use of the DWPT for sub-band division and energy detection, and then we analyze the proposed feature detection method. Section 3 presents the initial software simulation, while section 4 shows the hardware set-up and the test results using real recorded signals. Finally, in Section 5 we draw the conclusions and present the future work.

2   Energy Detection and Signal Classification

This two-stage approach is based on the work proposed initially in [6]. We are performing an energy detection based on DWPT sub-bands analysis, considering an initial band centered on the region occupied by the TV channels. The band is divided by means of a wavelet decomposition tree into sub-bands with a bandwidth specific to the various DTT standards (from 6.5 to 8 MHz). We are calculating the power level of the received signal in the wavelet domain by summing the corresponding squared wavelet coefficients for each sub-band. The resulting values are compared to opportune threshold values [7] to mark the channels for the frequencies corresponding to the TV channels of interest as free ("white") or occupied.

As known [2], the drawback of the energy detection method is the reliability of the power level thresholds. Therefore, in the second stage of the spectrum sensing, for all the channels that previously were identified as not "white", implicitly having a signal power surpassing the noise threshold mentioned in [7], we estimate whether they are occupied by PUs or SUs using a modulation classifier.

The modulation types used by the DTT broadcasting systems are standard, so a feature-based classifier for the modulation schemes typical for terrestrial communications can be used to classify a possible modulated signal. The proposed scheme supports the classification of QPSK, 16QAM and 64QAM modulations, specific for the European DVB-T standard. Similar to the methods proposed in [8], we are starting from the normalized histogram generation of wavelet-transformed coefficient with N_i the samples in the particular process.

The first-order moment of the statistical process is the mean given by

$$\mu_1(x) = \sum_{i=0}^{N-1} x_i \, p(x_i) \ . \tag{1}$$

The second-order moment of the DWPT represents the variance, given by

$$\mu_2 = \frac{1}{N} \sum_{i=0}^{N-1} |c_i|^2 - \left[\frac{1}{N} \sum_{i=0}^{N-1} |c_i| \right]^2, \tag{2}$$

where c_i are the wavelet coefficients in each single sub-band.

The constellation type, circular (MPSK) or in quadrature (MQAM), can be detected by comparing the mean with a first threshold T_1, defined as

$$T_1 = \frac{\mu_1 MQAM \cdot \mu_2 MPSK + \mu_1 MPSK \cdot \mu_2 MQAM}{\mu_2 MQAM + \mu_2 MPSK}. \tag{3}$$

If the modulation is a MPSK we can compare the variance with a second threshold T_{p2} to find if it is a QPSK or a different PSK order modulation. If the modulation is a QAM we can detect if it is a 16QAM or a 64QAM with the variance and a third threshold T_{q4}.

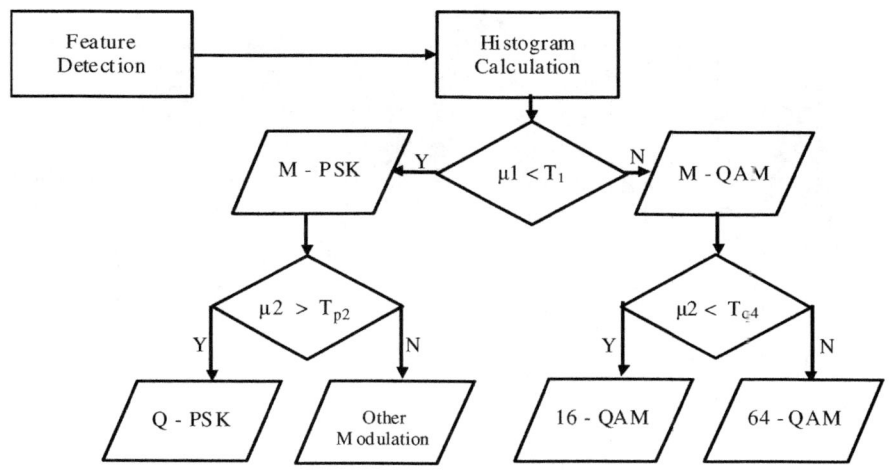

Fig. 1. Flowchart of the functionality of the proposed feature detection method

Opposed to the work presented in [9] we are considering the coefficients in each of the S sub-bands for calculating and using N/S coefficients from the original N signal samples.

Figure 1 presents the flowchart of the proposed feature detection method. If the channel is identified as being occupied by a PU, the corresponding channel is definitively marked as "black", meaning it is undoubtedly used by PUs and therefore not suitable for transmission. If the statistical analysis fails to identify a known type of modulation (QPSK, 16QAM, 64 QAM), we categorize the channel as being "grey", which means that there is no broadcaster transmitting, but still the channel is occupied, most probably by another SU. Therefore, the channel is not completely discarded, being a potential candidate to be analyzed again after a certain amount of time in order to be re-evaluated and eventually included in the white list.

The channels marked as "black" are not suitable for transmission and therefore, after the first energy and feature detection, we have to consider only the "grey" and "white" channels, thus reducing the number of operations and making the algorithm suitable for the use with real, live signals. Furthermore, the wavelet transformation has to be performed only once for both the two stages of the spectrum sensing scheme, the coefficients used for energy detection and signal classification being the same.

3   Software Simulations

The proposed spectrum sensing method is implemented using a two-stage algorithm. We perform an energy detection based on discrete wavelet packets sub-bands analysis followed by a feature detection stage as shown in section 2.

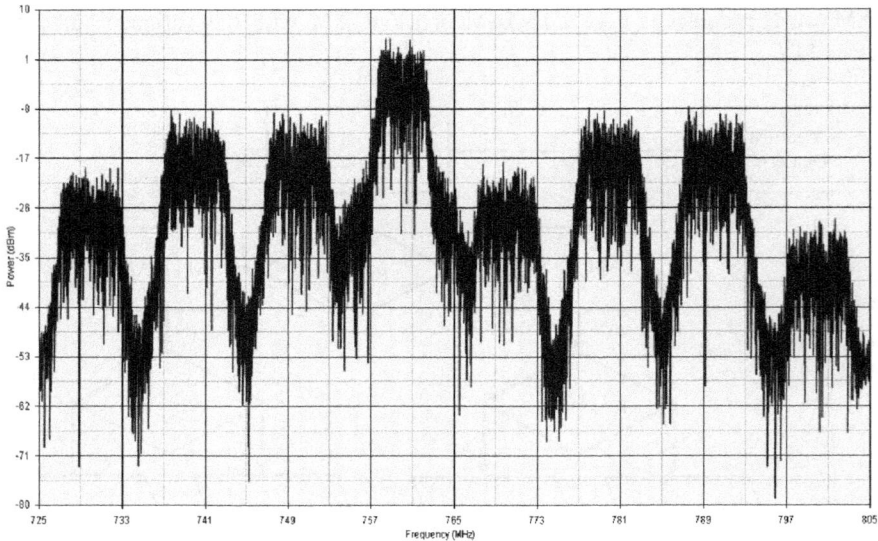

Fig. 2. Test signal composed by 8 DVB-T signals with different standard modulations

The algorithm was implemented in Simulink / Matlab using a 4-level DWPT block with 15th order Daubechies [9] wavelet filters.

The test signals for the system's functionality were generated using the Agilent SystemVue Software: 8 signals, spaced evenly at 8 MHz on 7.61 MHz - wide channels, with constant additive white Gaussian noise (AWGN), were modulated on carrier in the upper UHF band in order to simulate the behavior of a real terrestrial DVB-T system. Figure 2 depicts one of the test signals. Different modulation schemes were chosen even for adjacent signals in order to test the capacity of the system to correctly compute the thresholds values needed for the signal classification.

The first step consisted in the calculation of the threshold of the energy detection stage based also on the results from [7]. The next step implied calculating the reference thresholds for the modulation classifier based on the simulated input signals with known modulation type and signal-to-noise ratios.

Based on these thresholds, pre-set in the simulator, we tested the reliability of the proposed application scenario. A series of test consisting in changing the amplitude of the signals, their SNR and the modulation type has been performed. A total set of 200 different test signals has been fed to the simulator. For SNR values higher than 5 dB and specifically for the signals in the 700 MHz band, the proposed method equaled the best methods presented in the literature [8], [9]: for the typical DVB-T standard modulations the signal classifier had an identification percentage of 96.5 %.

4   Hardware Set-Up and Tests

After testing the functionality of the software implementation and calculating the appropriate threshold values for both the energy detector and the modulation classifier, we validated these results with real DVB-T signals acquired using RF hardware.

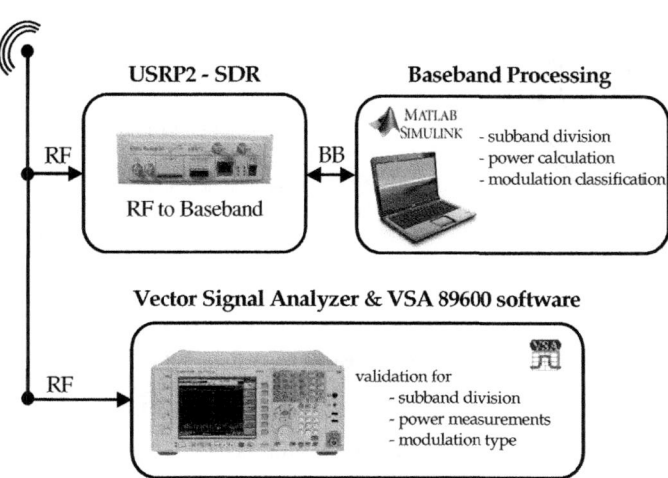

Fig. 3. Flowchart of the functionality of the proposed feature detection method

The hardware set-up consisted of USRP2 software radio boards equipped with WBX wideband daughterboards covering a spectrum range from 50 MHz to 2.2 GHz. The software radios were connected to a PC running a Simulink model that commands the RF hardware and implements the entire baseband processing. Appropriate antennas for the frequency band of interest were used during the tests. The measurements on the power of the DVB-T signals and on their modulation type were crosschecked using as reference the Agilent EXA9020A Vector Signal Analyzer, the 89600VSA software and the instrument's onboard software (figure 3).

A set of 50 real DVB-T signals, each one 30 seconds long, with different signal characteristics (modulation, symbol rate, FEC, SNR) were recorded to the Matlab environment using the RF hardware and an appropriate antenna. The signals were contemporary fed to the Agilent Vector Signal Analyzer for identifying their features.

The Simulink / Matlab set-up presented in the previous section was tested with the real recorded signals in terms of detection reliability. Figure 4 presents the receiver operating characteristics (ROCs) of our combined spectrum sensing approach, for both simulated and real DTT signals. While the system performed well for simulated signals with a SNR as low as 10 dB, we noticed a degradation of the detection curve for real signals. It can be seen that the ROC for a real signal with a 10 dB SNR is worse than the ROC curve for simulated signals with the same SNR. The discrepancy is due to the calculation of the initial thresholds values of the feature detection stage, done based on simulated signals.

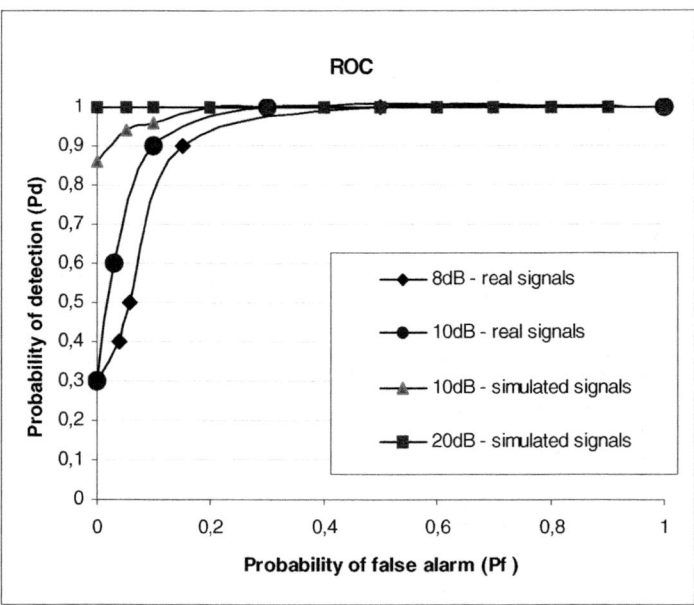

Fig. 4. Receiver operation characteristics for the proposed two-stage spectrum sensing method

5   Conclusions and Future Work

In this paper we propose a two-stage spectrum sensing approach for white spaces exploitation in the 470 - 790 MHz band, combining energy and feature detection methods in the wavelet domain, continuing the work initially presented in [6]. The current work is specifically adapted for PU signals compliant to the European DVB-T standard.

Different from other approaches we are looking at the energy detection and signal classification in a combined way, both at system level and at computational level. First we implemented the proposed sensing scheme in a fully software simulator and calculated with these simulated signals the thresholds for both energy detection and the signal classification stages. The simulations validated our design for the next step, i.e. the tests with real DVB-T signals. Consequently we tested our spectrum sensing approach in a functional system consisting of an RF hardware front-end (Ettus Research USRP2 Platform) connected to a computer running a Matlab / Simulink model. Real signals were acquired and fed offline to the computational model in order to test the system's behavior in real conditions. The receiver operation characteristics were slightly inferior to those obtained with the simulated signals.

The initial tests results and the obtained ROC curves showed us the need of improving the reliability of the sensing method for much lower SNR. A first step will be the calculation of new, more accurate threshold values using a large set of real DVB-T signals with various characteristics. Further steps will include also tests with different wavelet filters.

The hardware set-up presented in figure 3 will be extended to be used as an indoor short-range distribution system for video signals, with immediate implementations as home entertainment or infotainment systems. A central device (implemented by the USRP2 SDR platform), performing as server, can distribute video signals to clients in different areas of a building (e.g. apartment or hotel rooms). The aim is to implement a full cognitive transmission system to be deployed in the 470-790 MHz bands.

Acknowledgments. This paper is supported by the Sectoral Operational Programme Human Resources Development (SOP HRD), financed from the European Social Fund and by the Romanian Government under the contract number POSDRU/6/1.5/S/6 and by the Young Researchers Grant funded by Region of Sardinia, PO Sardegna FSE 2007/2013, L.R.7/2007, "Promotion of scientific research and technological innovation in Sardinia".

References

[1]   CEPT ECC Reprot 159 - Technical And Operational Requirements for the Possible Operation of Cognitive Radio Systems in the 'White Spaces' of the Frequency Band 470-790 Mhz (October 2010)

[2]   Budiarjo, I., Lakshmanan, M.K., Nikookar, H.: Cognitive Radio Dynamic Access Techniques. Wireless Personal Communications 45(3) (May 2008)

[3] Benko, J., Cheong, Y.C., Cordeiro, C., et al.: A PHY/MAC Proposal for IEEE 802.22 WRAN Systems Part 1: The PHY. IEEE 802.22-06/0004r1 (February 2006)

[4] Yue, W., Zheng, B.: A Two-Stage Spectrum Sensing Technique in Cognitive Radio Systems Based on Combining Energy Detection and One-Order Cyclostationary Feature Detection. In: Proceedings of the 2009 International Symposium on Web Information Systems and Applications (WISA 2009), Nanchang, P. R. China, May 22-24, pp. 327–330 (2009)

[5] Maleki, S., Pandharipande, A., Leus, G.: Two-Stage Spectrum Sensing for Cognitive Radios. In: Proceedings of the ICASSP 2010 Conference, Dallas, USA (March 2010)

[6] Fadda, M., Murroni, M., Popescu, V.: Spectrum Sensing in the DVB-T Bands using Combined Energy Detection and Signal Classification in the Wavelet Domain. In: Proceedings of the 13th International Symposium on Wireless Personal Multimedia Communications (WPMC), Recife, Brazil (October 2010); ISBN: 978-85-60307-02-9

[7] Yoon, Y., Leon, H., Lung, H., Lee, H.: Discrete wavelet packet transform based energy detector for cognitive radios. In: IEEE Vehicular Technology 2007, Dublin, Ireland (2007)

[8] Prakasam, P., Madheswaran, M.: M-ary Shift Keying Modulation Scheme Identification Algorithm Using Wavelet Transform and Higher Order Statistical Moments. Journal of Applied Science 8(1), 112–119 (2008)

[9] Linfoot, S.: A Study of Different Wavelets in Orthogonal Wavelet Division Multiplex for DVB-T. IEEE Transactions on Consumer Electronics 54(3) (August 2008)

A Spectrum Sensing Algorithm for White Spaces Detection Validated in Real Environments

Irati Lázaro, Maurizio Murroni, Iratxe Redondo, Mikel Sánchez, and Manuel Vélez

Escuela Técnica Superior de Ingeniería de Bilbao,
Alameda Urquijo s/n, 48013 Bilbao, Spain
{ilazaro006,iredondo010}@ikasle.ehu.es, murroni@diee.unica.it,
mikel.sanchez@tecnalia.com, manuel.velez@ehu.es

Abstract. Cognitive Radio Systems have been proposed as the solution to spectrum scarcity, and Spectrum Sensing a good way to detect which frequencies are being used by primary users and avoid interferences. When primary signals are unknown, energy detection is the best while easiest technique for the sensing process. In this paper, we consider energy detection based spectrum sensing for narrowband signals in the TV wideband. Simulations are performed to obtain ROC curves. Designed detector has been validated both with signals generated in the laboratory and with real signals captured from the radio space.

Keywords: Energy detection, Spectrum Sensing, Cognitive Radio.

1 Introduction

Nowadays, due to spectrum scarcity, there is a big interest in a more efficient use of the spectrum by using Cognitive Radio systems. Spectrum Sensing techniques provide very useful information about spectrum occupancy and the presence of white spaces.

Many spectrum sensing methods have been proposed and theoretically analyzed in the literature. Matched filter has demonstrated good performance when primary user's signal is known [1]. Cyclostationary detection is based on the cyclic characteristics of primary signals to obtain good results [2]. Energy detection is a very used technique to detect the presence of unknown primary signals [3]. It is very flexible because it can detect many types of signal and it is not necessary to have any knowledge about them.

Up to now, there have been many theoretical studies of algorithms for finding spectrum holes, but it arises necessary to validate the good operation of this kind of algorithm with signals received in real environments and to obtain empirical results.

The main goal of this project is to develop an energy detection algorithm on the Matlab environment and to validate it with real signals. The algorithm is valid for detecting signals of different bandwidths along TV wideband.

L. Atzori, J. Delgado, and D. Giusto (Eds.): MOBIMEDIA 2011, LNICST 79, pp. 73–77, 2012.
© Institute for Computer Sciences, Social Informatics and Telecommunications Engineering 2012

2   Energy Detector

In order to detect the presence of narrowband signals in a wideband, frequency domain energy detector has been developed. Consequently detection process results as follows [4]:

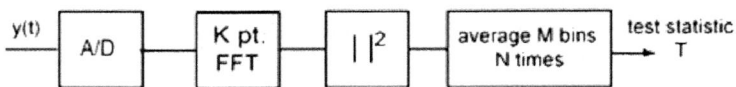

Fig. 1. Block diagram of energy detector in frequency domain

After digitizing, the time domain signal $y[n]$ is transformed into the frequency domain signal $Y[k]$ by applying a K-point FFT. The number of points of the FFT (K) is a function of the desired frequency resolution.

The detection is based on the test of two following hypotheses:

$$H_0: \; Y[k] = W[k] \qquad \text{signal absent.}$$
$$H_1: \; Y[k] = X[k] + W[k] \quad \text{signal present.} \tag{1}$$
$$k = 1 \dots K; \text{ where K is the number of points of the FFT.}$$

$W[k]$ is additive white Gaussian noise with zero mean and variance σ_w^2. $X[k]$ can also be considered as a Gaussian distributed variable with zero mean and variance equal to signal average power P_x.

To measure energy, M continuous K-point FFTs are done and, then, power is averaged over each frequency bin. The test statistic is:

$$T[k] = \frac{1}{M} \sum_{m=1}^{M} \left| Y_m[k] \right|^2 . \tag{2}$$

where $Y_m[k]$ is the m-th FFT of $y[n]$.

The decision test consists on deciding that primary signal is present at bin k when $T[k]$ is greater than a threshold γ. Otherwise, it should decide that primary signal is absent.

Detection probability (P_d) and false alarm probability (P_{fa}) are given by:

$$P_{fa} = Q\left(\sqrt{M} \, \frac{\gamma - \sigma_w^2}{\sigma_w^2} \right) . \qquad P_d = Q\left(\sqrt{M} \, \frac{\gamma - (\sigma_w^2 + P_x)}{(\sigma_w^2 + P_x)} \right) . \tag{3}$$

For a desired false alarm probability, threshold value (γ) can be set without the knowledge of primary signal power by P_{fa} equation. Once threshold value is obtained, P_d can be calculated by substituting the threshold in P_d equation.

For fixed P_{fa} and P_d, it is possible to calculate the minimum number of K-point FFTs (M) required for achieving that pair of probabilities. Representing signal to noise ratio as $SNR = P_x / \sigma_w^2$:

$$M = \left[\left(Q^{-1}\left(P_{fa}\right) - Q^{-1}\left(P_d\right)\right) \cdot SNR^{-1} - Q^{-1}\left(P_d\right)\right]^2 . \qquad (4)$$

3   Experimental Results

Experiments have consisted of a comparison between measuring noise power by two different methods. In *method 1* it is measured on a previously known free channel and in *method 2* equipment's noise power is considered.

First, we have used signals generated in the laboratory with the Alitronika AT2780 modulator for generating a DVBT signal of 5 MHz bandwidth. Then, we have used the Anritsu MS2690A Vector Signal Analyzer in order to save the IQ samples with a sampling frequency of 50 Msamples/sec and an observation interval of 50 msec. We have added Gaussian noise to get a desired SNR. We have realized the minimum number of averages necessary to achieve a desired P_d and P_{fa} (equation (4)). By doing 5000 experiments we have obtained the results of figure 2 and 3. These results have been obtained for a desired $P_d = 0.9$, a desired $P_{fa} = 0.1$, spectral resolution of 0.2MHz and a SNR = -15 dB.

As seen in figures 2 and 3, it is possible to obtain good results by using only the minimum number of averages. It is appreciable that better performance is achievable when using noise power of a previously known free channel (method 1).

The algorithm has been validated with field trials carried out in Bilbao (Spain). Channels from 58 to 63 (766-814 MHz) have been recorded at different locations and with different antennas. Previously mentioned methods for estimating noise power have been compared.

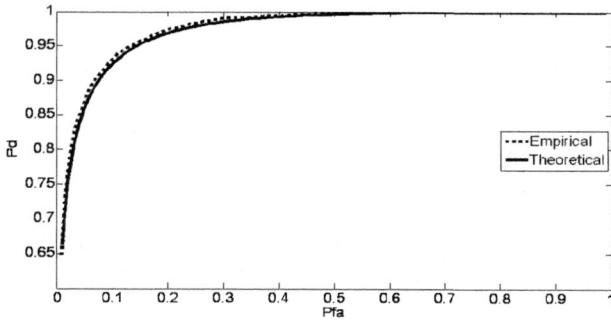

Fig. 2. ROC for method 1 (noise power in free channel)

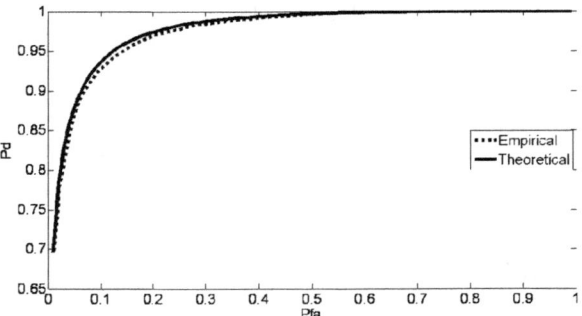

Fig. 3. ROC for method 2 (equipment's noise power)

Real TV broadcasting signals usually have high SNRs. Therefore, few averages are necessary to obtain good performance (high P_d and low P_{fa}). For example, for a *SNR* of 0 dB, a P_d of 0.99 and a P_{fa} of 0.01, only 49 averages are necessary (by equation (4)). In this case, central limit theorem (CLT) [5] can't be applied. Hence, we have realized the minimum number of averages to satisfy the CLT, $M = 125$. In that minimum sensing time the algorithm has decided that the frequency is occupied or free. After realizing this process many times, a percentage of occupancy has been obtained.

In figure 4, a comparison between both methods of measuring noise power can be observed, for one location where the primary signal has been captured with a log-periodic antenna located in a window of a building facing a transmitter. Channels 59, 61 and 63 are really occupied and 58, 60 and 62 are unoccupied.

For the representation, white color has been used for 'occupied' detected frequencies and black color for 'unoccupied' detected frequencies. It can be seen the evolution over time for all the frequencies. In the last column, on the right of each image, the percentage of occupancy of each frequency bin has been represented following the legend.

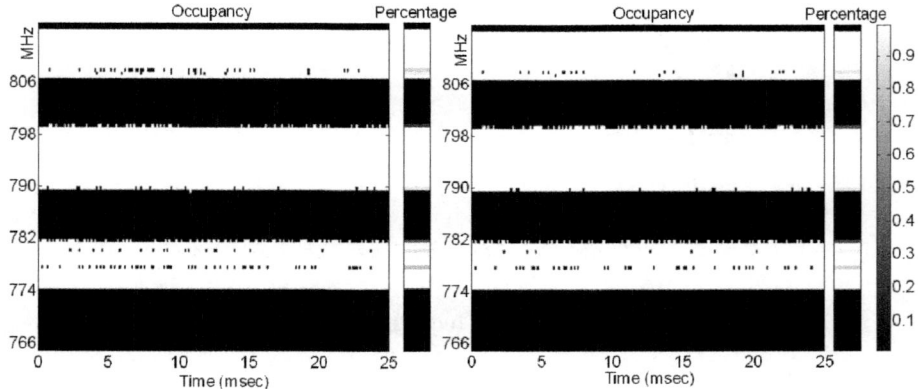

Fig. 4. Algorithm's decision results using Method 1 (left) and Method 2 (right)

It can be observed that with method 2, P_d is higher than with method 1, but also P_{fa} increases with this method. However, there aren't significant differences. So it can be said that the equipment's noise power is a good approximation for estimating noise power and calculating the decision threshold.

More experiments have been carried out, analyzing the influence of measuring the noise power in an unoccupied channel with different bandwidths (1MHz, 4MHz and 8MHz). Results demonstrate that better performance is achievable when using a bandwidth of 8MHz. This is because when using the whole unoccupied channel, we are also considering the power introduced by the tails of adjacent channels. Therefore, a lower P_{fa} is obtained.

When using a higher sensing time (product of K and M), we can achieve better performance. Nevertheless, with a higher sensing time, the algorithm is slower and loses the capability to be used for 'real-time' applications.

4   Conclusions

For real TV broadcasting signals, equipment's noise power is a good estimation of real noise in the sensing channel. At these frequencies, atmospheric noise is very low and the unique measurable noise is equipment's noise.

When measuring noise power in an unoccupied channel it achieves better performance by considering the full channel (8 MHz) instead of considering a narrower bandwidth.

By increasing the number of averages (M) it is possible to get better results but at the expense of losing speed.

References

1. Proakis, J.G.: Digital Communications, 4th edn. McGraw-Hill (2001)
2. Cabric, D., Mishra, S.M., Brodersen, R.W.: Implementation Issues in Spectrum Sensing for Cognitive Radios. In: Conference Record Thirty-Eighth Asilomar Conference on Signals, Systems and Computers (2004)
3. Li, G., Fang, J., Tan, H., Li, J.: The Impact of Time-Bandwidth Product on the Energy detection in the Cognitive Radio. In: 3rd IEEE International Conference on Broadband Network and Multimedia Technology, IC-BNMT (2010)
4. Yan, Y., Gong, Y.: Energy Detection of Narrowband Signals in Cognitive Radio Systems. In: International Conference on Wireless Communications & Signal Processing (2009)
5. Urkowitz, H.: Energy detection of unknown deterministic signals. Proc. IEEE 55, 523–531 (1967)

Cooperative Spectrum Sensing
for Geo-Location Databases

Mauro Fadda[1], Maurizio Murroni[1], Vlad Popescu[1], and Vlad Cristian Stoianovici[2]

[1] Department of Electrical and Electronic Engineering,
University of Cagliari, Piazza d'Armi, 09123 Cagliari, Italy
{mauro.fadda,murroni,vlad.popescu}@diee.unica.it
[2] Department of Electronics and Computers,
University of Transilvania, Str. Politehnicii 1-3 500019 Brasov, Romania
stoianovici.vlad@unitbv.ro

Abstract. Spectrum sensing techniques are the key components for identifying and exploiting unused radio spectrum resources in the perspective of the cognitive radio paradigm. Based on the centralized cooperative sensing techniques, vast generalized databases named geo-location databases (GL-DB) can be deployed in order to centralize sensing and general radio spectrum data for the benefit of secondary cognitive radio users. The authors propose a hybrid conceptual application that involves spectrum sensing and a Human Network Interaction (HNI) model with the purpose of perceptually representing, in an immersive way, the available GL-DB information from a specific location for a natural user perception and interaction with the area of interest.

Keywords: Spectrum sensing, Geo-location Databases, Cognitive Radio, Spectrum Management, 3D Virtual Environments.

1   Introduction

The general trend of the current regulations and standardization efforts for the cognitive radio (CR) paradigm is the deployment of large geo-location databases (GL-DBs). In the US, the FCC has already commissioned the creation of geo-location databases, which can be accessed by any CR device without the use of additional resources. The database entries will provide, for a certain location (geographical coordinates), the list of available channels and the allowable maximum effective isotropic radiated power (EIRP) useful to transmit without providing harmful interference [2]. Even if these GL-DBs are upgraded on a daily basis, the values corresponding to a specific geographical point are still the results of calculations based on a traditional signal propagation models and estimated power levels. Due to this static (for short term at least) approach, the provided data might be inaccurate for different reasons such as variable atmospheric conditions or multipath and fading phenomena [3, 4].

A possible approach to alleviate the above-mentioned shortcomings could be the implementation of a cooperative spectrum sensing architecture [1], based on sensor networks. This scenario is applicable both for outdoor and indoor applications.

L. Atzori, J. Delgado, and D. Giusto (Eds.): MOBIMEDIA 2011, LNICST 79, pp. 78–83, 2012.
© Institute for Computer Sciences, Social Informatics and Telecommunications Engineering 2012

Cooperative spectrum sensing is typically divided into operational networks, handling cognitive transmissions, and sensing networks. The latter would involve a set of sensors deployed in an area of interest, which would sense the spectrum and would relay the process' results to a Cognitive Radio Controller (CRC) [5, 6]. The CRC further processes the collected data and sends the sensed area of interest's spectrum occupancy information to a GL-DB, to which it is connected, in a transparent way, through Future Internet typical infrastructure [7]. The database centralizes all the sensing information from its attached CRCs and serves as a general register that secondary users, who no longer require their own dedicated sensing equipment, can inquire for accessing sensing information for their particular area of interest.

The focus of this paper is a proof-of-concept on the use of spectrum sensing for populating a GL-DB by implementing and deploying a sensor network – based sensing architecture and extending the functionality of the GL-DB towards human users. This approach features the concepts of cluster radio mapping and natural sensing information perception through 3D Virtual Reality (VR) representations of the GL-DB relevant information for the benefit of a spectrum manager or developer.

The structure of the paper is divided into 5 chapters. The first chapter is an introductory one that sets the tone for the current survey and states the goal of our implementation to a conceptual application that integrates Spectrum Sensing with Virtual Reality - specific methods and equipments for the end result to be an immersive Dynamic Radio Spectrum Management application dedicated to spectrum managers and developers. Chapter two presents the general idea of a radio configuration management system that centralizes the spectral information. The third chapter shows how the previously mentioned functionality can be implemented on the basis of a deployed wireless sensor node, in a specific area of interest and the interpretation of the gathered data with the help of a Virtual Environment design tool. The fourth chapter quantifies the results as a consequence of the functional implementation from the previous chapter. The final chapter contains the quantified results, portraying an appropriate context and expressing possible future developments and research opportunities.

2   The Dynamic Spectrum Management Model

Radio Mapping techniques are employed in an attempt to predict and graphically represent network coverage on the basis of a number of connection measurements from locations in an area of interest. A cluster is defined by an area where there is an active CRC and a number of deployed spectrum-sensing sensors. This translates into a real-time electromagnetic profile of the specific area where the sensing sensors are deployed. This profile serves for the design and development of radio architectures over the considered area, and reveals such data as optimum transmission pathways, radio propagation obstacles and, especially, sensing information.

What the authors intend is to implement an architecture that will centralize the real-time statistical sensing information, normally intended for secondary cognitive transmitting users who no longer perform the sensing stage, from different areas of interest, in a GL-DB. Also, we will employ novel perceptual representation in order to provide a radio spectrum manager with a way of perceiving and assimilating this statistical information in a natural and efficient way.

In order to validate our above-mentioned approach we have implemented an indoor functional proof-of-concept prototype, able to capture, represent and transmit sensing information towards a GL-DB. Instead of offering this information only to the secondary users, our application will interpret and represent it in 3D Graphic User Interface (GUI) and illustrate it for the benefit of a radio spectrum manager in order to assist, as a development tool, in the optimization of radio spectrum allocation. Basically we will translate the statistic sensing data that a cognitive radio makes use of, into a perceivable and understandable representation.

3   Conceptual Application Implementation

The sensing sensor network employed for the functional implementation of the conceptual application is a Crossbow ZigBee Wireless Sensor Network (WSN) that uses wireless sensor nodes know as MICAz Motes [8]. Although it is limited to central frequencies between 2.405 and 2.485 GHz, with low throughput, ZigBee does have sixteen 5 MHz channels that we used for testing the sensing algorithm. In other words, our testing scenario involves limiting the concept of radio frequency spectrum to the 2.405 GHz - 2.485 GHz domain, and its 16 channels. All the sensors of the WSN have in their transmission stream's frames a Received Signal Strength Indicator (RSSI) slot, which reveals a numerical value of the gateway's signal power as perceived by that particular network node, which we can interpret as a power measurement, equivalent to the energy detection spectrum sensing method. The RSSI is a naturally available resource when dealing with wireless nodes, and can be used to implement obstacle and position detection and estimation algorithms, in dealing with both primary and secondary users of a CR Network [9].

The total number of utilized wireless nodes is 192. The area of interest is split into 1.5 m side squares, disposed as 16 in length and 12 in width. Each square is the sensing area of a specific sensor, positioned in the middles of the square at 0.6 meters from the floor. The position of each node represents an increment of the measurement step, of 1.5 meters. Theoretical values of the detected signal power (and implicitly of the sensed signal RSSI) can be found out by utilizing the logarithmic correlation between received signal strength and distance as was previously done in [10].

It is at this point that we will consider a theoretical division of the WSN Gateway into two distinct functional entities. The first will be considered as an entity that gathers the sensing information from the WSN nodes, and therefore, also performs the CRC characteristic functionality of GL-DB update, while the other will handle the WSN Gateway's transmission and will be considered a typical CR primary transmitting user, using one of the typical ZigBee channels.

To sum up, the measurement and validation scenario enforces the following suppositions:

-    the radio spectrum is the ZigBee standard frequency domain with its 16 channels;
-    the WSN Gateway is a primary transmitter;
-    the sensors are secondary users who employ energy detection sensing (RSSI);
-    the WSN Gateway's data gathering and GL-DB update is transparent;

The primary user (WSN Gateway) is placed in the corner correspondent to row 0 and column 0 of the senor grid in the area of interest, at 0.6 m from floor level.

All the sensing information gathered by the sensing sensors is real-time processed and forwarded by the WSN Gateway entity, playing the CRC role, towards the GL-DB, to be made available for secondary users or spectrum managers.

As previously stated in the definition of the concept, our functional implementation aims at modeling and developing a real-time natural perception and interaction GUI that brings additional functional uses for the GL-DB concept. The data, contained in the GL-DB, that was originally intended for cognitive users can be employed by spectrum and network managers in order to better understand, develop and utilize available channels and spectrum resources.

For perceptually representing the gathered GL-DB data, the authors implemented a 3D VR Environment build upon traditional desktop equipment, which portrays the available sensed information. Inside the GUI, along with information perception, the user can interact with relays and switches that control actuators from the area of interest, in so enabling the reconfiguration of the sensing architecture to better suit the user's informational needs. The consequence of the interaction inside the GUI and its implicit sensing sensor pattern reconfiguration is a real-time change in the sensed information and accordingly in its representation. Also inside the 3D GUI, there is a navigation menu that allows the user to move inside the virtual environment in order to gain better perspective and perception of the relevant information.

The deployed 3D Virtual Environment is supported by the VR Media's XVR virtual reality framework [11]. The virtual environment is a 3D replica of the real world area of interest, starting from the RSSI information interpretation, upon which, obstacle and position detection and estimation algorithms were based, can be implemented. A view of the 3D representation of the gathered data is presented in figure 1.

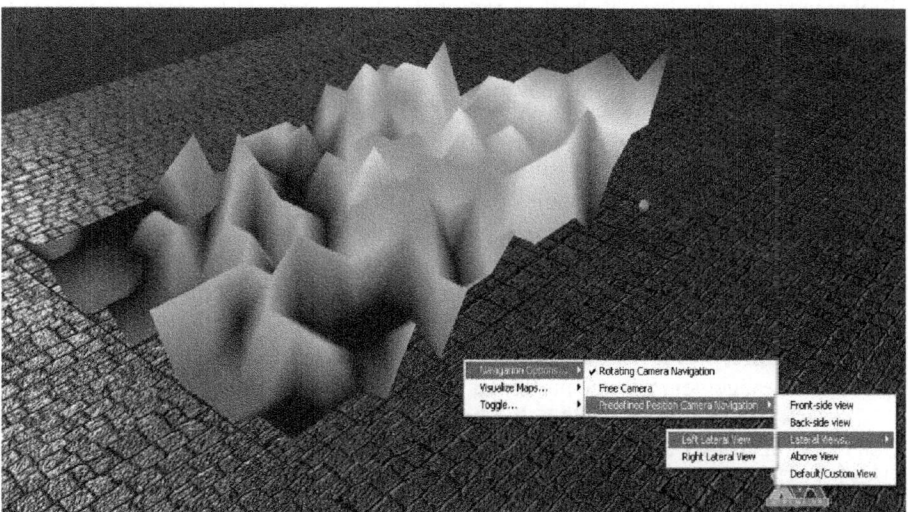

Fig. 1. Radio Map representation of the signal power distribution with smooth transition between colors, inside the area of interest (form Red the highest value to Violet the lowest); The green sphere represents the WSN Gateway primary user

4   Results

The graphic representation describes a Cognitive Radio Network primary user's (WSN Gateway) signal power distribution as received by a sensing WSN. This radio signal is on a typical ZigBee frequency channel, having 5 MHz of bandwidth.

The represented values are expressed in [dBm] and they are subject to the propagation constraints provided by typical electromagnetic indoor obstacles, disposed in the area of interest. The highest RSSI values are color-coded red, and are given a high value on the Y axis (the orthogonal direction from the wireless sensor node arrangement geometry represented as a blue grid plane), in the representation, while the lowest are color-coded violet, and have a value of 0 (null) on the Yaxis. Between the two extremes the values are interpolated, for a smooth transition.

The green sphere marks the location of the transmitting primary user, and, as expected, has the highest RSSI value. While the signal power distribution profile is loosely consistent with Friis' model equations, the inherent anomalies signify electromagnetic obstacles, typical to indoor environments. The measured Gateway RSSI values were in between a -60 dBm and -100 dBm. The WSN's PER (Packet Error Rate) was established to be 3.4%.

The main result is the added value derived from the hybrid implementation of the Radio Spectrum Management field with Virtual Reality representation and multimodal interaction methodologies.

5   Conclusions and Future Developments

Sensing sensor networks can be extended to the paradigm of Centralized Coordinated Techniques that involve CRCs, if operational networks would be implemented on top of sensing networks (collocated) and additional functionality (from the point of view of the transmission and processing power) would be passed from the CR Controller to the sensor nodes, which can be implemented by employing SDR platforms (USRP2) [12]. This is not an evolution of the Sensing sensor networks approach but rather a parallel alternative for a better-suited purpose scenario, both collocated and separated architectures approach having their pros and cons.

Because of the immersive nature, high interactivity and powerful sense of presence, the authors' 3D GUI complies perfectly with the 3D Internet [13] component of the Future Internet that offers users an augmented interaction and navigation metaphor. Also, the wireless nodes network features a functionality that emulates Internet of Things specific scenarios, while the whole conceptual application offers a radio spectrum management service particular to the Internet of Services.

The VR environment could be further developed by employing advanced visualization, sound and haptic devices specific to immersive VR applications (i.e. CAVE [14]).

The first tests, performed in order to validate the proof-of-concept application, using a single frequency, showed that merging exponentially developing domains such as Virtual Reality and its characteristic techniques and devices with the field of spectrum sensing and, generally, dynamic spectrum management, results in added functionality and significant added value for the latter. These initial tests will be continued by analyzing a more extended range of frequencies and by implying more intelligent sensor nodes, for example software-defined radios.

Acknowledgments. The authors would like to thank professor Marcello Carrozzino, from Scuola Superiore Sant'Anna / PERCRO, for his support and guidance. This paper is supported by the Sectoral Operational Programme Human Resources Development (SOP HRD), financed from the European Social Fund and by the Romanian Government under the contract number POSDRU/6/1.5/S/6 and by the Young Researchers Grant funded by Region of Sardinia, PO Sardegna FSE 2007/2013, L.R.7/2007, "Promotion of scientific research and technological innovation in Sardinia".

References

1. Thanayankizil, L., Kailas, A.: Spectrum Sensing Techniques (II): Receiver Detection and Interference Management (2008),
   http://www.personal.psu.edu/bxg215/spectrum%20sensing.pdf
   (retreived on April 22, 2011)
2. Ko, G., Antony Franklin, A., You, S.-J., Pak, J.-S., Song, M.-S., Kim, C.-J.: Channel Management in IEEE 802.22 WRAN Systems. IEEE Communications Magazine (September 2010)
3. FCC Second Memorandum Opinion and Order, FCC 10-174 (September 2010), http://www.fcc.gov
4. Arslan, H.: Cognitive Radio, Software Defined Radio and Adaptive Wireless Systems. Springer, Heidelberg (2007); ISBN: 1402055412
5. IEEE SCC 41 - White Paper Sensing techniques for Cognitive Radio - State of the art and trends (April 2009)
6. Shankar, N.S., Cordeiro, C., Challapali, K.: Spectrum agile radios: utilization and sensing architectures. In: DySPAN 2005, Philips Res. Briarcliff Manor, CA, USA (2005)
7. European Commission Information Society and Media, The future of the internet. A compendium of European projects on ICT research supported by the EU 7th framework programme for RTD, ftp://ftp.cordis.europa.eu/pub/fp7/ict/ docs/ ch1-g848-280-future-internet_en.pdf (retrieved on December 08, 2010) ISBN: 978-92-79-08008-1
8. Crossbow Technology, XServe Gateway Middleware, http://www.xbow.com/ Technology/GatewayMiddleware.aspx (visited on November 03, 2009)
9. Srinivasan, K., Levis, P.: RSSI is under appreciated. In: Proc. of the Third Embedded Networked Sensors (EmNets 2006), pp. 15–20 (May 2006)
10. Stoianovici, V.C., et al.: A Virtual Reality Based Human-Network Interaction System for 3d Internet Applications. In: 12th International Conference on Optimization of Electrical and Electronic Equipment, OPTIM 2010, Brasov, Romania (2010)
11. VRmedia Italy, XVR, http://www.vrmedia.it/Xvr.htm (accessed November 2009)
12. Brodersen, R., et al.: CORVUS: A Cognitive Radio Approach for Usage of Virtual Unlicensed Spectrum. Berkeley Wireless Research Center (BWRC) White paper (2004)
13. Alpcan, T., et al.: Towards 3D Internet: why, what, and how? In: IEEE Int. Conf. on Cyberworlds (Cyberworlds 2007), Hannover, Germany (October 2007)
14. Ihren, J., Frisch, K.: The fully immersive cave. In: Proc. 3rd International Immersive Projection Technology Workshop, pp. 59–63 (1999)

Spectrum Occupancy and Hidden Node Margins for Cognitive Radio Applications in the UHF Band

Miren Alonso[1], Irati Lázaro[1], Maurizio Murroni[2], Pablo Angueira[1], Manuel Vélez[1], J. Morgade[1], Mikel Sánchez[3], and Pablo Prieto[3]

[1] Bilbao Faculty of Engineering, UPV/EHU, Bilbao, Spain
[2] DIEE University of Cagliari, Cagliari, Italy
[3] Telecom Unit, Tecnalia Research Centre, Zamudio, Spain
{malonso038,pablo.angueira}@ehu.es, murroni@diee.unica.it

Abstract. This paper presents the study of the spectrum occupancy in the UHF frequency band (470-870MHz) in Bilbao area, Spain. The study has been performed at three different sites and at different height to determinate the hidden node margin, the main problem of the cognitive radio. The objective of the paper is to determinate by signal power measuring if the cognitive device will be able to detect and distinguish the empty and occupied channel to carry out the communication. The results from the spectrum measurements taken in all the sites have been analyzed and compared to the official spectrum regulation. The study reveals that the spectrum occupancy is minimum, thus there are a lot of white spaces in this band.

Keywords: Cognitive Radio, White Space, Hidden Node Margin, Spectrum Measurement, Spectrum Occupancy.

1   Introduction

The feasibility of the idea of "cognitive radio" [6] for the UHF band is strongly dependent on the border conditions of the UHF band occupancy. Also, and not in a less relevant position, the potential propagation channel and attenuation of the signal radiated from cognitive devices are also key parameters to be analyzed before a sensible discussion on the feasibility of the technology is raised. This paper aims at providing field measurement data that could serve as a reference for the technical discussions around the possible use of White Spaces in the UHF band. This paper does not aim at providing an "a priori" opinion on the use of this spectrum management approach.

In principle, Cognitive Radio proposes that user devices perform a dynamic adaptation of their physical layer to the radio environment on a secondary use of the spectrum basis. The Cognitive Devices (CD) would use information gathering techniques to make intelligent choices for their radio resources: choosing the operating frequency, maximum ERP, modulation parameters etc. These techniques are based on Spectrum Sensing and on geolocation databases that will provide each

L. Atzori, J. Delgado, and D. Giusto (Eds.): MOBIMEDIA 2011, LNICST 79, pp. 84–88, 2012.
© Institute for Computer Sciences, Social Informatics and Telecommunications Engineering 2012

device a reference of the use of the spectrum in the location where the device is operating. With this technology, unlicensed services could temporally use the white spaces of a certain broadcast service area.

Spectrum Sensing techniques are based on signal detection by different signal processing methods. In any of the techniques a problem called Hidden Node Margin (HNM) is a major issue. The HNM represents the difference of the channel usage estimation by the CD and the actual spectrum picture at the receiving antenna of the primary service (in our case Digital TV). Due to this HNM an occupied channel would be identified as free (White Space) by a CD and a potential interference to the primary (to be protected) service would occur.

Previous work on this topic can be found in [2][3][4][5][7]. A theoretical study from the BBC [1] has proposed a reference value of 40 dB for outdoor HNM (10 m to 1.5 m margin) and an additional 20 dB attenuation for the indoor case. Other studies have obtained values that range from 16.6 dB to 33 dB on channel 22. Additional work has based the results on the field strength simulated values provided by different algorithms, including results with 3D ray tracing and conventional UHF empirical and semi empirical methods (ITU-R P.1546). In this case, the values range from 4 to 46 dB.

This work tries to provide consistent values from a carefully planned measurement campaign.

This paper is organized as follows. Section 2 contains a description of the methodology and scenarios. Section 3 contains a discussion on the channel occupancy decision threshold. Section 4 presents a sample of the results that are expected at the end of the project and a discussion based on the first results. Finally, Section 5 contains the conclusions at this stage of the work.

2   Methodology and Scenarios

The first step of this survey was to elaborate an environment classification that would allow extrapolation of the measurement results and conclusions to a number of real cases as wide as possible.

Table 1 shows the features of the different scenarios that have been considered for the measurement campaign. The same table contains the first three measurement locations of the trial that have been used as the reference data for this paper. A higher number of locations have been already planned for a next phase of the research work.

Outdoor roof and indoor measurements were carried out in the three environments. Indoor measurements were performed in each floor of the building under tests, using a dipole antenna at 1.5 m. above the ground level.

At each location, the man-made-noise was evaluated in two frequencies, 408MHz (radioastronomy frequency) and 520MHz (free frequency) to determinate the channel occupancy decision threshold. After obtaining the reference noise value, 1MHz 100 power measures were taken starting from 470MHz up to 870MHz. A second measurement round was carried out at a selected list of channels using a 100KHz resolution bandwidth in order to obtain more accurate data at frequencies where potential doubts of occupancy could appear.

Table 1. Environment Description

	Urban	Suburban	Rural
Characteristics	Buildings with more than 3 floors Heavy traffic	Low-rice buildings and duplex Little traffic	Isolated houses. Usually surrounded by vegetation or in open areas
Place	Bilbao 43°15'40.46''N 2°56'54.80''O	Lezama 43°16'25.15''N 2°49'46.59''O	Maruri 43°23'02.78''N 2°52'19.50''O

The measurement system consisted of an ESPI test receiver (Rhode and Schwarz) connected to a PC and a set of antennas. A calibrated dipole was used for 1MHz measurements (ETS-LINDGREN 3121-D). A logperiodic antenna was used for taking measurements with a 100 kHz resolution. The system was controlled from a PC where data files were also recorded for further analysis. The Test Receiver was configured with a pre-amplifier for higher sensitivity.

Fig. 1. Measurement System. Examples of suburban and urban environments.

3   Decision Threshold

The choice of threshold is an important decision. Its value sets up the condition for channel occupancy decision. If the power measured is above the threshold, the device would decide that the channel is being used. In the same way, if the power is lower than the threshold, the device will decide that the channel is empty.

The election of this parameter has been a difficult choice. Low values lead to noise values higher than the threshold and may mislead the device. On the opposite case, if the threshold is too high, when a signal is extremely low, the device can decide that the channel is empty when it's not, and create interferences.

Based on other researches all the measurements were analyzed with a threshold 3 and 4dB above the power measures. A usual value for this threshold in the literature is

5 dB [4]. It was found that the Radio Noise in rural and suburban environments is below the receiver internal noise (-105 dBm/1MHz, -115 dBm/100kHz). In the case of urban spots, the Radio Noise was 3 dB higher than the equipment threshold.

4   Measurements

The power values measured along the frequency band from 470MHz to 870MHz were compared with the decision threshold. The results obtained are shown in Table 2, for band occupancy calculation. The figures represent the percentage of occupied channels as detected at each location.

Table 2. Spectrum Occupancy

Site	Rural Mean percentage	Suburban Mean percentage	Urban Mean percentage
Roof	%27.00	%32.25	%28.25
4rd floor			%21.50
3rd floor			%26.75
2nd floor	%18.00	%17.50	%25.00
1st floor	%20.00	%20.00	%25.00
Street		%22.25	
Basement		%3.00	%7.50

The spectral occupancy decision is higher at the roof of each building than the register. These channels present at the roof but not detected at lower locations are signals coming from transmitters outside the service area of the potential interfered transmitters. In any of the cases the spectral occupancy never exceeded the %32 of the band. If a higher threshold is assumed, 4dB above the measurements, the results are very similar and the percentage is reduced by %1.

The hidden node margin has been calculated as the difference of the signal power measured at the roof and all the floors. To show some results we have chosen the channel 22.

Table 3. Hidden node margin

Site	Rural HNM(dB)	Suburban HNM(dB)	Urban HNM(dB)
4rd floor			17.53
3rd floor			8.21
2nd floor	17.14	16.30	8.85
1st floor	23.50	15.76	12.35
Street		17.91	
Basement		28.86	38.34

The results obtained agree with the references available in the literature. Some theoretical estimations mention values close to 60 dB if indoor reception is considered [1] and in [2] HNM values were obtained in real measurements for different scenarios providing a 28.5 dB difference for outdoor measurements in a 90% of the locations. The values range from 18 dB for rural cases to 37 dB for dense urban spots. Considering that the trials of this paper are indoor, an average building penetration losses need to be added to our values. In that case, the value obtained in the trial ranges from 8 dB to 38 dB (rural and dense urban cases respectively).

5   Conclusion

The present paper investigated the spectral occupancy of the UHF band allocated for TV broadcasting trough real measurements performed in urban, sub urban and rural scenarios at roof, middle floor and basement. Results have shown that a top occupancy of 32% of the bandwidth is achieved in the roof and that the hidden node margin obtained range from 8 to 38 dB on channel 22 depending on the environment and lead to the conclusion that cognitive communications to be performed in the UHF TV band need the joint use of geolocation databases and spectrum sensing technique to avoid harmful interference to the primaries services of the broadcasters.

References

1. Waddell, M.: Compatibility Challenges for Broadcast Networks and White Space Devices. BBC Research White Paper, WHP 182 (January 2010)
2. Randhawa, B.S., Wang, Z., Parker, I.: Report Title: Analysis of hidden node margins for cognitive radio devices potentially using DTT and PMS spectrum (January 2009)
3. Contreras, S., Villardi, G., Funada, R., Harada, H.: Report Title: An investigation into the spectrum occupancy in Japan in the context of TV White Space Systems. In: 6th International ICST Conference on Cognitive Radio Oriented Wireless Networks and Communications, Osaka, Japan, June 1-June 3 (2011)
4. Chiang, R.I.C., Rowe, G.B., Sowerby, K.W.: A queantitative Analysis of Spectral Occupancy Measurements for Cognitive Radio. IEEE Research, 3016–3020 (2007) ISSN:1550-2252
5. López-Benítez, M., Umbert, A., Casadevall, F.: Report Title: Evaluation of Spectrum Occupancy in Spain for Cognitive Radio Applications, Dept. Signal Theory and Comn. IEEE, Barcelona (2009)
6. Mitola III, J., Maguire Jr., G.Q.: Article Title: Cognitive Radio: Making Software Radios More Personal. IEEE Personal Communications 6(4), 13–18 (1999)
7. Barbiroli, M., Carciofi, C., Guidotti, A., Guiducci, D.: Evaluation and analysis of the hidden node margin for cognitive radio system operation in a real scenario. In: Proceeding of the 5th European Conference on Antennas and Propagation

Delay Model for Super-Frame Based Resource Reservation in Distributed Wireless Networks

Xiaobo Yu, Pirabakaran Navaratnam, and Klaus Moessner

University of Surrey, Guildford, GU2 7XH, UK
{x.yu,p.navaratnam,k.moessner}@surrey.ac.uk

Abstract. This paper proposes an analytical framework for evaluating the delay performance of super-frame (SF) based MAC schemes with distributed resource reservation in IEEE 802.11e enhanced distributed channel access (EDCA). SF-based resource reservation (RR) schemes divide the airtime into service intervals (SIs) with contention-free period (CFP) for providing guaranteed QoS for RTSNs and contention access period (CAP) for pledging fairness toward other sessions. The proposed analytical framework models the delay performance of RTSNs that obtain dedicated resources in a distributed manner. In addition, the optimization of system parameters, such as size of transmission opportunity (TXOP) and SI are studied in order to enhance the overall network capacity. The accuracy of the analytical framework is verified through numerical simulation and analytical results, which also suggest that the optimum resource allocation and SI can be found for improving the network capacity.

Keywords: QoS, IEEE 802.11e, resource reservation.

1 Introduction

Nowadays, IEEE 802.11-based wireless communication technology pervades in various areas such as Wi-Fi hot spots, city wide mesh networks, vehicular communication, and similar application areas. Most personal communication devices such as laptop computers as well as mobile phones are armed with 802.11a/b/g adapters or 802.11-compliant entities. Despite of the general application, there are still lots of issues that pose difficulties in providing Quality of Service (QoS) in 802.11-based distributed wireless networks.

So far, many research works have been focused on providing QoS for real-time sessions (RTSNs) in IEEE 802.11-based distributed wireless networks. Since the legacy distributed coordination function (DCF) can not differentiate the services between RTSNs and non-real-time sessions (NRTSNs), an enhancement of DCF named enhanced distributed channel access (EDCA) has been standardised in IEEE 802.11e [1]. Its fundamental QoS support is proved helpful for QoS support but its enhancement is still limited. For further improving the QoS for EDCA, some of the contributions [2,3] are made to enhance the probability of channel access for RTSNs by tuning the parameters of deferral and back-off algorithms.

L. Atzori, J. Delgado, and D. Giusto (Eds.): MOBIMEDIA 2011, LNICST 79, pp. 89–104, 2012.
© Institute for Computer Sciences, Social Informatics and Telecommunications Engineering 2012

Optimizations of queueing algorithm [4–6] for improving the QoS for RTSNs are also achieved. Although QoS advancements toward RTSNs can be implemented by using these approaches, the transmissions of RTSNs are still affected under interference environment and the deferral and back-off, as the channel access overheads, are unavoidable.

To solve the aforementioned issues, one of the most effective solutions is the super-frame (SF) based MAC scheduling mechanisms with resource reservation (RR), which can partition the channel airtime into contention-free period (CFP) for providing guaranteed QoS for RTSNs and contention access period (CAP) for the fairness toward other types of traffic sessions. The RTSNs can get periodic and dedicated resources through this distributed RR scheme so that their QoS requirements can be met. Following the idea of this distributed SF-based RR method, several MAC protocols [7,8] have been proposed and simulations have been conducted for validating the effectiveness of these schemes. However, the analysis as well as the optimization of these distributed SF-based RR schemes are still open issues.

This paper mainly proposes an analytical framework for modelling the delay performance of QoS guaranteed RTSNs in the distributed SF-based mechanisms ,that were devised for IEEE 802.11e EDCA. The analytical framework is capable of predicting QoS performance of RTSNs on both saturated and unsaturated traffic conditions. Based on the guaranteed QoS, the enhancement of network capacity (i.e. the maximum amount of RTSNs) is studied through the optimization of bandwidth allocation for RTSNs as well as the system parameter such as service interval (SI) in order to accommodate more RTSNs in CFP. Note that the network capacity in this paper implies the maximum amount of RTSNs that are allowed to reserve transmission opportunities (TXOPs) in CFP.

The rest of this paper is organized as follows. Section 2 depict the SF based RR mechanism and its derivative protocols - EDCA/RR and EDCA/DRR. Section 3 specifies the analytical model for delay performance. The optimization study is presented in Section 4. Simulation and analytical outcomes are shown in Section 5. Finally, section 6 concludes this paper.

2   Overview of Super-Frame Based Resource Reservation in IEEE 802.11e Networks and Its Derivative Protocols

As mentioned before, the SF-based RR schemes utilize SI to partition the services between admitted RTSNs and other sessions. Fig. 1 shows an example of the SF-based RR scheduling, a QoS guaranteed RTSN will obtain a dedicated bandwidth called transmission opportunity (TXOP) during which multiple frames of the corresponding RTSN can get transmitted provided that the dedicated duration is adequate. If the residual time can not afford a further data transmission, the corresponding RTSN will wait until the next TXOP.

To implement this distributed RR, EDCA/RR [7] proposes a signalling process. An add traffic stream (ADDTS) request frame is broadcasted from the source of the RTSN if RR is required. The signalling frame takes the traffic

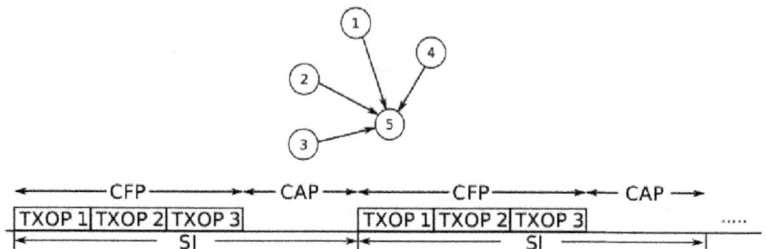

Fig. 1. Super-frame based resource reservation scheduling

specification (TSPEC) which contains the parameters such as service start time, delay bound, etc. Upon the receipt of the request frame, destination will decide whether to accept the RTSN and reserve bandwidth for it. If the residual bandwidth in CFP can support the QoS demand of the RTSN, destination will confirm the reservation request through replying an ADDTS response frame. Otherwise, it will send back the response frame to reject the RTSN. The reservation request is also validated by contending nodes within the transmission range. They will confirm the new request given that their dedicated resources are not offended. Otherwise, they decide to reject the new RTSN and inform the source by sending signalling messages.

Although EDCA/RR is able to successfully implement the SF-based RR, it ignores the dynamic resource allocation for the TXOPs that become idle after their corresponding RTSNs stop transmitting. This will incur the wastage of bandwidth in CFP and degrade the network capacity. In EDCA/DRR [8], a dynamic resource allocation scheme is proposed for addressing this problem. Arrival priority (APR) is introduced for differentiating the precedence of the rejected RTSNs that are made to be transmitted in CAP. The rule is that the earlier the RTSN accesses the CAP, the higher its APR is. An adaptive admission control is devised for monitoring and controlling the transmissions in CFP. If there are RTSNs being transmitted in CAP when idle resources appear in CFP, the idle resources will be assigned to the RTSN with the highest APR and then all the other rejected RTSNs shift their priority accordingly. If no rejected RTSN exists, the idle resources will be allocated to the CAP.

3   Analytical Model

In this section, an analytical framework for modelling the delay performance of RTSNs with dedicated resources is proposed for distributed SF-based RR mechanisms. To enhance the efficiency of transmission time and reduce the channel deferral time, MAC service data unit (MSDU) is formed by aggregating several frames of a session [9]. This can help improve the throughput of the session. During CFP, the amount of MSDUs for $RTSN_i$ that are permitted to be transmitted within its dedicated TXOP depends on the duration of SI denoted by ΔSI, mean

MSDU size $\overline{s_{DATA_i}}$, and required scheduling rate which is represented by $\lambda_i{}^1$. Therefore, we can obtain

$$n_{t,i} = \lceil \frac{\Delta SI \times \lambda_i}{s_{DATA_i}} \rceil \tag{1}$$

where $n_{t,i}$ is the amount of MSDUs of RTSN$_i$ that are able to be accommodated by a TXOP. Fig. 2 shows an example of a scheduled TXOP which can be used for multiple data transmissions. The duration of a TXOP is expressed by

$$t_{TXOP_i} = n_{t,i}(E[t_{DATA_i}] + t_{ACK}) + 2n_{t,i} \cdot t_{SIFS} \tag{2}$$

where t_{ACK} and t_{SIFS} denote the duration of ACK and SIFS, respectively. $E[t_{DATA_i}]$ in the above equation stands for the average duration cost by transmitting an MSDU of RTSN$_i$. The analytical model for delay performance of QoS guaranteed RTSNs is specified as follows.

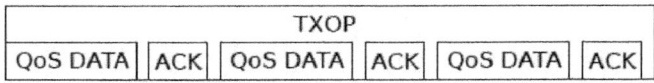

Fig. 2. A scheduled TXOP

3.1   Delay Model for RTSNs with TXOPs

In this subsection, we analyse the delay performance of RTSNs with TXOPs in CFP. The average delay $d_{ave,i}$ of MSDUs for RTSN$_i$ is equivalent to the average duration from the instant that its MSDU buffers in the queue to the moment that it successfully completes transmission. In general, the delay is comprised of channel access delay $d_{ca,i}$, queueing delay $d_{q,i}$ as well as transmission delay $d_{tr,i}$. The detail analyses are shown below.

Channel Access Delay. The channel access delay is defined as the time from an MSDU reaches the head of the interface queue to the instant that it starts accessing the channel. Owning dedicated bandwidth in CFP, each MSDU needs to wait for its time-slots in CFP to get transmitted. The instant when the MSDU arrives the head of queue determines its channel access delay. As shown in Fig. 3, for each RTSN, time can be regarded to be composed of periodic TXOP for data transmission and non-TXOP time during which its MSDUs have to wait. Based on the relationship between traffic load of a RTSN and the size of its allotted TXOP. Three conditions can be defined: (i). Unsaturated condition which indicates that the allocated resources can not be used entirely by the RTSN. (ii). Saturated condition which implies that the RTSN can exactly feed

---

[1] In this paper, it is assumed that application rate is equivalent to required scheduling rate.

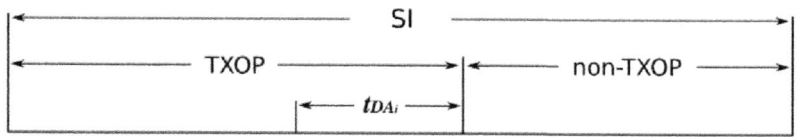

Fig. 3. TXOP, t_{DA_i} and non-TXOP

all the duration of TXOP. (iii). Over-saturated condition which indicates that the maximum transmission capability of reserved bandwidth is not sufficient for accommodating the traffic load of the RTSN. The third condition results in buffer overflow and thus devastates the performance of RTSN. Since the analytical model is aimed at evaluating the delay performance of QoS guaranteed RTSNs, the unsaturated and saturated conditions will be the focus hereafter.

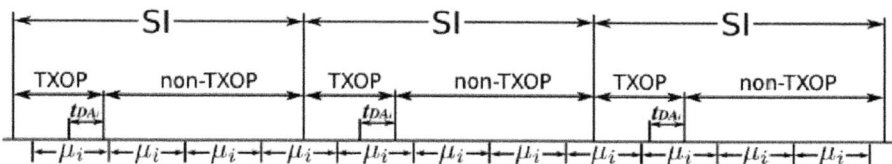

Fig. 4. SI, sending interval and TXOP

As a key parameter for analysing the channel access delay, the MSDU arrival time is mainly affected by the required scheduling rate λ_i of RTSN$_i$ and the duration of allocated TXOP t_{TXOP_i}. To formulate the channel access delay using the above parameters, sending interval is introduced to map the required scheduling rate into the periodic SI. Let μ_i denote the sending interval of RTSN$_i$. We can obtain

$$\mu_i = \frac{\overline{s_{DATA_i}}}{\lambda_i} \tag{3}$$

As shown in Fig. 4, the relationship between the required scheduling rate and the TXOP can be easily indicated if sending interval is employed. Assumed that the MSDUs of RTSNs regularly generate, it is able to figure out whether the instant that an MSDU arrives at the queue belongs to its TXOP or its non-TXOP. Since the channel access delay is determined by the arrival instant of each MSDU, a function $f = \delta(x)$ is given for representing normalized offset of each arrival within a SI. It can be expressed by

$$\delta(x) = x - [x] \tag{4}$$

where $[x]$ is used for taking the integer part of variable x. An example of a normalized offset of MSDU arrival time is shown in Fig. 5. To express the offset duration of each arrived MSDU within its SI, $\eta(j)$ is introduced and expressed by

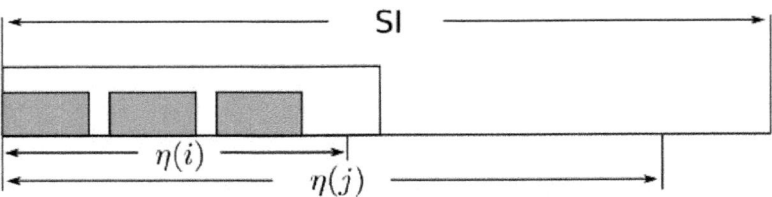

Fig. 5. Arrival offset

$$\eta(j) = \delta(\frac{j \cdot \mu_i}{\Delta SI}) \cdot \Delta SI \tag{5}$$

where j stands for the jth MSDU of the RTSN. If a head-of-line MSDU arrives and finds out that the residual time is not sufficient for another data transmission, it can not be transmitted within this TXOP but has to wait for the TXOP in the next SI. Let t_{DA_i} stand for the entire duration for a successful data transmission for RTSN$_i$, which is given by

$$t_{DA_i} = E[t_{DATA_i}] + t_{SIFS} + t_{ACK} + t_{SIFS} \tag{6}$$

If an MSDU arrives at the head of queue within the last t_{DA_i} of a TXOP, it is not allowed to get transmitted until experiencing another non-TXOP duration. Fig. 3 shows the relationship among TXOP, t_{DA_i} and non-TXOP. Under unsaturated and saturated conditions, if the MSDU arrives inside the first $t_{TXOP_i} - t_{DA_i}$ of a TXOP, it can get transmitted within the current TXOP. This is because when considering the unsaturated and saturated conditions, there is no MSDU buffered in the queue at the instant of $t_{TXOP_i} - t_{DA_i}$ of each TXOP. The reason is that transmission capability of the reserved bandwidth in these two cases is at least not less than actual traffic load of the corresponding RTSN. For simplify the equations hereafter, the duration of the first $t_{TXOP_i} - t_{DA_i}$ within a TXOP is represented by Φ_i. The first MSDU that arrives since the time of Φ_i will reach the head of the line and then it waits for the consequent dedicated bandwidth. In this case, the channel access delay is the duration between the moment that the MSDU reaches the head of the line and the beginning of the next TXOP. If an MSDU generates after the time when the first MSDU reaches the head of queue within the interval $[\Phi_i, \Phi_i + \mu_i]$ of the current SI, it will buffer in the queue and reach the head of the line until its prior MSDUs finish their transmissions in the subsequent reserved time-slots. As a result, these MSDUs have no channel access delay. The channel access delay for MSDUs of RTSN$_i$ in different conditions can be formulated by

$$dc_{j,i} = \begin{cases} 0, & if \ 0 \leq \eta(j) \leq \Phi_i \\ (1 - \frac{\eta(j)}{\Delta SI}) \cdot \Delta SI, & if \ \Phi_i < \eta(j) \leq \mu_i + \Phi_i \\ 0, & if \ \eta(j) > \mu_i + \Phi_i \end{cases} \tag{7}$$

To simplify the computation of average channel access delay for a QoS guaranteed RTSN, a period can be identified for $\eta(j)$. The proof is shown as follows.

Proof: Assumed that there is an integer P, which denotes the subsequent Pth MSDU that arrives after the jth MSDU. The offset duration of the Pth MSDU can be derived by

$$\eta(j + P) = \delta(\frac{(j + P) \cdot \mu_i}{\Delta SI}) \cdot \Delta SI$$
$$= \delta(\frac{j \cdot \mu_i}{\Delta SI} + \frac{P \cdot \mu_i}{\Delta SI}) \cdot \Delta SI \tag{8}$$

The SI can be deemed as a fixed set of $SlotTime$ σ, which is denoted by

$$\Delta SI = K \cdot \sigma, \quad if \ K \in \mathbf{N}^+ \tag{9}$$

Similarly, we can obtain

$$\mu_i = K' \cdot \sigma, \quad if \ K' \in \mathbf{N}^+ \tag{10}$$

where sending interval is expressed by an integer amount of σ. Using (9) and (10), the variance of (8) can be derived by

$$\eta(j + P) = \delta(\frac{j \cdot \mu_i}{\Delta SI} + \frac{P \cdot K'}{K}) \cdot \Delta SI \tag{11}$$

Note that K and K' are both taken as integer values. Consequently, a minimum value of integer P can be found in order to make the value of $\frac{P \cdot K'}{K}$ equal to a positive integer. Due to the property of $\eta(x)$, the term $\frac{P \cdot K'}{K}$ can be ignored. Thus,

$$\eta(j) = \eta(j + P) \tag{12}$$

This verification suggests that the offset value of an arbitrary MSDU will periodically reappear after a certain duration which can be viewed as a period. Therefore, the average channel access delay can be obtained through computing the average value of channel access delay for all the MSDUs arrived within a period. The average channel access delay is expressed by

$$\overline{d_{ca,i}} = \frac{\sum_{j=1}^{p_i} dc_{j,i}}{p_i} \tag{13}$$

where p_i denotes the minimum period for the MSDUs of $RTSN_i$.

Queueing Delay. The queueing delay is measured from the moment that an MSDU pumps into the interface queue to the instant that it reaches the head of the line. Considering unsaturated and saturated conditions, the buffered MSDU during a SI can be completely transmitted using the TXOP in the subsequent SI. As a result, there is no MSDU buffered at the time of Φ_i in each TXOP. As shown in Fig. 6 and Fig. 7, there are three conditions of queueing delay for an MSDU.

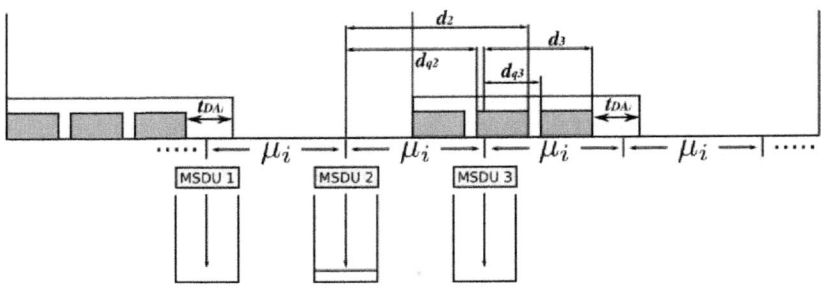

Fig. 6. Queueing delay on different time-slots

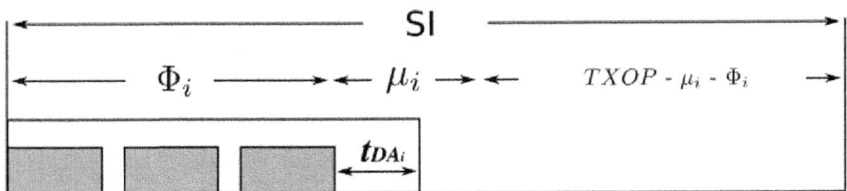

Fig. 7. Queueing delay on different conditions

First, the MSDU arrives inside the interval $[\Phi_i, \Phi_i + \mu_i]$ of the current SI. Second, the MSDU arrives within the interval $[\Phi_i + \mu_i, \Delta SI]$ of the current SI. Third, the MSDU arrives during the first Φ_i of the TXOP within the current SI.

In the case of the first situation, the newly arrived MSDU directly becomes the head-of-line MSDU and defers until the start time of its next TXOP. Thus, the MSDU has no queueing delay. For the MSDU following the second situation, it will buffer in the queue and get transmitted in the subsequent TXOP. Therefore, its queueing delay is equal to the deferral time of non-TXOP duration plus the transmission time of the prior MSDUs buffered in the queue. The amount of MSDUs that arrive prior to the tagged MSDU is $\lfloor \frac{\eta(j) - \Phi_i}{\mu_i} \rfloor$. They will cost the transmission time of $\lfloor \frac{\eta(j) - \Phi_i}{\mu_i} \rfloor \cdot t_{DA_i}$, which is part of the queueing delay of the tagged MSDU. The rest part of its queueing delay is the non-TXOP duration which is $\Delta SI - \eta(j)$. For the third situation, the MSDU will buffer in the queue and get transmitted in the current TXOP. As a result, its queueing delay is the transmission time of the remained MSDUs buffered before plus the residual transmission time of the MSDU which is being transmitted at the moment. In order to figure out the queueing delay in this situation, the amount of accumulated MSDUs from the instant Φ_i of last TXOP to the arrival time $\eta(j)$ of the tagged MSDU needs to be figured out. Since there is no MSDU at the timestamp Φ_i of the last TXOP, the number of MSDUs that still buffer in the queue is $\lfloor \frac{\eta(j) + \Delta SI - \Phi_i}{\mu_i} \rfloor - \lfloor \frac{\eta(j)}{t_{DA_i}} \rfloor - 1$. It excludes the current transmitting MSDU, which requires the time of $\lfloor \frac{\eta(j)}{t_{DA_i}} + 1 \rfloor \cdot t_{DA_i} - \eta(j)$ in order to finish its transmission. The amount of MSDUs that have already been sent is $\lfloor \frac{\eta(j)}{t_{DA_i}} \rfloor$. Finally, the queueing delay for the third situation can be derived as $(\frac{\eta(j)+\Delta SI-\Phi_i}{\mu_i} \rfloor - \lfloor \frac{\eta(j)}{t_{DA_i}} \rfloor - 1) \cdot t_{DA_i} + (\lfloor \frac{\eta(j)}{t_{DA_i}} + 1 \rfloor \cdot t_{DA_i} - \eta(j))$. The queueing delay for all the situations can be denoted by (14). Using the periodicity property of

$$dq_{j,i} = \begin{cases} (\lfloor \frac{\eta(j)+\Delta SI-\Phi_i}{\mu_i} \rfloor - \lfloor \frac{\eta(j)}{t_{DA_i}} \rfloor - 1) \cdot t_{DA_i} + (\lfloor \frac{\eta(j)}{t_{DA_i}} + 1 \rfloor \cdot t_{DA_i} - \eta(j)), & if \ \eta(j) < \Phi_i \\ 0, & if \ \Phi_i \le \eta(j) \le \mu_i + \Phi_i \\ \lfloor \frac{\eta(j)-\Phi_i}{\mu_i} \rfloor \cdot t_{DA_i} + \Delta SI - \eta(j), & if \ \eta(j) > \mu_i + \Phi_i \end{cases}$$

(14)

the $\eta(j)$, the average queueing delay of RTSN$_i$ is expressed by (15).

$$\overline{d_{q,i}} = \frac{\sum_{j=1}^{p_i} dq_{j,i}}{p_i} \tag{15}$$

Transmission Delay. Transmission delay is equal to the duration from the instant that an MSDU begins accessing the channel to the moment it is successfully transmitted. The average transmission delay can be denoted by

$$\overline{d_{tr,i}} = E[t_{DATA_i}] + t_{SIFS} + t_{ACK} + t_{SIFS} \tag{16}$$

where $E[t_{DATA_i}]$ stands for the average transmission time of an MSDU for RTSN$_i$.

4 Optimization Study Based on Delay Bound for SF-Based RR Scheme

In this section, we study the optimization of system parameters such as SI and the size of allocated TXOP for each RTSN. It can be implied from the previous analysis that guaranteed QoS of a RTSN can be achieved under unsaturated and saturated conditions in which required scheduling rate of the RTSN does not exceed the maximum transmission capability of its reserved TXOP. According to (1), the size of TXOP allocated for a RTSN lies with the amount of MSDUs that are allowed to be transmitted. Considering the delay bound, the duration of TXOP for a RTSN is closely associated with the size of SI, required scheduling rate and the average size of an MSDU. RTSNs with different required scheduling rates need to obtain distinct amount of resources (i.e. TXOPs) in order to ensure their guaranteed QoS. The delay of a RTSN with large required scheduling rate may be bound through allocating sufficient bandwidth. However, it will make less RTSNs reserve TXOPs in CFP. The trade-off between the optimal amount of RTSNs accommodated in one SI and the guaranteed delay for these RTSNs is an open issue.

On the other hand, another trade-off exists between a small and a large SI. Small SI can enhance the maximum transmission capability of each allocated

TXOP so that it is capable of accommodating RTSNs with a higher required scheduling rate. Small SI also reduces the delay for each admitted RTSN because of its short non-TXOP duration. However, the amount of RTSNs that can reserve the bandwidth in CFP decreases if small SI is employed. Using a large SI, more resources can be allocated to CFP. But each RTSN may require more bandwidth for satisfying their delay bound due to the degraded performance caused by large SI. To balance this trade-off, the optimization study is a necessity.

The aim of the optimization study is to accommodate maximum amount of RTSNs in an optimum SI given that the QoS of each RTSN is guaranteed. To investigate the optimization, the requirement of the reserved bandwidth for guaranteeing QoS toward RTSNs needs to be identified first. In fact, each dedicated TXOP has its own maximum transmission capability. In order to formulate the maximum transmission capability of a reserved TXOP, the reserved scheduling rate $\lambda_{r,i}$ for RTSN$_i$ is introduced. It can be given by

$$\lambda_{r,i} = \frac{n_{t,i} \times \overline{s_{DATA_i}}}{\Delta SI} \qquad (17)$$

The reserved scheduling rate stands for the maximum transmission capability of the allocated TXOP. The prerequisite of the optimization is that each RTSN need to obtain satisfactory QoS which can be embodied by the requirement of delay bound. Thus, the QoS demand can be defined by

$$d_{ave,i} \leq d_{rmax,i} \qquad (18)$$

where $d_{rmax,i}$ stands for the delay bound for RTSN$_i$. Using the definition of delay in section 3.1, the variance of (18) can be expressed by

$$\overline{d_{ca,i}} + \overline{d_{tr,i}} + \overline{d_{q,i}} \leq d_{rmax,i} \qquad (19)$$

It has been indicated that the guaranteed QoS can only be achieved under the unsaturated and saturated conditions. This argument can be converted to the relationship between the reserved scheduling rate and the required scheduling rate, which is denoted by

$$\lambda_{r,i} \geq \lambda_i \qquad (20)$$

The above argument suggests that QoS demand of a RTSN can only be satisfied when the reserved scheduling rate exceeds the required scheduling rate. In order to find out the connection between the required scheduling rate and the size of TXOP which is denoted by TXOPlimit, the relationship between the reserved scheduling rate and the TXOPlimit needs to be investigated. The size of TXOP can be expressed by

$$t_{TXOP_i} = \frac{\overline{s_{DATA_i}} \cdot n_{t,i} + O_{t,i}}{R} \qquad (21)$$

which indicates that an entire duration of a TXOP for RTSN$_i$ is consumed by transmissions of MSDUs, ACK, MAC header and deferral time SIFS. The transmissions of ACK, MAC header and the deferral time SIFS are deemed as the overhead denoted by $O_{t,i}$. It is expressed by

$$O_{t,i} = n_{t,i}(s_{ACK} + O_{mac}) + 2n_{t,i} \cdot t_{SIFS} \cdot R \qquad (22)$$

where O_{mac} represents the MAC header. Using (17), the derivative of (21) is obtained by

$$t_{TXOP_i} = \frac{\lambda_{r,i} \cdot \Delta SI + O_{t,i}}{R} \qquad (23)$$

Considering the requirement defined by (20), (23) can be transformed into

$$t_{TXOP_i} \geq \frac{\lambda_i \cdot \Delta SI + O_{t,i}}{R} \qquad (24)$$

which stands for the relationship between the TXOPlimit and required scheduling rate under saturated and unsaturated conditions. Since multiple RTSNs can reserve bandwidth in CFP, the average TXOP $\overline{t_{TXOP}}$ is introduced and given by

$$\overline{t_{TXOP}} = \frac{1}{N} \sum_{i=1}^{N} t_{TXOP_i} \qquad (25)$$

Provided that the size of CFP as well as the SI are fixed, the optimum network capacity can be obtained if the existing reserved TXOPs occupy the minimum duration in CFP given that the delay bound of each RTSN is strictly satisfied. Converting this argument to the average TXOP, the network capacity can be maximized if the $\overline{t_{TXOP}}$ gets the minimum value. The expression is derived by

$$f(\lambda_{r,i}) = min\{\overline{t_{TXOP}}\}, \ \ if \ \ d_{ave,i} \leq d_{rmax,i}$$

$$= min\{\frac{\sum_{i=1}^{N} t_{TXOP_i}}{N}\}, \ \ if \ \ d_{ave,i} \leq d_{rmax,i} \qquad (26)$$

Using (23), the equation of optimum resource allocation can be finalized by

$$f(\lambda_{r,i}) = min\{\frac{\sum_{i=1}^{N}(\lambda_{r,i} \cdot \Delta SI + O_{t,i})}{N \cdot R}\}, if \ d_{ave,i} \leq d_{rmax,i} \qquad (27)$$

where the reserved scheduling rate is the variable. It can be indicated from (27) that under the delay bound and fixed SI, the optimum resource allocation as well as the optimum network capacity can be achieved once the reserved scheduling rate of each RTSN is taken as the minimum value. It can be derived from (24) that if delay bound is sufficed, the lowest value of (27) is $\frac{\sum_{i=1}^{N}(\lambda_i \cdot \Delta SI + O_{t,i})}{N \cdot R}$ when $\lambda_{r,i}$ is equal to λ_i.

After figuring out the optimum resource allocation, the optimization of the SI can be investigated. Given the amount of QoS guaranteed RTSNs, if the size of SI changes, the reserved TXOPs of these RTSNs will vary accordingly in order to satisfy distinct level of delay bound. The optimum SI is obtained when the proportion of the reserved TXOPs for these RTSNs in CFP takes the minimum value. Therefore, the optimum SI can be formulated by

$$g(SI) = min\{\frac{\sum_{i=1}^{N} t_{TXOP_i}}{\Delta SI}\}, \ \ if \ d_{ave,i} \leq d_{rmax,i} \qquad (28)$$

where SI is the variable of the optimization. When the optimal SI is achieved, maximum duration can be left in SI for accommodating more RTSNs. Thus, maximizing the network capacity.

5   Performance Evaluation

In this section, we apply the analytical model to predict the delay performance of RTSNs in two SF-based RR schemes which are EDCA/RR and EDCA/DRR. Simulation and analytical outcomes are compared in order to verify the accuracy of the proposed mathematic model. Optimization results are also shown and discussed. Tab. 1 recaps all the parameters used in the evaluation. Several senders with a RTSN per-node as well as a receiver comprise the network. All the nodes are randomly deployed within each other's transmission range.

Table 1. Simulation parameters

Parameter(units)	Value
SIFS(μs)	10
Slot time(μs)	20
ACK size(bytes)	28
MAC header(bytes)	36
Channel capacity(Mbps)	11
Interface queue size(packets)	50
Transmission range (m)	250
Traffic application	CBR over UDP

Fig. 8 shows the simulation and analytical results of delay performance for QoS guaranteed RTSNs in EDCA/RR. The analytical results well-match the simulation outcomes. In order to study the impact of SI as well as the required scheduling rate on delay, three distinct size of SIs (i.e. 10ms, 15ms, and 20ms) are used and the required scheduling rate varies from 300kb/s to 1500kb/s. In this case, the allocated resource is made equal to the required bandwidth. It can be concluded that given a SI, delay of RTSNs with TXOPs become higher along

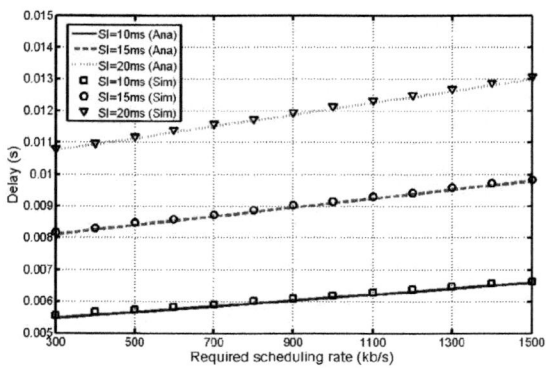

Fig. 8. Delay of RTSNs in EDCA/RR

with the increment of the required scheduling rate. The reason is that average queueing delay increases once the number of MSDUs allowed to be transmitted in a TXOP becomes larger. It also can be seen from Fig. 8 that compared with smaller SI, delay leaps dramatically in the face of larger size of SI. This is because the increasing channel access delay contributed by the non-TXOP duration plays the major impact on the average delay.

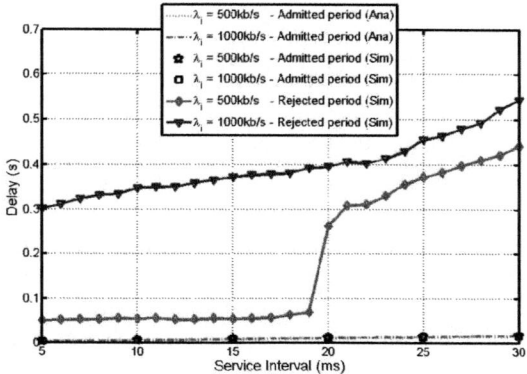

Fig. 9. Delay of RTSNs in EDCA/DRR

Fig. 9 shows the delay performance of RTSNs in EDCA/DRR. Two different required scheduling rates (i.e. 500kb/s and 1000kb/s) are used and the delay of RTSNs is tested under different size of SIs. Sufficient bandwidth is assigned to CFP (i.e. maximum CFP duration is equal to 0.75SI) in order to accommodate more RTSNs. On the other hand, a certain proportion of duration is allocated to CAP for the fairness toward other types of sessions or the rejected RTSNs which can not obtain dedicated resources in CFP. Simulation result implies that the rejected RTSN suffers from a degrade performance. This attributes to the channel contention from other NRTSNs and the limited duration for contending the chance of channel access. On the other hand, in the contention-based environment, rejected RTSN with higher required scheduling rate receives a worse delay performance than the rejected RTSN with lower required scheduling rate. This is because the buffered MSDUs in the interface queue accumulates faster in terms of the RTSN with higher required scheduling rate. In contrast to their performance in CAP, the delay drastically decreases if the RTSNs are re-admitted by the admission control algorithm in EDCA/DRR and transmitted during CFP when certain bandwidth in CFP becomes idle. In addition, a good agreement between the analytical outcomes and simulation results can be seen in Fig. 9.

As analysed in the previous section, the reserved scheduling rate of a RTSN needs to be at least equal to the required scheduling rate so that its QoS demand such as the delay bound can be possibly satisfied. Even if the above requirement is met, a RTSN may need a higher reserved scheduling rate compared with the

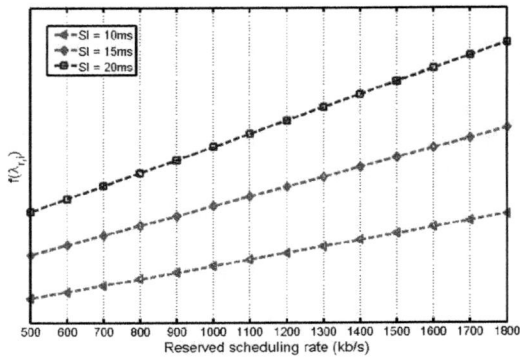

Fig. 10. Optimum scheduling rate

required scheduling rate if it is desired by a lower delay bound. Delay reduces if
more bandwidth can be reserved by the corresponding RTSN. Consequently, to
consider the optimization of system parameters such as the allocated resource
(i.e. TXOP) and SI, the bandwidth reserved for each RTSN need to be the minimum value that can exactly meet the delay bound under certain value of SI.
Fig. 10 shows the optimum scheduling rate of RTSNs with required scheduling
rate of 500kb/s. It suggests that under a relatively loose delay bound, the optimum scheduling rate is equal to the required scheduling rate. From Fig. 10,
we can also conclude that the more resources allocated to these RTSNs, the
worse the network capacity becomes. However, reserved scheduling rate needs
to be adjusted according to the QoS requirement (i.e. delay bound and required
scheduling rate). To investigate the optimum SI under different QoS demand, a
set of required scheduling rates (i.e. 200kb/s, 800kb/s, 1400kb/s and 2000kb/s)
are given under the delay bound of 25ms. The SI is tested from 5ms to 100ms

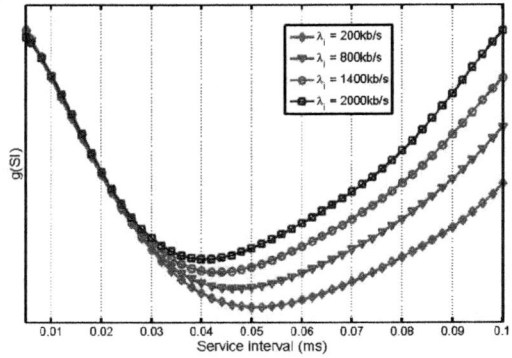

Fig. 11. Optimum SI

in order to identify the optimum value. Fig. 11 shows the results that have been figured out for all the different conditions. It can be seen that the optimum SI tends to be larger if RTSNs with lower required scheduling rates are employed. The reason is that under a light traffic load, RTSNs can reserve a relatively small size of TXOP which can satisfy their QoS. As the SI is increasing, the requested bandwidth will not leap dramatically so that the optimum value can be achieved at a larger value. The results shown in Fig. 10 and Fig. 11 validate the feasibility of the proposed optimization study.

6 Conclusion

Resource reservation plays an important role in QoS provisioning for multimedia applications in IEEE 802.11-based wireless networks. This paper provides an analytical model and an optimization study for SF-based distributed RR mechanisms. The proposed model has been proved accurate for evaluating the delay performance of RTSNs with TXOPs under different traffic conditions. In addition, optimization of system parameters such as the size of allocated TXOPs and the SI has been conducted for maximizing the network capacity, making more RTSNs obtain dedicated bandwidth in CFP. Simulation and analytical results have verified the accuracy of the analytical model. In addition, the analytical results for optimization have shown optimum parameters under different situations, which has proved the feasibility of the optimization study.

Acknowledgement. This work was performed in the project QoSMOS which has received research funding from the EU FP7 framework.

References

1. IEEE 802.11-2007: IEEE Standard for Wireless LAN Medium Acess Control (MAC) and Physical Layer (PHY) Specification (June 2007) (revised version)
2. Ghazizadeh, R., Fan, P.: Dynamic Priority Scheduling Mechanism through Adaptive InterFrame Space. In: Proc. IEEE WiCOM 2007, Shanghai, China (September 2007)
3. Naoum-Sawaya, J., Ghaddar, B., Khawam, S., Safa, H., Artail, H., Dawy, Z.: Adaptive Approach for QoS Support in IEEE 802.11e Wireless LAN. In: Proc. IEEE WiMOB 2005, Montreal, Canada (August 2005)
4. Kim, M.S., Shrestha, D.M., Ko, Y.B.: EDCA-TM: IEEE 802.11e MAC Enhancement for Wireless Multi-Hop Networks. In: Proc. IEEE WCNC 2009, Budapest, Hungary (April 2009)
5. Feng, Z., Wen, G., Zou, Z., Gao, F.: RED-TXOP Scheme for Video Transmission in IEEE 802.11e EDCA WLAN. In: Proc. IEEE ICCTA 2009, Beijing, China (October 2009)
6. Zhang, Y., Foh, C.H., Cai, J.: An On-Off Queue Control Mechanism for Scalable Video Streaming over the IEEE 802.11e WLAN. In: Proc. ICC 2008, Beijing, China (May 2008)

7. Hamidian, A., Körner, U.: Extending EDCA with Distributed Resource Reservation for QoS Guarantees. J. Sel. Telecommun Syst. 39(3-4), 187–194 (2008)
8. Yu, X., Navaratnam, P., Moessner, K.: Distributed Resource Reservation Mechanism for IEEE 802.11e-Based Networks. In: Proc. IEEE VTC 2010, Ottawa, Canada (September 2010)
9. Lin, Y., Wong, V.W.S.: Frame Aggregation and Optimal Frame Size Adaptation for IEEE 802.11n WLANs. In: Proc. IEEE GLOBECOM 2006, San Francisco, USA (December 2006)

A Method for Detection/Deletion via Network Coding for Unequal Error Protection of Scalable Video over Error-Prone Networks

Michele Sanna and Ebroul Izquierdo

School of Electronic Engineering and Computer Science,
Queen Mary, University of London, E1 4NS, London, UK
{michele.sanna,ebroul.izquierdo}@eecs.qmul.ac.uk

Abstract. The development of universal systems for video streaming needs transmission strategies that exploit the characteristics of the transmission medium such as a wireless network. Scalable video coding allows partial decoding of the video for multiple demands or under severe reception conditions. Network coding increases the transmission rate and provides error control at network level. We propose a detection/deletion system for error reduction in presence of channel noise. We combine the error detection capabilities of the network code with erasure decoding and unequal error protection to improve the visual quality of the video.

Keywords: scalable video coding, network coding, network error correction, multicast, error detection, detection/deletion, erasure decoding.

1 Introduction

Due to the diffusion of connectivity and multimedia services, multimedia communication in the future will try to reach several kind of communication platforms. In universal video systems, a wide landscape of users are connected via different physical means to a service provider. On the one hand, a backbone infrastructure connects heterogeneous platforms to the streaming server. On the other hand, users display requirements, reception conditions and Quality of Service (QoS) demands can be deeply different between the platforms. Systems like Digital Video Broadcasting (DVB) differentiate the physical interface between terrestrial (DVB-T), satellite (DVB-S), and Handled (DVB-H) connected by a common backbone network.

Flexibility is required both from the backbone network and the coded data carried along. Scalable Video Coding (SVC) is a video coding paradigm that embraces Layered Coding (LC) applied to video, exploiting the variety of the transmission channels and users requirements [1]. SVC allows partial decoding of the video stream by extracting parts of the stream if needed. Decoding at reduced resolution, frame rate or quality from the same source bitstream is possible in case of different display requirements or impossibility to decode part of the stream due to channel errors.

L. Atzori, J. Delgado, and D. Giusto (Eds.): MOBIMEDIA 2011, LNICST 79, pp. 105–120, 2012.

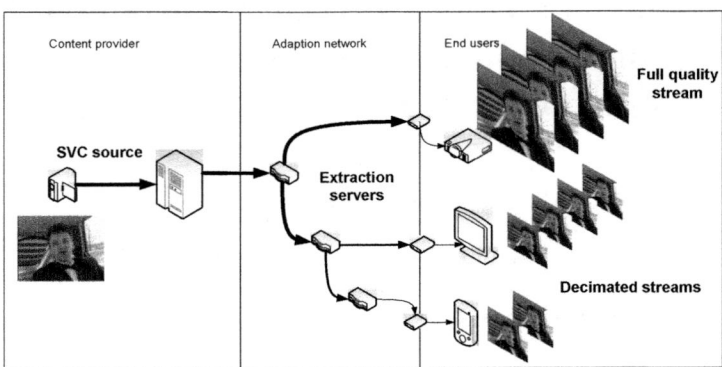

Fig. 1. An SVC application scenario with stream adaption for differentiated services

Multirate content delivery is possible in a structured scenario like the one in Fig. 1. Mobile devices with low computational power and small displays need a compact stream with reduced quality and resolution, whereas home users usually make use of big displays and a large bandwidth and big screens and demand a superior quality.

When it comes to transmission with channel noise, like in wireless links, SVC has an intrinsic mechanism to combat information loss and errors. Noise has different and independent impact on the layers. Errors occurring on the higher layers have less impact in the visual quality than those in the base layers. They are independently encoded and decoded, thus separating the impact of the error. Unequal Error Protection (UEP) increases the visual quality with respect to equal error protection and to the conventional non-scalable codecs [2,3]. Loss (e.g., in packet networks) or corruption of the information in the highest layers do not nullify the decoding of lower layers, thus offering a continuous service. Unequal Loss Protection (ULP) and Multiple Description Coding (MDC) techniques have been proposed to boost the performance in lossy packet networks [4,5].

In this paper we use an MDC-like coding technique based on ULP-FEC codes to protect against channel errors. We use an erasure decoding method with Detection/Deletion (D/D) via Network Coding. Network Coding (NC) was introduced by Ahlswede *et al.* in 2000 [6]. This paradigm allows intermediate nodes to decode the incoming packets and re-transmit to the other nodes a function of the received information. NC overpasses the traditional forwarding at intermediate nodes, which can be regarded as a special case of network coding. Conventional networking can not cope with bottlenecks, thus it is not always possible to serve all receivers at the same time in a multicast setting as in Fig. 2 (a). With network coding the information can be combined in the bottlenecks and communicated to all receivers, which under determinate conditions are able to decode the source information (Fig. 2 (b)). Network coding has found potential application possibilities in Content Delivery Networks (CDN) and Peer-to-Peer (P2P) networks [7]. NC increases the network throughput [6] and allows error control against link failures at network level thanks to Network Error Correction [8,20]. Further

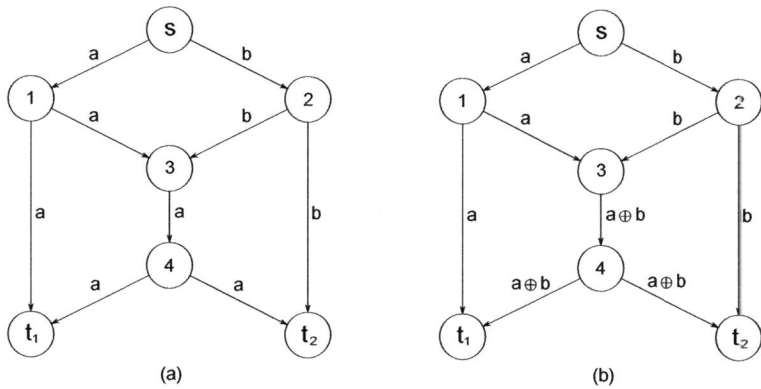

Fig. 2. The two-sources two-sinks butterfly network without (a) and with (b) network coding

benefits are found in using network coding in wireless networks, in which physical superposition of signals can be used opportunistically [9,10].

We consider the multicast transmission of scalable video on error prone networks and we aim at decreasing the impact of channel errors on the visual quality. ULP techniques have been reproposed in various works to produce multiple descriptions of data [11,5]. Multiple Description Coding (MDC) produces a number of dependent representations of the source information and allows partial decoding with increasing quality as more chunks of coded data are recovered. We use a similar technique to combat errors instead of packet losses, by modifying the construction of FEC-based MDC. We produce data blocks coded with erasure codes with differentiated persistence via UEP. The streams are transmitted in independent paths through a network performing network coding and error detection is performed at the receivers. Errors are detected via Network Error Correction and data blocks with errors are deleted. Erasure correction on the temporal direction is performed on the remaining data blocks of the streams. The spatial dimension of error detection is separated from the temporal dimension of erasure correction. Orthogonal coding redundancy achieves superior error protection performance of the scalable video with respect to the cases of separated FEC coding and network error correction only.

This system benefits from increased throughput in multicast scenario thanks to the network coding. Although MDC-derived, our detection/deletion scheme is designed for robustness against channel errors, rather than packet losses, for the wireless radio channel. Packet losses can be treated with statistical decoding of NEC-coded packets.

The organization of the paper is as follows: Section 2 presents the scalable video coding paradigm and the traditional FEC-based MDC scheme. In Section 3 we discuss the implementation of the transmission system with network coding and we introduce the detection/deletion method of MDC derivation. In Section 4 the performance of the transmission system is evaluated. Section 5 concludes the paper.

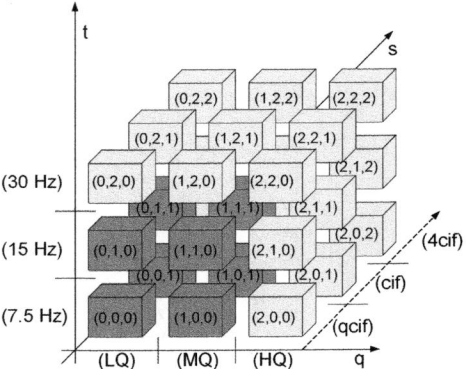

Fig. 3. Representation of atoms of an SVC stream with 3 levels of temporal, spatial and quality scalability [12]

2   Scalable Video Coding

Classic video codecs are designed to work under definite display requirements and channel rate. The variable nature of the wireless channel and the variety of transmission possibilities in large networks requires flexibility in terms of source coding. An Scalable Video Coding (SVC) approach supports multiple decoding configurations from the same embedded bitstream for different display and channel requirements, avoiding overloading the network and decreasing the coding computational load [1]. Full scalability allows reducing spatial resolution, frame rate and quality (SNR) if needed. The H.264/SVC extension of the video compression standard will support full scalability with a DCT block-based approach. Other alternative solutions have been proposed, like the W-SVC codec which has also full scalability but it is based on wavelet transforms [12].

The SVC approach to video coding produces an embedded bitstream of elementary units (atoms) containing the coded information of an enhancement level of time/space (T/S) resolution or quality for every Group of Pictures (GOP), as shown in Fig. 3. Atoms are organized in hierarchy where each layer always depends on the lower ones for decoding. Partial decoding can be performed by bitstream parsing according to a desired display configuration, by extracting and assembling in a reduced bitstream the required atoms to decode at a desired display configuration. SVC is designed to cope with differentiated demands as well as unpredictable channel conditions. When particularly adverse channel conditions nullify the decoding of parts of the stream, partial decoding at reduced time/space resolution or quality provides a continuous service.

Boosting the decoding performance is possible with proper channel coding of SVC. The hierarchy of the coding layers implies that lower layers have a higher impact on video quality. Also, higher layers are useless if the underneath layers are not decoded successfully. Unequal Error Protection (UEP) differentiates the redundant bits allocated to protect different layers under particular rate

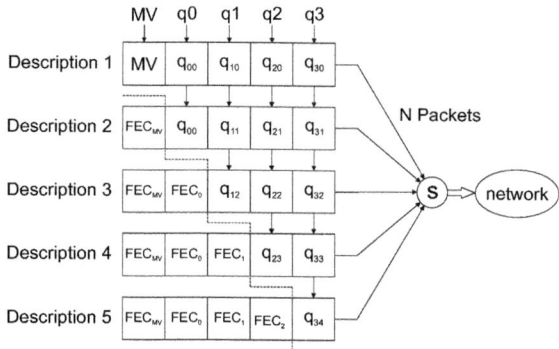

Fig. 4. Scheme for Multiple Description Coding via Forward Error Correction and for unequal packet-loss protection

constraints. Traditional channel coding techniques are applicable to differentiate FEC coding among layers, achieving superior quality and reducing the percentage of undecoded GOPs than equal protection or standard non-scalable video coding approaches [2,3].

2.1 FEC-Based Multiple Description Coding for Scalable Data

We now describe a FEC-based Multiple Description Coding (MDC) technique, whose structure is adopted in our Deletion/Detection method to protect against channel errors. MDC has been proposed to allow partial decoding of source data under random loss of information, e.g. in packet networks [13]. MDC can exploit the independence of the coding layers of SVC and yield to opportunistic recovery of information under severe channel conditions. The traditional method for generating multiple descriptions of a scalable stream is resumed in Fig. 4 [14]. In case of quality scalability, the quality layers $Q_i, i = 1, 2...$, are grouped in non overlapping sets and organized in independent streams, as:

$$Stream_0 = \{Q_0, Q_1, \ldots, Q_{n_1}\},$$
$$Stream_1 = \{Q_{n_1+1}, \ldots, Q_{n_2}\},$$
$$\ldots$$
$$Stream_N = \{Q_{n_{i-1}+1}, \ldots, Q\},$$

and with source rates (without error-control coding):

$$R_{stream_i} = \sum_{i=n_{i-1}+1}^{n_i} R_i. \tag{1}$$

where R_i is the rate of the i-th video layer.

Coded blocks of equal length are obtained by applying erasure codes with variable rate to each stream. Portions of k_i symbols are taken from each i-th layer and organized into FEC-coded packets of $n = k_i r_{FEC}^{(i)}$ symbols, in an

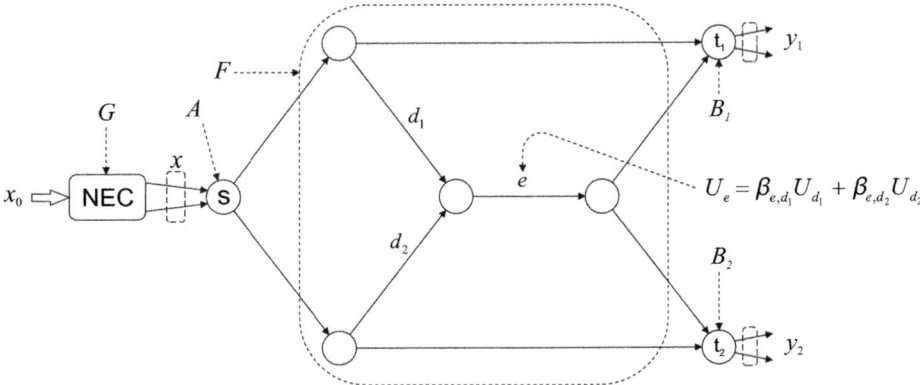

Fig. 5. Algebraic modeling of network coding with local linear encoding

interleaving-like manner, as show in Fig. 4. Smaller messages from the base layers are encoded with a stronger erasure code, with rate $r^{(0)}_{FEC} < r^{(1)}_{FEC} < \ldots < r^{(N)}_{FEC}$, so that

$$R_{stream_0}/r^{(0)}_{FEC} = R_{stream_1}/r^{(1)}_{FEC} = \ldots = R_{stream_N}/r^{(N)}_{FEC}. \qquad (2)$$

The loss of l packets, with $n - k_i \geq l > n - k_j, i < j$, allows erasure decoding of the streams up to the i-th, from the $n - l$ remaining packets. The recovered packets being not enough to decode the other layers, partial decoding of the video is possible, with increasing quality as more packets are received.

This scheme is useful in transmission affected by packet losses. We use this criterion to perform opportunistic deletion of coded symbols across the packets upon detection of errors via network coding. Our scheme protects against channel errors occurring in an error-prone network, as explained in Sec. 3.

3   Network Coding for Error Detection

Network coding transmission has the property of modifying the information like a coding operator. The motivation for employing network coding is the possibility of reaching the network capacity in multicast settings and the possibility of performing error correction/detection on the spatial dimension at network level. We consider the single-source multicast transmission on an acyclic network, i.e., without closed cycles, where linear coding and decoding operations in a vector space over a finite field are implemented at the nodes. [15].

Consider a network model as a directed graph $\mathcal{G} = (V, E)$, where V is a set of vertices and E is a set of edges. A vertex is designated as the source node s and the receivers are grouped in the nodes set T. Every edge has unitary capacity. Links with capacity multiple of the unit are treated as multiple parallel edges. Given a non-source node i with input edges $d \in In(i)$ and output edges $e \in Out(i)$, then the message U_e transmitted on an edge e can be expressed as:

$$U_e = \sum_{d \in In(e)} \beta_{d,e} U_d. \tag{3}$$

where $\beta_{d,e}$ constitute the *local coding kernels* of the edges $e, \forall e \in E$. Blocks of m bits are considered as symbols in a finite field $\mathbb{F}$ of size q, e.g., a Galois Field of size $q = 2^m$.

Generation-based transmission is considered, where each nodes stores in a buffer all the packets belonging to the same generation before transmitting their linear combination. Unitary coding vectors are put before the payload in the packets so that the receivers can deduce the network code [16]. This corresponds to a transmission delay equal to the delay of the longest path.

For modeling the network transmission, an algebraic model can be employed, where the network transform characteristic can be modeled by a transfer matrix M_t from the source to the receiver t [17]. The source data is arranged in vectors $\mathbf{x} = [x_1, x_2, \ldots, x_h]$. If the network supports a rate of h symbols through h edge-disjoint paths, the sink t receives a vector $\mathbf{y}_t = [y_1, y_2, \ldots, y_h]$ obtained by the network transformation as:

$$\mathbf{y}_t = \mathbf{x} M_t + \mathbf{z} F_t \tag{4}$$

which also models random errors and erasures by means of a $1 \times |E|$ additive error vector $\mathbf{z}$.

3.1 Error Control via Network Error Correction

Consider a network with a max-flow h to each destination. With Network Error Correction (NEC), the set of source messages $\mathbf{x}$ belongs to an ω-dimensional vector space $\mathcal{C} \subseteq \mathbb{F}^h$. The codebook $\mathcal{C}$ generated by means of traditional block codes with $\omega \times h$ generation matrix G, is a Minimum Distance Separable (MDS) code with minimum distance $d_{min,s} = h - \omega + 1$. Every network codeword is generated by means of

$$\mathbf{x} = \mathbf{x}_0 G, \tag{5}$$

where $\mathbf{x}_0 = [x_{0,1}, x_{0,2}, \ldots, x_{0,\omega}]$. The coding space at the receivers $t \in T$ can have by construction minimum distance $d_{t,min} = h - \omega + 1$ and the redundant linear combinations can be used at the receiver for error control purposes [8]. The receivers are able to correct up to $\lfloor \frac{d_{min,t}-1}{2} \rfloor$ and detect up to $d_{min,t} - 1$ errors.

A distributed approach for the construction of the network code can be assumed. A randomized choice of the coding kernels achieves an MDS code with correct minimum distance with increasing probability as the field size grows. For codebooks with low redundancy ($d_{min} = 0, \ldots, 2$), a reasonable choice of field size is $q = 2^8, 2^{12}, 2^{16}$, which achieve success probability above 90% [18]. Packetized transmission is necessary to communicate the chosen kernels to the receiver. This approach is more suitable in wireless ad-hoc networks, where the nodes occasionally join a mesh network and take part to the transmission.

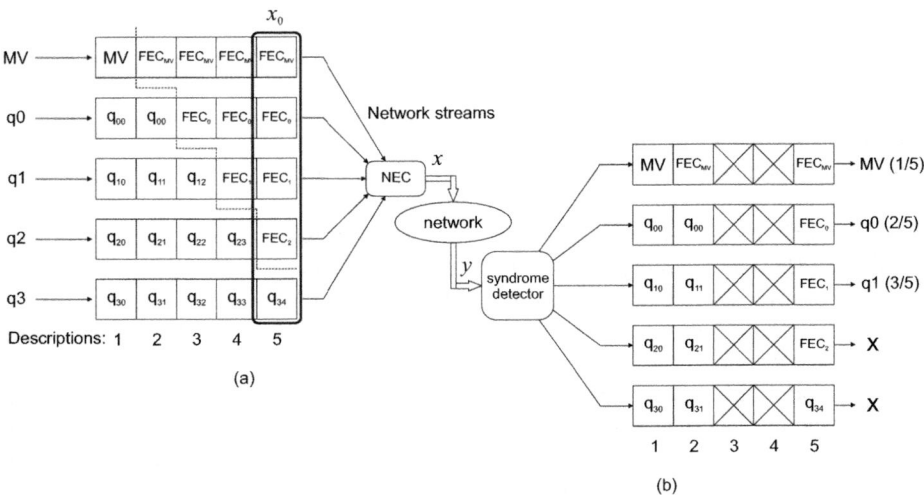

Fig. 6. (a) Scheme for Multiple Description Coding via Forward Error Correction and modified transmission criterion for detection via Network Error Correction, deletion and erasure decoding. (b) An example of partial decoding of MDC in case of erroneous decoding of network codewords 3 and 4.

3.2 The Detection/Deletion Scheme

The detection/deletion (D/D) system for robustness against random errors has packetization scheme transversal to the MDC scheme.

In a network coding transmission system, the loss of a packet in one link is not perceived from the receiver as with traditional networking. A packet loss is perceived as an alteration of the coding space, by $\mathbf{z}Ft$ as in Eq. (4), rather than an actual loss of information. This impedes identification of losses and thus traditional erasure correction, because losses are analogous to channel errors.

The D/D system detects error by means of the NEC code and applies erasure decoding on the remaining symbols. The system described in the following is not formally an MDC system, but achieves superior resistance to random channel errors which impair wireless transmissions.

Coding. The coding scheme is shown in Fig. 6 (a). As for traditional MDC, ω streams are build from the scalable data. On the temporal dimension every stream is independently erasure-encoded. FEC coding rates are such that the resulting datablocks have length equal to the network packet lenght, as in Eq. (2). The source streams are arranged in vectors $\mathbf{x_0} = [x_{0,1}, x_{0,2}, \ldots, x_{0,\omega}]$, where every $\mathbb{F}$ symbol from the i-th stream stays at the corresponding position $x_{0,i}$ of $\mathbf{x_0}$. Then the streams are NEC coded with the chosen generation matrix and sent in parallel on the network paths (Fig. 6).

Detection/Deletion and Decoding. All symbols in a packet share the same coding at the intermediate network nodes. On the other hand random errors affect network codewords $\mathbf{y}_t$ independently. Traditional decoding of network coding is based on complete decoding, i.e., $\lfloor \frac{d_{min,t}-1}{2} \rfloor$ errors are corrected at network level.

At the receiver the network codewords are generated by means of a system of linear equations with core M_t. Since the number of equations h is more than the number of variables ω, if the system is undetermined the codewords are detected as not belonging to the target codebook at the receiver. A syndrome detector can be used for this purpose [19].

The received codeword is detected as erroneous if such word has a minimum distance from the rest of the receiver's codewords higher than the number of errors occurred during the transmission slot. All the symbols at the corresponding position in the ω streams are flagged as erroneous and discarded (Fig. 6). Of all the symbols of a packet, those that are not recognized as erroneous can be used for erasure decoding. If l network codewords belonging to the same data block are flagged, with $n - k_i \geq l > n - k_j, i < j$, the blocks of the streams up to the i-th are correctly decoded from the remaining symbols.

This technique decouples the errors happening on the temporal dimension with a detector on the spatial dimension that detects up to $d_{min,t} - 1$ errors on the same time slot. E.g., when using a $(h - 1, h, 1)$ code ($d_{min,t} = 1$) error patters of weight equal to 1 (i.e., $|\mathbf{z}| \leq 1$) are always detected. Patterns with higher number of errors are partially detectable if the base field is large enough (Fig. 7). In a multirate setting, the sinks can receive differentiated flows, with differentiated detection capabilities.

Code construction in decentralized setting is performed in a randomized manner. The intermediate nodes retransmit random linear combinations of the packets that they have in the buffer. The coding distance of the codewords at the receiver $d_{min} = h - \omega$ is random. The probability of having a minimum coding distance for all codewords is higher for small distances and decreases for values close to the coding limit [18]. Small coding distances are required for the task of only detecting errors, i.e., a distance $d_{min,t} = 2, t \in T$ is required for detection, which is more probable than a distance $d_{min,t} = 3, t \in T$ for complete NEC decoding. In a randomized network code the distance between the codewords can be irregular. Codewords are more probable to have small distance (usable for detection purposes) among them. Detection performance of a random code is shown in Fig. 7 and commented in the next section. In order to cope with packet losses, a traditional method for statistical decoding can be used [20]. Even with small coding distances, e.g. $d_{min,t} = 2$, due to the fact that all the symbols in a packet share the same network code, a possible erasure pattern and error vector can be found for all the slots in a packet with high probability, and the packet loss corrected.

We perform network detection at receivers by means of a syndrome-based detector [19]. Consistently with classic syndrome decoding, a parity check matrix H_t is built from the codebook generator matrix G and the system matrix M_t. If the systematic form of the generator matrix of the code at the receiver t is

$$GM_t = [I_\omega, P],$$ (6)

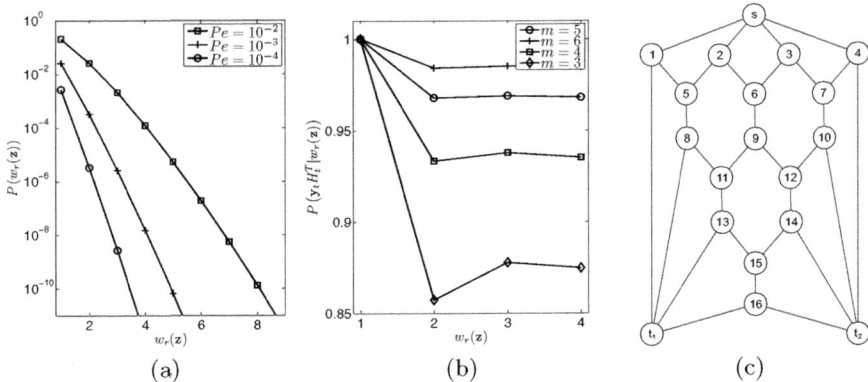

Fig. 7. (a) Probability of successful detection of errors with weight $w_r(\mathbf{z})$ (field size $q = 2^m$). Network code is $(\omega, h, d_{min,t}) = (3, 4, 2), t in T$ and the network is (c). (b) Probability of occurrence of error patterns with weight $w_r(\mathbf{z})$. (c) Network used for graphics in (a) and (b).

then the parity check matrix for the source code projected at receiver t is

$$H_t = \left[P^T, I_{h-\omega} \right]. \tag{7}$$

For all network codewords transmitted without errors:

$$\mathbf{x}M_t \cdot H_t^T = \mathbf{0}, \quad \mathbf{x} \in \mathcal{C} \tag{8}$$

Codewords with syndrome not equal to $\mathbf{0}$ are flagged as erroneous.

In the next section we present the simulation results of the detection/deletion system with scalable video transmission.

4   Simulation Results

When the network allows, the network code can be constructed deterministically to achieve the desired minimum distance at each receiver $d_{min,t} = h - \omega + 1, t \in T$. This is in general not easy to implement in wireless networks unless a centralized authority controls and manages the connections. Randomized code construction can be performed in decentralized manner. The packets are mixed at the intermediate nodes with random coefficients which are attached to the packet header.

We consider the detection capabilities of the network code. A network code with minimum distance $d_{min,t}$, is designed to detect all errors with pattern weight:

$$w_r(\mathbf{z}) = \text{rank}(\rho_\mathbf{z}) = d_{min,t} - 1, \tag{9}$$

where the error pattern $\rho_\mathbf{z}$ is a $1 \times E$ vector with unitary components in the positions corresponding to the non-zero components of $\mathbf{z}$. Its rank is the number

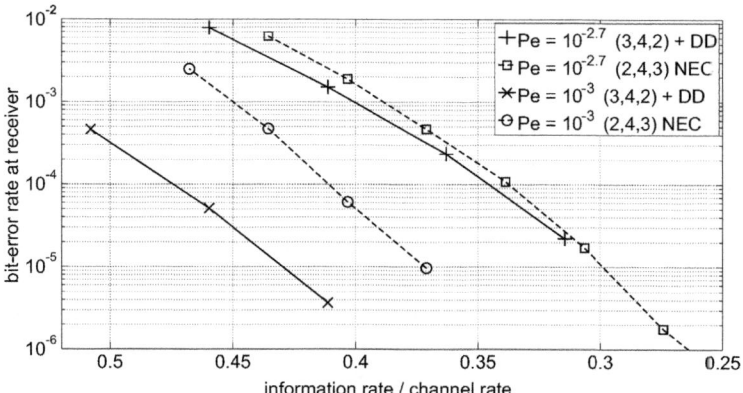

Fig. 8. Comparison of bit-error rate at the receivers at different FEC coding rates with traditional network error correction and with the detection/deletion method (D/D) under bit-error probability at the intermediate links is 10^{-3} and $10^{-2.7}$. The field size is $q = 2^6$ and block size for symbol-based FEC code $N = 186$ bits.

of non-zero components. Although a network code with distance $d_{min,t}$ allows detection of up to $d_{min,t} - 1$ link errors per time slot of a symbol at receiver t, the following considerations motivate the use of a code with source redundancy $\delta = h - \omega = 1$. Such codewords can achieve a distance $d = 2, t \in T$ with higher probability than a code with higher redundancy [18]. Especially for small field sizes, whose choice benefits the computational loads at the intermediate nodes and at the decoder, small coding distances among the codewords (even if the minimum distance is smaller) are more probable than higher distances, e.g., $d_{min,t} = 3, t \in T$ for detecting errors with $w_r = 2$ or correcting $u_r = 1$ [18]. Error patterns with $w_r > 1$ are often dominated by patterns with $w_r = 1$. Such capability of detecting error patterns beyond the limit is shown in Fig. 7 (a) for the network in Fig. 7 (c). The probability of detecting error patterns with weights beyond > 1 with a code with $d_{min,t} = 1$ is over 95% when using a code in a Galois Field with size $q = 2^m$ with m higher than $m = 6$. Additionally, error patterns with increasing weight appear with decreasing probability as shown by Fig. 7 (b). Such results suggest that the missed detection of errors happens for a small percentage of patterns and thus with small probability. In order to cope with packet losses, a traditional method for statistical decoding can be used [20].

The erroneous codewords detected via network coding are flagged as erroneous and erasure decoding is performed at the receiver based on the symbols received correctly (Fig.6). Symbol-based erasure codes are used, such as Reed-Solomon codes, with generator polynomial in a field of dimension 2^m. We test the transmission on the network in Fig. 7(c), with random coding at intermediate nodes for every generation of packets, i.e., every h packets generated at the source at the same time are synchronously coded throughout the network. Detection is performed by means of a $(\omega, h, d_{min,t}) = (3, 4, 2)$ network code $t = 1, 2$. We compare the detection/deletion (D/D) transmission technique with a

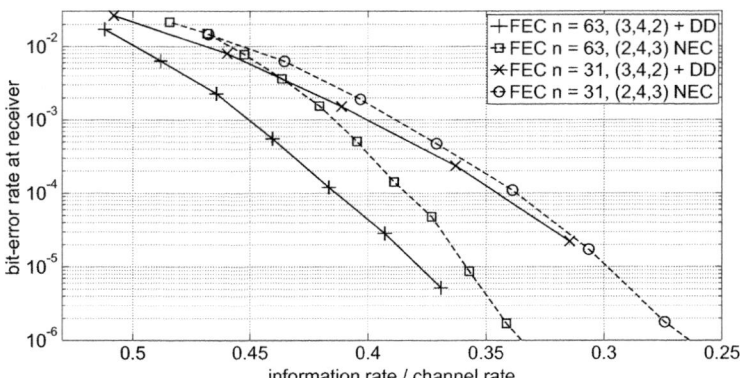

Fig. 9. Comparison of bit-error rate at the receivers at different FEC coding rates with traditional network error correction and with the D/D method and block sizes for symbol-based FEC code $N_1 = 31 \times m$, $N_2 = 63 \times m$. Bit-error probability at the intermediate links is $Pe = 10^{-2.7}$, field size is $q = 2^6$.

transmission with a network error correction code $(\omega, h, d_{min,t}) = (2, 4, 3)$. This code can correct all single errors at network level. Since the information rate is 2/3 of the other case, error-control codes with lower protection are used. Differentiated local FEC codes are applied at the source, but globally the allocation of the error protection is less flexible because the NEC code protects equally the video layers.

Fig. 8 shows the error protection performance of the D/D method and the normal NEC transmission under two different link error rates. For low error rates the performance of the D/D method is always higher than the traditional NEC for all chosen coding rates. The ratio between information rate and channel rate is calculated considering the rate of the erasure codes and the NEC code rate.

The choice of large block length for the erasure code can sensibly reduce the impact of link errors on the bitstream decoded by the receiver. This is shown in Fig. 9. The error protection performance of the D/D method is higher than NEC with the tested block lengths. As the block length increases the NEC system performs similarly. The use of small block sizes can be motivated by the fact that an erroneous decoding of a whole datablock propagates the errors to a large part of the bit stream and may lead to undecodability of a GOP.

Fig. 10 shows the error protection performance of the D/D method and the normal NEC transmission with different field size. The distance between the codewords has a higher probability of being respected with a larger field size. The increase in the field size increases more the performance of the traditional NEC rather than the detection deletion method. On the other hand the datablock size is reduced in order to accommodate the same length in terms of number of bits, thus reducing the correction capability of the erasure code and yielding, for the tested setting, to a higher error rate at the receiver.

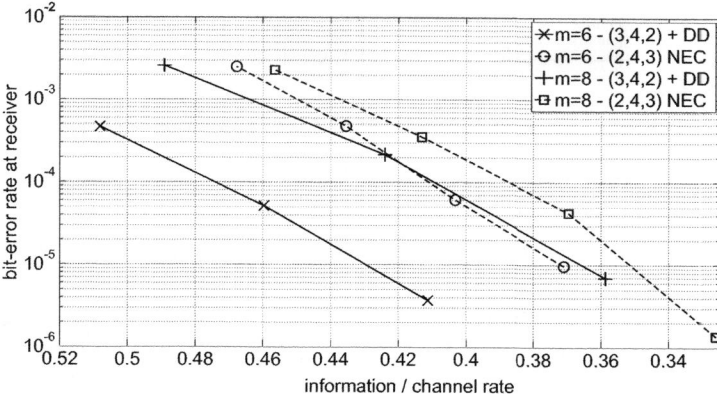

Fig. 10. Comparison of bit-error rate at the receivers at different FEC coding rates with traditional network error correction and with the D/D method and field sizes $q_1 = 2^6$ and $q_2 = 2^8$. Bit-error probability at the intermediate links is $Pe = 10^{-3}$ and block size for symbol-based FEC code $N = 378$ bits.

We test the transmission of video coded with the W-SVC codec with 3 levels of quality scalability, at the bitrates of 288 kbps for the base layer and 480 kbps and 800 kbps for the refinement layers. Following the D/D scheme in Fig. 6 we apply unequal error correction (UEP) of the video layers by means Reed-Solomon erasure codes. We compare the video quality by means of Peak Signal-to-Noise Ratio (PSNR) of the decoded video, shown in Fig. 11 under variable error rate at the intermediate links. The error-rates of the same streams are compared in Fig. 12. The error correction performance and the video quality are higher with the D/D system in the shown configuration. The error rate with the D/D method is equal to zero up to a certain threshold. This characteristic allows to perform perfect decoding of the video up to this value. Even if NEC allows a gentle increase of the error, even few errors often nullify the decoding of a GOP, thus it's preferable to have zero errors for a wider range of link-error rates. The nature of the scalable video, coupled with the UEP, allows to decide to drop a video layer when this becomes uncorrectable due to the undecodable errors. This happens for an error rate at the receiver of around 10^{-4}, which is reached earlier by the NEC method. This in general interests first the enhancement layers and then the lower ones as the link-error rate increase. The NEC system overcomes the D/D system for higher error rates. This error rates are though not acceptable for decoding the video. The D/D method performs better at low or controllable error rates, which makes necessary some kind of link-based error control system. It also performs better with small field sizes which also yield to less computational load for the intermediate nodes. Much less complexity weights on the receiver, due to the simple detection method (rather than the statistical solution of overranked linear systems) and the need of a smaller field size.

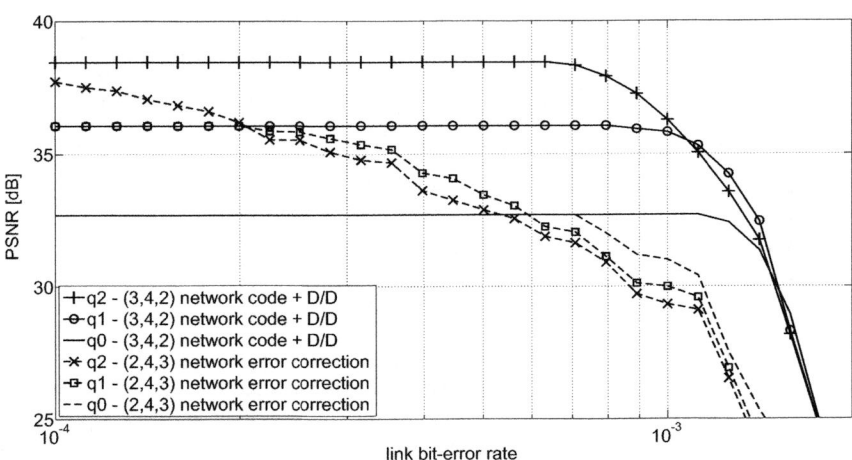

Fig. 11. Average Peak Signal-to-Noise ratio for *city* sequence at CIF resolution on the network in Fig. 7(c). Detection/Deletion by means of a network code $(\omega, h, d_{min,t}) = (3, 4, 2)$ is compared with a traditional $(2, 4, 3)$ NEC code.

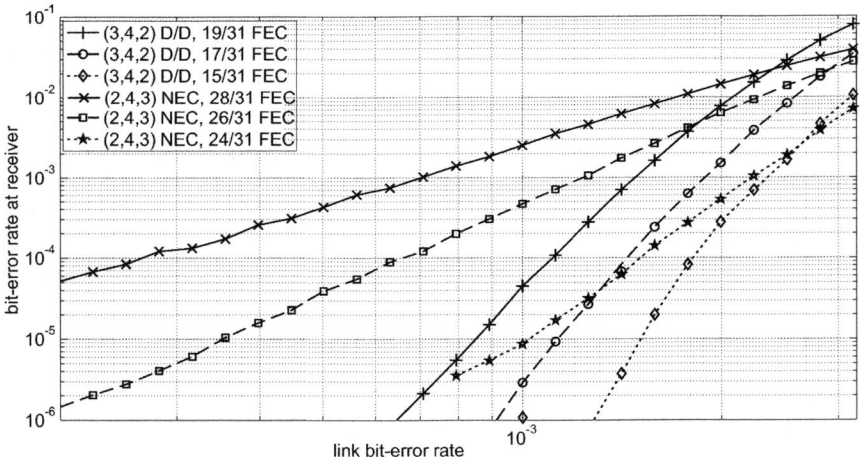

Fig. 12. Bit-error rates for the three video streams of the example in Fig. 11

5   Conclusions

We presented a system for error detection and deletion via Network Error Correction codes and erasure correction for scalable data. This technique allows for better use of transmission rate for error control purposes. Detection and deletion of errors allows zero-error decoding of the video up to higher levels of noise than traditional network error correction coding and normal forward error correction codes, exploiting the scalable nature of the video for higher error rates and yielding to an untouched video quality and decodability.

Acknowledgments. This research was partially supported by the European Commission under contract FP7-247688 3DLife and FP7-248474 SARACEN.

References

1. Schwarz, H., Marpe, D., Wiegand, T.: Overview of the scalable video coding extension of the h.264/avc standard. IEEE Trans. Circuits Syst. Video Techn. 17(9), 1103–1120 (2007)
2. Ramzan, N., Wan, S., Izquierdo, E.: Joint source-channel coding for wavelet-based scalable video transmission using an adaptive turbo code. J. Image Video Process. (2007) (January 2007)
3. van der Schaar, M., Radha, H.: Unequal packet loss resilience for fine-granular-scalability video. IEEE Transactions on Multimedia 3, 381–394 (2001)
4. Albanese, A., Blomer, J., Edmonds, J., Luby, M., Sudan, M.: Priority encoding transmission. IEEE Transactions on Information Theory 42, 1737–1744 (1996)
5. Mohr, A., Riskin, E., Ladner, R.: Unequal loss protection: graceful degradation of image quality over packet erasure channels through forward error correction. IEEE Journal on Selected Areas in Communications 18, 819–828 (2000)
6. Ahlswede, R., Cai, N., Li, S.-Y., Yeung, R.: Network information flow. IEEE Transactions on Information Theory 46, 1204–1216 (2000)
7. Jain, K., Lovász, L., Chou, P.: Building scalable and robust peer-to-peer overlay networks for broadcasting using network coding. Distributed Computing 19, 301–311 (2007), doi:10.1007/s00446-006-0014-9
8. Yeung, R.W., Cai, N.: Network error correction. I: Basic concepts and upper bounds, and II: Lower bounds. Commun. Inf. Syst. 6(1), 19–54 (2006)
9. Chou, P.A., Wu, Y.: Network Coding for the Internet and Wireless Networks. tech. rep., Microsoft Research (June 2007)
10. Katti, S., Rahul, H., Hu, W., Katabi, D., Médard, M., Crowcroft, J.: Xors in the air: Practical wireless network coding. IEEE/ACM Transactions on Networking 16, 497–510 (2008)
11. Goshi, J., Mohr, A., Ladner, R., Riskin, E., Lippman, A.: Unequal loss protection for h.263 compressed video. IEEE Transactions on Circuits and Systems for Video Technology 15, 412–419 (2005)
12. Sprljan, N., Mrak, M., Zgaljic, T., Izquierdo, E.: Software proposal for wavelet video coding exploration group. tech. rep., M12941, 75th MPEG Meeting, ISO/IEC JTC1/SC29 /WG11/MPEG2005, Bangkok, Thailand (January 2006)
13. Goyal, V.: Multiple description coding: compression meets the network. IEEE Signal Processing Magazine 18, 74–93 (2001)
14. Puri, R., Ramchandran, K., Lee, K.W., Bharghavan, V.: Forward error correction (fec) codes based multiple description coding for internet video streaming and multicast. Signal Processing: Image Communication 16(8), 745–762 (2001)
15. Li, S.-Y., Yeung, R., Cai, N.: Linear network coding. IEEE Transactions on Information Theory 49, 371–381 (2003)
16. Yunnan, P.C., Chou, P.A., Wu, Y., Jain, K.: Practical network coding. In: Allerton Conference in Communication, Control and Computing, Monticello, IL (October 2003)

17. Koetter, R., Médard, M.: An algebraic approach to network coding. IEEE/ACM Transactions on Networking 11, 782–795 (2003)
18. Balli, H., Yan, X., Zhang, Z.: On randomized linear network codes and their error correction capabilities. IEEE Transactions on Information Theory 55, 3148–3160 (2009)
19. Bahramgiri, H., Lahouti, F.: Robust network coding against path failures. IET Communications 4, 272–284 (2010)
20. Zhang, Z.: Linear network error correction codes in packet networks. IEEE Transactions on Information Theory 54, 209–218 (2008)

Multiple Description Coded Video Streaming with Multipath Transport in Wireless Ad Hoc Networks

Yuanyuan Xu and Ce Zhu

School of Electrical & Electronic Engineering,
Nanyang Technological University, Singapore
{xuyu0004,eczhu}@ntu.edu.sg

Abstract. Multiple description coding (MD coding or MDC) generates multiple decodable bitstreams for a single source to combat packet loss, which is suitable for video streaming in error-prone wireless ad hoc networks. In this paper, two problems are investigated for MD coded video streaming in wireless ad hoc networks. The first problem addresses multipath selection for balanced two-description coded video streaming. We formulate an interference-aware MDC multipath routing for single-radio networks by employing a time-division link scheduling method to eliminate wireless interference, and ultimately obtain an optimal path selection corresponding to the minimum achievable distortion. A heuristic solution is developed for the interference-aware multipath routing, by defining a path metric taking into account interference, link bandwidth and link "up" probability. The second problem addresses MDC redundancy control according to varying channel conditions of multiple paths. We design an unbalanced redundant slice based two-description video coding, which optimally selects the amount of inserted redundancy for each description. Simulation results demonstrate the effectiveness of the proposed MDC multipath routing scheme and the unbalanced MD video coding approach over heterogeneous paths.

Keywords: Multiple description video coding, multipath routing, wireless ad hoc networks, wireless interference.

1 Introduction

Wireless ad hoc networks consist of wireless nodes, which exchange data without infrastructures support. The lack of infrastructure support poses a great challenge for multimedia applications in wireless ad hoc networks, especially for delay-sensitive and high bandwidth-demanding video streaming services. Video traffic may suffer from great packet losses due to the varying network topology and fragile wireless links. Retransmission of loss packets is sometimes undesirable due to the mobility of source node, delay constraint or congestion control. To provide continuous video delivery over fragile links without retransmission, multiple description coding (MD coding or MDC) [1] can be used to generate

L. Atzori, J. Delgado, and D. Giusto (Eds.): MOBIMEDIA 2011, LNICST 79, pp. 121–135, 2012.
© Institute for Computer Sciences, Social Informatics and Telecommunications Engineering 2012

different decodable bitstreams for a single source, where each bitstream is called a description. Each description is delivered along its own path and can provide a coarse version of source independently, while more descriptions result in a finer quality of reconstruction. By transmitting multiple descriptions over multiple paths, bandwidth of wireless paths is aggregated. Unless all the paths fail simultaneously, MDC coupled with multipath transport can provide continuous delivery of source content.

The performance of MD coded video streaming in wireless ad hoc networks depends on both MD coding efficiency and MDC path selection. Extensive efforts have been devoted to MDC codec design [2, 3, 4, 5, 6, 7, 8, 9]. Recently, a multiple description video coding (MDVC) technique [10] makes use of redundant slice representation option of H.264/AVC video coding standard [11], and generates two balanced descriptions by interlacing primary slices and redundant slices of each frame. This MDVC approach is fully compatible with the H.264/AVC standard. The two descriptions for two paths are equally protected with the same amount of inserted redundancy. However, multi-hop paths in wireless ad hoc networks are heterogeneous. Packets transmitted in different paths are likely to encounter different packet loss rates.

Compared with works on MDC codec design, there is relatively less work on MDC path selection. Without support of fixed base stations or wired backbone networks, routing becomes a crucial issue for the design of wireless ad hoc networks. Studies of multimedia-centric MDC multipath routing in wireless ad hoc networks are presented in [12, 13, 14, 15], where video quality is optimized via routing operations with respect to available link bandwidth, link "up" probabilities and loss burst length on links. In those works, wireless links are modeled as point-to-point channels in wired networks for simplicity and convenience, where interference which is a fundamental issue of wireless networks is either implicitly or inadequately considered. There are a few network-centric approaches addressing the inference issue, e.g. WCETT [16], MIC [17], iAWARE [18]. Nevertheless, these works are designed for single path routing in multi-radio wireless networks, and focus on network criteria rather the received video quality.

In this paper, we focus on two problems of MD coded video streaming in wireless ad hoc networks, which are multipath routing for balanced two-description coded video and MDC redundancy control for heterogenous paths. For the first problem, we formulate an interference-aware MDC multipath routing for video unicast services. Conflicting wireless links due to interference are scheduled to be active in different time fractions. We adopt the concept of conflict graph in [19] to obtain a schedulable maximum flow rate on each path subject to interference constraint. Based on maximum flow rates and estimated "up" probabilities of multiple paths, optimal path selection with minimum achievable distortion of reconstructed video can be obtained for a given wireless network. We also develop a heuristic solution for interference-aware MDC multipath routing. A new path metric is defined considering both video streaming characteristics and network-centric criteria. For the second problem, an unbalanced MDVC scheme is designed for multiple paths with varying channel conditions. MDVC inserts different amounts of redundancy into two descriptions according to varying channel conditions of their selected paths.

The remainder of this paper is organized as follows. The background of MDC and MDC multipath routing is provided in Section 2. In Section 3, we discuss and formulate interference-aware MDC multipath routing, and develop a heuristic solution. In Section 4, we consider the unbalanced MDVC design. Section 5 presents simulation results of MDC multipath routing, and evaluates the performance of unbalanced MDVC. Section 6 draws a conclusion.

2   Background and Preliminaries

2.1   Multiple Description Coding

MDC is a promising technique to combat transmission error by generating multiple decodable descriptions for a single source. Reconstructed quality can be refined as the number of received descriptions increases. Various MDC methods can be classified into preprocessing-based MDC, encoding-based MDC and postprocessing-based MDC, depending on which stage the one-to-multiple mapping occurs at [9]. In the preprocessing-based MDC, the original source is split into multiple subsources before encoding, and then subsources are encoded separately to generate multiple descriptions, e.g., subsampling based MDC in the temporal and spatial domains [7]. For encoding-based MDC, the one-to-multiple mapping is performed by dedicated coding techniques, such as MD scalar quantization [4], MD lattice vector quantization [6, 8], and MD correlating transform [5]. The postprocessing-based MDC realizes the one-to-multiple mapping in the compression domain by transforming an encoded bit stream into multiple streams, for instance, FEC-based MDC [3]. The redundant slice based MDVC in [10] is a kind of postprocessing-based MDC.

Generally the decoding of one or partial descriptions is known as side decoding corresponding to side distortion d_s, while the decoding of all the descriptions is central decoding resulting in a smallest central distortion d_c. If none of the descriptions is received, distortion d_{null} can be calculated according to a specific error concealment technique ($d_{null} > d_s > d_c$). We focus on two-description coding in this paper. Till now, the entire achievable MD rate-distortion region is only known for two-description coding with quadratic Gaussian source [20]. The distortions of zero-mean Gaussian source satisfy [21]

$$d_{s1} \geq \sigma^2 2^{-2R_{s1}}, \tag{1}$$

$$d_{s2} \geq \sigma^2 2^{-2R_{s2}}, \tag{2}$$

$$d_c \geq \frac{\sigma^2 2^{-2(R_{s1}+R_{s2})}}{1 - (\sqrt{\Pi} - \sqrt{\Delta})^2}, \tag{3}$$

where $\Pi = (1 - \frac{D_{s1}}{\sigma^2})(1 - \frac{D_{s2}}{\sigma^2})$, and $\Delta = (\frac{D_{s1}D_{s2}}{\sigma^4}) - 2^{-2(R_{s1}+R_{s2})}$. R_{si}, d_{si} are bitrate and side distortion corresponding to description i, respectively. Rate-distortion region for a specific MDC codec can be estimated or obtained empirically.

2.2   MDC Multipath Routing

Given a source and a destination, an MDC multipath routing generates two paths for MDC descriptions. More than two paths may exist between the source and the destination. The MDC multipath routing computes paths that minimize expected distortion of two-description coded video. MDC multipath routing in wireless ad hoc networks can be described as follows.

$$
\begin{aligned}
\text{Minimize } D^s(R_{s1}, R_{s2}) = {} & (1 - p^s_{path1})(1 - p^s_{path2})d_{null} \\
& + p^s_{path1}(1 - p^s_{path2})d_{s1}(R_{s1}) \\
& + (1 - p^s_{path1})p^s_{path2}d_{s2}(R_{s2}) \\
& + p^s_{path1}p^s_{path2}d_c(R_{s1}, R_{s2}), \tag{4}
\end{aligned}
$$

where $D^s(R_{s1}, R_{s2})$ is expected distortion of a unicast session s, and p^s_{pathi} is successful delivery probability of ith path P^s_i. As MDC introduces redundancy between descriptions, it is usually coupled with best-effort transmission which implies no retransmissions. A description is successfully delivered to the destination, if all wireless links along the transmission path are "up", $p^s_{pathi} = \prod_{l_{jk} \in P^s_i} p_{jk}$, where p_{jk} is "up" probability of link l_{jk}. Link may be "down" due to mobility of nodes, energy shortage or connection errors. Different paths differ in their "up" probabilities and maximum supporting flow rates R_i, which cause different expected distortions at the destination as shown in (4).

3   Interference-Aware MDC Multipath Routing

In this section, we focus on MDC multipath routing in single-radio wireless ad hoc networks. With special consideration of wireless interference, an interference-aware multipath routing problem is formulated. A heuristic solution is developed at the end of this section.

3.1   Wireless Interference and Link Scheduling

Assume that each node in a wireless ad hoc network is supported by an omnidirectional antenna and communicates only via wireless medium at the same frequency band. These nodes are distributed in the physical space. They assume to stay at their own locations during a video streaming session. A communication range R_c can be defined for each wireless node. Two nodes can communicate directly through a wireless link, if they are within each other's communication ranges. Each link l_{ij} is characterized by its maximum capacity b_{ij} and link "up" probability p_{ij}. Packet losses on a link are assumed to be independent.

Due to broadcast nature of wireless medium, transmission on one link interferes with transmissions on its neighboring links. While accurately measuring and estimating wireless interference is a complicated issue, we use a protocol model in [22] assuming interference to be an all-or-nothing phenomenon.

An interference range R_f is defined for each wireless node ($R_c \leq R_f$). An actively transmitting node causes severe interference within its interference range R_f and no interference outside range R_f.

To eliminate interference in a shared wireless medium, transmissions on wireless links in the neighborhood have to be scheduled in a time-division manner. A direct transmission from node i to node j is considered to be successful if: (1) a wireless link exists between node i and node j, i.e. $d_{ij} \leq R_c$, where d_{ij} is the distance from node i to node j; (2) transmissions from other node k causing interference at node j, i.e. $d_{kj} \leq R_f$, are not happening, where $k \neq i$ [19].

Conflict graph in [19] can be used to model interference between links, whose vertices correspond to individual links. An edge exists between two vertices l_{ij} and l_{pq} in the conflict graph, if two links l_{ij} and l_{pq} cannot be active simultaneously due to interference, i.e. $d_{iq} \leq R_f$ or $d_{jp} \leq R_f$. Maximal independent sets $I_1, .., I_n$ can be found in the conflict graph, where no edge exists between any two of the vertices in an independent set. By including every active link in one of the maximal independent sets, and having each individual set I_i active at its own time fraction λ_i, wireless links are scheduled to eliminate interference [19]. The maximum output flow rate of node i is subject to

$$f_i \leq \sum_n \lambda_n b_{ik}, l_{ik} \in I_n, \tag{5}$$

$$\sum_{i=1}^{n} \lambda_i \leq 1. \tag{6}$$

Note that, although we use protocol model of wireless interference for simplicity, the proposed scheme can be generalized to a more practical physical model [19] of interference as well. With the physical model, signal strength and signal to interference plus noise ratio (SINR) of each transmission are calculate at the receiver. If the receiver receives the signal at or above its sensitivity level, the transmission is successful. Similarly, conflict graph and maximal independent sets can be obtained using the physical model, and wireless links are scheduled to eliminate interference.

3.2   Problem Formulation

With the aid of conflict graph, wireless links delivering multiple descriptions to the destination are carefully scheduled to eliminate interference. An interference-aware multipath routing for balanced MDC can be formulated as follows:

$$
\begin{aligned}
\text{Minimize } D^s(R) = {}& (1 - p^s_{path1})(1 - p^s_{path2})d_{null} \\
& + p^s_{path1}(1 - p^s_{path2})d_s(R) \\
& + (1 - p^s_{path1})p^s_{path2}d_s(R) \\
& + p^s_{path1}p^s_{path2}d_c(R),
\end{aligned} \tag{7}
$$

subject to

$$f_i \leq \sum_n \lambda_n b_{ik}, l_{ik} \in I_n, \tag{8}$$

$$\sum_{i=1}^{n} \lambda_i \leq 1, \tag{9}$$

where

$$f^s(P) = R, \tag{10}$$

$$f_i = \sum_{s=i,l_{ij}\in P_i^s} f^s(P) + \sum_q \sum_{l_{qi}l_{ij}\in P_i^s} f^s(P) - \sum_{dest(s)=i,l_{qi}\in P_i^s} f^s(P), q \neq j, \tag{11}$$

$$p_{pathi}^s = \prod_{l_{jk}\in P_i^s} p_{jk}. \tag{12}$$

The optimization problem aims to minimize expected distortion of a two-description coded video unicast session initiated from source s to destination $dest(s)$ with respect to link capacities and "up" probabilities as shown in (7). Maximum output flow rate f_i of a node depends on active time and capacities of its output links b_{ik} as given in (8). Interference between links is eliminated using a time-division scheduling method that gives a constraint on time fractions (9). One of the two balanced description with bitrate R has to be delivered on each path as indicated in equation (10). Equation (11) guarantees that the amount of output flow at node i equals the sum of flows initiated from node i and transit flows at node i, excluding traffic flows ending at node i, where $l_{qi}l_{ij} \in P_i^s$ means l_{qi} and l_{ij} are two consecutive links along path P_i^s.

The solution of this problem is a pair of paths from source s to destination $dest(s)$ for two balanced descriptions. These two paths may be disjoint or partly shared. Given any multipath routing decision, maximal independent sets of its conflict graph can be obtained, and wireless links along these paths are scheduled accordingly to eliminate interference. The maximum path flow rate for each routing decision is subject to linear constraints (8-11), thus can be obtained using linear programming. Therefore a schedulable maximum flow rate on each path within interference constraint can be obtained. The maximum flow rate is achieved with an ideal media-access-control protocol, which finely controls and schedules transmissions at the individual nodes. The design of such a media-access-control protocol in wireless ad hoc networks is very challenging, and is out of the scope of our research. With maximum flow rates and path "up" probabilities, we can obtain the corresponding distortion. A global optimal solution with the minimum expected distortion can be found by exhaustively searching all the combinations of two qualified path candidates.

3.3   Heuristic Routing Solution

In this subsection, a heuristic routing solution is developed for solving the aforementioned interference-aware multipath routing problem. For computational purpose, we consider the subset of MD achievable region of zero-mean Gaussian source where three inequalities in (1) are all active. Thus rate-distortion of Gaussian source can be approximated as

$$D_{s1} = \sigma^2 2^{-2R_{s1}},$$
$$D_{s2} = \sigma^2 2^{-2R_{s2}},$$
$$D_c = \frac{\sigma^2 2^{-2(R_{s1}+R_{s2})}}{2^{-2R_{s1}} + 2^{-2R_{s2}} - 2^{-2(R_{s1}+R_{s2})}}. \tag{13}$$

Substituting (13) into (7), we can have $\frac{\partial D}{\partial R_{s1}} \leq 0, \frac{\partial D}{\partial R_{s2}} \leq 0, \frac{\partial D}{\partial p^s_{path1}} \leq 0, \frac{\partial D}{\partial p^s_{path2}} \leq 0$ [15]. Therefore, D is non-increasing with R_{s1} and R_{s2}. Assuming packet losses are independent, D is non-increasing with p^s_{path1} and p^s_{path2}.

We define a conflict factor $n^{f_n}_{ij}$ for a link l_{ij} delivering flow f_n, where the value of $n^{f_n}_{ij}$ equals the number of active links that are conflicting with the delivery of f_n on l_{ij}. Note that the same link l_{ij} carrying another flow f_m $(m \neq n)$ increases $n^{f_n}_{ij}$ by 1. Conflict factor measures both intra-path interference and inter-path interference. Larger value of $n^{f_n}_{ij}$ indicates severer interference, probably leads to shorter active time. However, short active time can be compensated by large link bandwidth. Therefore, a path metric $I(P, f_i)$ can be defined for a path P carrying flow f_i as follows,

$$I(P, f_n) = (1 - \alpha) \prod_{l_{ij} \in P} p_{ij} + \alpha \sum_{l_{ij} \in P} \frac{b_{avg}}{b_{ij}} \frac{n^{f_n}_{ij}}{N}, \tag{14}$$

where b_{avg}, N are the approximate average link bandwidth in the network and number of all the active links, respectively. The value of α is tunable, where $0 \leq \alpha \leq 1$. Larger α helps to find paths with greater flow rate, while lower α results in more reliable paths.

The proposed path metric $I(P, f_n)$ is non-isotonic, which is not suitable for link-state routing protocols to find loop-free and minimum weighted paths. However, on-demand distance vector or source routing protocols can use non-isotonic metrics to find efficient paths [17]. To find two paths P^s_1 and P^s_2 for one MDC video streaming session, we search all paths connecting source node s and destination $dest(s)$. The path with the minimum $I(P, f_1)$ is selected as P^s_1. Then $I(P, f_2)$ path metric for f_2 is updated counting in interference caused by P^s_1. We search in the rest of paths and choose the one with the minimum updated $I(P, f_2)$ to be P^s_2.

4   Unbalanced Multiple Description Video Coding

In [10], redundant slices based MDVC aims to produce two balanced descriptions for a video sequence. These descriptions are of the same size and correspond to the same side distortion. However, there are hardly any identical multi-hop paths between a source and a destination in wireless ad hoc networks. Descriptions transmitted in different channels encounter different degrees of packet losses, thus should be unequally protected. In this section, we present an unbalanced redundant slice based MDVC technique, which adjusts the amount of inserted redundancy in each description according to channel condition of its own path.

4.1   Unbalanced MDVC Scheme

The concept of redundant slice defined in the H.264/AVC standard [11] represents an alternative representation of a picture. When normally coded primary slice cannot be decoded correctly, the decoder will replace it with the corresponding correctly decoded redundant slice. By interlacing primary slices and redundant slices, two descriptions can be generated as shown in Fig. 1. As in [10], only primary slices are used as references for motion predication of subsequent pictures. The primary and redundant representations of the same slice are transmitted in different channels to provide robustness against loss. While MDVC in [10] codes redundant slices in a picture with the same quality, we tune the quality of these redundant slices according to their allocated channels. The quality of slices can be controlled by employing different values of quantization parameter (QP).

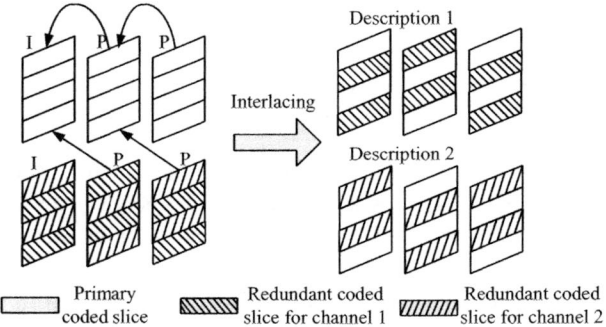

Fig. 1. Unbalanced redundant slices based MDVC

4.2   Optimal Redundancy Allocation

When redundant slice replaces primary slice at the decoder, an error is introduced in prediction loop due to the mismatch between primary and redundant representations. This mismatch error propagates to the subsequent frames. If we consider transmitting primary representation of kth slice in path 1 with successful probability p^s_{path1}, and redundant representation in path 2 with p^s_{path2}. According to [10], expected distortion of slice k can be evaluated as

$$d_{k,1} = p^s_{path1}d_{p,k} + p^s_{path2}(1 - p^s_{path1})\phi_i d_{r,k,2} + (1 - p^s_{path1})(1 - p^s_{path2})d_{0,k}$$
$$\approx p^s_{path1}d_{p,k} + p^s_{path2}(1 - p^s_{path1})\phi_i d_{r,k,2}, \tag{15}$$

where slice k belongs to ith picture $i = M(k)$, and $d_{p,k}$, $d_{r,k,2}$, $d_{0,k}$ represent encoding distortion of kth primary slice, encoding distortion of kth redundant slice allocated to channel 2, and distortion of kth slice losing both representations, respectively. In (15), mismatch and propagation errors due to redundant slice are captures by ϕ_i, which is the summation of power transfer functions $\phi_i = \sum_{n=0}^{N-M(k)} e^{-\alpha n} = (1 - e^{-\alpha(N-i+1)})/(1 - e^{-\alpha})$. The term $(1 - p_{path1}^s)(1 - p_{path2}^s)d_{0,k}$ can be ignored at a low packet loss probability.

$$d_{k,1} \approx p_{path1}^s d_{p,k} + p_{path2}^s(1 - p_{path1}^s)\phi_i d_{r,k,2}. \tag{16}$$

If primary slice is transmitted in path 2, the expected distortion becomes

$$d_{k,2} \approx p_{path2}^s d_{p,k} + p_{path1}^s(1 - p_{path2}^s)\phi_i d_{r,k,1}. \tag{17}$$

We aim to minimize expected distortion of N frames by adjusting the amount of inserted redundancy, where bitrate of each description is constrained by maximum flow rate of its path. The allocation problem can be formulated as

$$
\begin{aligned}
\text{Minimize} \quad & \sum_{i=1}^{N} \sum_{k:M(k)=i} d_k = \sum_{i=1}^{N} \left(\sum_{k:M(k)=i,(k+i)\%2=1} d_{k,1} + \sum_{k:M(k)=i,(k+i)\%2=0} d_{k,2} \right) \\
\approx & \sum_{i=1}^{N} \sum_{k:M(k)=i,(k+i)\%2=1} (p_{path1}^s d_{p,k} + p_{path2}^s(1 - p_{path1}^s)\phi_i d_{r,k,2}) \\
+ & \sum_{i=1}^{N} \sum_{k:M(k)=i,(k+i)\%2=0} (p_{path2}^s d_{p,k} + p_{path1}^s(1 - p_{path2}^s)\phi_i d_{r,k,1})), \tag{18}
\end{aligned}
$$

subject to

$$\sum_{i=1}^{N} \left(\sum_{k:M(k)=i,(k+i)\%2=1} R_{p,k} + \sum_{k:M(k)=i,(k+i)\%2=0} R_{r,k,1} \right) \leq R_1, \tag{19}$$

$$\sum_{i=1}^{N} \left(\sum_{k:M(k)=i,(k+i)\%2=0} R_{p,k} + \sum_{k:M(k)=i,(k+i)\%2=1} R_{r,k,2} \right) \leq R_2. \tag{20}$$

The optimization problem can be solved by minimizing the cost function

$$
\begin{aligned}
L = & \sum_{i=1}^{N} \sum_{k:M(k)=i} d_k + \lambda_1 \left(\sum_{i=1}^{N} \left(\sum_{k:M(k)=i,(k+i)\%2=1} R_{p,k} + \sum_{k:M(k)=i,(k+i)\%2=0} R_{r,k,1} \right) - R_1 \right) \\
& + \lambda_2 \left(\sum_{i=1}^{N} \left(\sum_{k:M(k)=i,(k+i)\%2=0} R_{p,k} + \sum_{k:M(k)=i,(k+i)\%2=1} R_{r,k,2} \right) - R_2 \right), \tag{21}
\end{aligned}
$$

where λ_1 and λ_2 is the Lagrangian multiplier. As we task for allocation of redundant slices, constant QP is used for all the primary slices ($QP_{p,k} = QP_p$). Using the standard H.264 R-D approximation $\frac{\partial D}{\partial R} = -0.85 \times 2^{(\frac{QP-12}{3})}$ [23], optimal QPs of redundant slices can be obtained:

$$QP_{r,k,1} = QP_p + 3\log_2{}^{(\frac{1}{(1-p^s_{path2})\phi_i})}, \tag{22}$$

$$QP_{r,k,2} = QP_p + 3\log_2{}^{(\frac{1}{(1-p^s_{path1})\phi_i})}. \tag{23}$$

5   Simulation Results

5.1   Performance of Interference-Aware MDC Multipath Routing

In the simulation, performance of MDC multipath routing is evaluated in randomly generated wireless ad hoc networks. In each network, wireless nodes are randomly placed with a uniform distribution in an area of 500×500 square. During a video session, these nodes remain at the same locations. Transmission range of wireless nodes R_c is a fixed value to ensure that average node degree is at least 3. Interference range R_f is a fixed value larger than transmission range R_c. If the distance between two nodes is within the transmission range, a wireless link exists between these nodes. Each wireless link l_{ij} has a fixed link "up" probability $p_{ij} = 0.995$, while its maximum capacity $b_{i,j}$ is randomly selected within the range of $[0.25, 1]$ Mb/s with a step size of 0.125 Mb/s. Source node and destination node for each MDC unicast session are randomly chosen among these nodes.

In Table 1, different multipath routing algorithms are used to select two paths for one MD coded video streaming session in two 15-node networks and two 30-node networks. The proposed interference-aware MDC multipath routing is compared with k shortest paths [24] and interference-oblivious MDC multipath routing algorithm ignoring wireless interference. The scheme of k shortest paths is a popular network-centric routing algorithm connecting a source and a destination with k paths that have minimum hops. The interference-oblivious MDC multipath routing minimizes average end-to-end distortion by finding paths with large bandwidth and high link "up" probability like those in wired networks, and the work in [14] adopts a similar approach.

Global optimal solution of interference-aware MDC multipath routing can be obtained by exhaustively searching all the combinations of two paths, where each path has a restricted number of hops. Maximum flow rate on each path is obtained under linear constraints using a linear programming function in MATLAB. As we can see from Table 1, the global optimal solution of proposed interference-aware outperforms the two existing algorithms. We also compare our heuristic solution of interference-aware routing with the global optimal solution, 2 shortest paths [24] and interference-oblivious routing. The heuristic solution is obtained by searching paths within certain hops to obtain one path with minimum value of path metric $I(P, f_1)$, and a second path with minimum updated $I(P, f_2)$. It can be observed from Table 1 that, the heuristic routing solution generally achieves better performance compared with two existing algorithms.

Table 1. Comparison of average distortion using different routing solutions

Topology	I (15-node)	II (15-node)	III (30-node)	IV (30-node)
2 shortest paths	0.804	0.687	0.776	0.799
Interference-oblivious	0.729	0.669	0.735	0.712
Global optimal	0.686	0.618	0.708	0.689
Heuristic solution	0.686	0.629	0.730	0.708

In Fig. 2, we compare reconstructed quality of individual frames of Foreman QCIF sequence using different routing solutions in Topology I. The redundant slice based MDVC in [10] is used to generate two balanced descriptions for each sequence. The video sequence is encoded at 30 frames per second using "IPP..P" coding structure with an intra period of 50 frames. Each coded slice is packed into a packet. Reconstructed quality of individual frames is measured in peak-to-noise ratios (PSNRs). The proposed solution achieves 3.96dB and 1.27dB gains in average over 2 shortest paths and its interference-oblivious counterpart, respectively. Reconstructed frame samples of Foreman sequence using different schemes in Topology I are shown in Fig. 3, while frame samples of Mobile QCIF sequence using different routing solutions in Topology II are shown in Fig. 4.

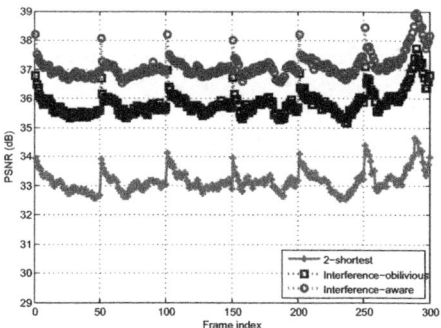

Fig. 2. Comparison of reconstructed "Foreman.qcif" using different routing solutions

5.2 Performance of Unbalanced MDVC

We compare the proposed unbalanced redundant slice based MDVC with balanced version of MDVC in [10]. The successful probabilities of two paths are set to 0.99 and 0.95, respectively. Each primary slice or redundant slice is packed into a packet for transmission. The proposed MDVC adds more redundancy to the description transmitted in the worse path, while balanced MDVC in [10] equally protects these description according to an average path successful probability. In Fig. 5, with the same QP for primary slices, the proposed MDVC generates

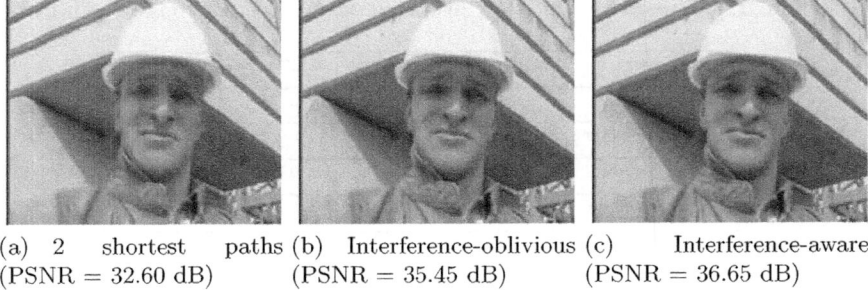

(a) 2 shortest paths (PSNR = 32.60 dB)   (b) Interference-oblivious (PSNR = 35.45 dB)   (c) Interference-aware (PSNR = 36.65 dB)

Fig. 3. Reconstructed 45th frames of "Foreman.qcif" using different routing solutions

(a) 2 shortest paths (PSNR = 28.67 dB)   (b) Interference-oblivious (PSNR = 29.39 dB)   (c) Interference-aware (PSNR = 30.26 dB)

Fig. 4. Reconstructed 135th frames of "Mobile.qcif" using different routing solutions

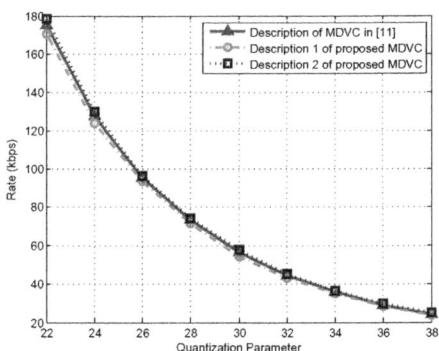

Fig. 5. Comparison of coding rate of Foreman QCIF video sequence

two descriptions whose bitrates are close to average bitrate per description of balanced MDVC. Decoded quality of individual frames using different MDVC schemes is compared with a same transmission trial in Fig. 6. It can be seen that the unbalanced MDVC gets better reconstructed images than balanced MDVC

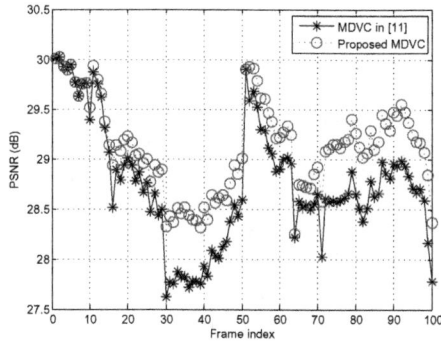

Fig. 6. Comparison of reconstructed Foreman.qcif in the same transmission trial

(a) MDVC in [10] (PSNR = 27.77 dB)   (b) Proposed MDVC (PSNR = 28.42 dB)

Fig. 7. Reconstructed 31th frames of Foreman.qcif using different MDVC schemes

with biased error protection. Fig. 7 shows the reconstructed frame samples using different MDVC schemes.

6 Conclusion

In this paper, we investigate two problems of two-description coded video streaming in wireless ad hoc networks, namely, multipath selection and MDC redundancy allocation. Firstly, we formulate an interference-aware MDC multipath routing in single-radio networks with a time-division link scheduling method. The proposed MDC multipath routing generates optimal paths that maximize average decoded quality. A heuristic solution is developed for the proposed interference-aware multipath routing, by defining a new path metric considering intra-path and inter-path interference. Secondly, we design an unbalanced MDVC scheme for multiple paths with different channel conditions, where the amount of redundancy of each description is optimally selected according to condition of its selected path. Simulation results verify the proposed MDC multipath routing scheme and the proposed unbalance MDVC approach.

References

[1] Goyal, V.: Multiple description coding: Compression meets the network. IEEE Signal Processing Magazine 18(5), 74–93 (2001)

[2] Vaishampayan, V.A.: Design of multiple description scalar quantizer. IEEE Transactions on Information Theory 39(3), 821–834 (1993)

[3] Puri, R., Ramchandran, K.: Multiple description source coding using forward error correction codes. In: Conference Record of the Asilomar Conference on Signals, Systems and Computers, pp. 342–346 (1999)

[4] Servetto, S., Ramchandran, K., Vaishampayan, V., Nahrstedt, K.: Multiple description wavelet based image coding. IEEE Transactions on Image Processing 9(5), 813–826 (2000)

[5] Goyal, V., Kovacevic, J.: Generalized multiple description coding with correlating transforms. IEEE Transactions on Information Theory 47(6), 2199–2224 (2001)

[6] Vaishampayan, V.A., Sloane, N.J.A., Servetto, S.D.: Multiple description vector quantization with lattice codebooks: Design and analysis. IEEE Transactions on Information Theory 47(5), 1718–1734 (2001)

[7] Wang, D., Canagarajah, C., Redmill, D., Bull, D.: Multiple description video coding based on zero padding. In: IEEE International Symposium on Circuits and Systems, vol. 2 (2004)

[8] Bai, H., Zhu, C., Zhao, Y.: Optimized multiple description lattice vector quantization for wavelet image coding. IEEE Transactions on Circuits and Systems for Video Technology 17(7), 912–917 (2007)

[9] Zhu, C., Liu, M.: Multiple description video coding based on hierarchical B pictures. IEEE Transactions on Circuits and Systems for Video Technology 19(4), 511–521 (2009)

[10] Tillo, T., Grangetto, M., Olmo, G.: Redundant slice optimal allocation for H. 264 multiple description coding. IEEE Trans. on Circuits and Systems for VideoTechnology 18(1), 59–70 (2008)

[11] Joint Video Team JVT of ISO/IEC MPEG and ITU-T VCEG: International Standard of Joint Video Specification (ITU-T Rec. H.264, ISO/IEC 14496-10 AVC) (March 2003)

[12] Mao, S., Hou, Y., Cheng, X., Sherali, H., Midkiff, S., Zhang, Y.: On routing for multiple description video over wireless ad hoc networks. IEEE Transactions on Multimedia 8(5), 1063–1074 (2006)

[13] Mao, S., Cheng, X., Hou, Y., Sherali, H.: Multiple description video multicast in wireless ad hoc networks. Mobile Networks and Applications 11(1), 63–73 (2006)

[14] Kompella, S., Mao, S., Hou, Y., Sherali, H.: Cross-layer optimized multipath routing for video communications in wireless networks. IEEE Journal on Selected Areas in Communications 25(4), 831–840 (2007)

[15] Mao, S., Cheng, X., Hou, Y., Sherali, H., Reed, J.: On joint routing and server selection for MD video streaming in ad hoc networks. IEEE Transactions on Wireless Communications 6(1), 338–347 (2007)

[16] Draves, R., Padhye, J., Zill, B.: Routing in multi-Radio, multi-Hop wireless mesh networks. In: MOBICOM (2004)

[17] Yang, Y., Wang, J., Kravets, R.: Designing Routing Metrics for Mesh Networks. In: WiMesh (2005)

[18] Subramanian, A.P., Buddhikot, M.M., Miller, S.: Interference aware routing in multi-radio wireless mesh networks. In: 2nd IEEE Workshop on Wireless Mesh Networks, pp. 55–63 (2006)

[19] Jain, K., Padhye, J., Padmanabhan, V., Qiu, L.: Impact of interference on multi-hop wireless network performance. Wireless Networks 11(4), 471–487 (2005)

[20] Ozarow, L.H.: On a source coding problem with two channels and three receivers. The Bell Syst. Tech. J. 59, 1909–1921 (1980)

[21] Chen, J., Tian, C., Berger, T., Hemami, S.S.: Multiple description quantization via Gram-Schmidt orthogonalization. IEEE Trans. Information Theory 52, 5197–5217 (2006)

[22] Gupta, P., Kumar, P.: The capacity of wireless networks. IEEE Transactions on Information Theory 46(2), 388–404 (2000)

[23] Wiegand, T., Schwarz, H., Joch, A., Kossentini, F., Sullivan, G.: Rate-constrained coder control and comparison of video coding standards. IEEE Trans. Circuits Syst. Video Technol. 13(7), 688–703 (2003)

[24] Eppstein, D.: Finding the k shortest paths. Journal on Computing Society for Industrial and Applied Mathematicas 28(2), 652–673 (1999)

Mobile Multipath Cooperative Network
for Real-Time Streaming

Viji Raveendran, Phanikumar Bhamidipati, Xun Luo, and Xiaolong Huang

Qualcomm Inc.
5775 Morehouse Dr, San Diego, California 92121, USA
{vraveend,pbhamidi,xunl,xhuang}@qualcomm.com

Abstract. Access links are often times the bottlenecks of wireless wide area networks (WWAN). The prevalent use of multimedia applications on mobile devices introduces an ever increasing traffic load on WWAN access links, leading to traffic congestion and unsatisfactory user experiences. In this paper, we introduce a mobile multipath cooperative network. In the system, multiple paths are dynamically established among cooperative devices over WWAN and WLAN so that multiple descriptions of a multimedia stream can be transported over distinct end-to-end paths between two mobiles. As a result, the capacities of multiple wireless access links can be utilized to enhance quality of experience of a multimedia application. We also introduce a MDC rate adaptation algorithm that jointly adapts the source coding rates among multiple paths. Our lab experiments show that real-time streaming with multiple description coding benefits significantly from the proposed cooperative network and subsequently enhance the quality of user experience for multimedia applications.

Keywords: Multiple description coding, mobile multimedia service, network architecture, overlay network, quality of experience, real-time streaming.

1   Introduction

The amount of digital information created and replicated in the world is expected to grow to 35 trillion gigabytes in the next 10 years [1]. The primary driver for this exponential growth is multimedia fueled by social media networks and the increasing desire of users to share content, be it a personal video of your cat or a digital photo taken at a social gathering. More than 70% of the data generated this year will be from individuals at home, at work or on the go [1]. Multimedia capture devices on mobiles play a major role in creating this user generated content (UGC) due to their inherent mobility. Mobiles are available anytime, anywhere, when the user wants to capture or access content.

The computing and multimedia processing capabilities in mobile devices has increased several fold over the past couple of years, particularly with respect to video formats (resolution, frame rate). For example, VGA at 30 fps video recording was prevalent until about 2010, while upwards of 8M pixel cameras, HD (up to 1080p60)

L. Atzori, J. Delgado, and D. Giusto (Eds.): MOBIMEDIA 2011, LNICST 79, pp. 136–151, 2012.

and 3D stereo video capture is becoming common going forward. In addition, the increasing memory capacity leads to a large amount of personal content to be stored on the phones. Consequently, mobiles become the primary sources of UGC with no efficient means to share the content instantly at its captured quality.

The primary means of sharing this content today is through enterprise servers. IDC (International Data Corporation) data shows that nearly 75% of the digital world today is a copy. The double digit annual growth in the amount of data exacerbated by high fidelity capture devices not only burdens the enterprise servers but makes this conventional means of sharing unsustainable.

QoE (Quality of Experience) is further compromised by limited capacity on radio access links in wireless networks. This makes it difficult for users to instantly share their mobile multimedia contents over cellular networks. As shown in Fig. 1, the uplink capacity of WWANs (both 3G and 4G) is significantly lower than that of the downlink. The capacity of a cellular access link is also subject to constraints, such as the maximum bit rate allowed for a user according to its service subscription, the number of simultaneously active users, and channel condition fluctuation.

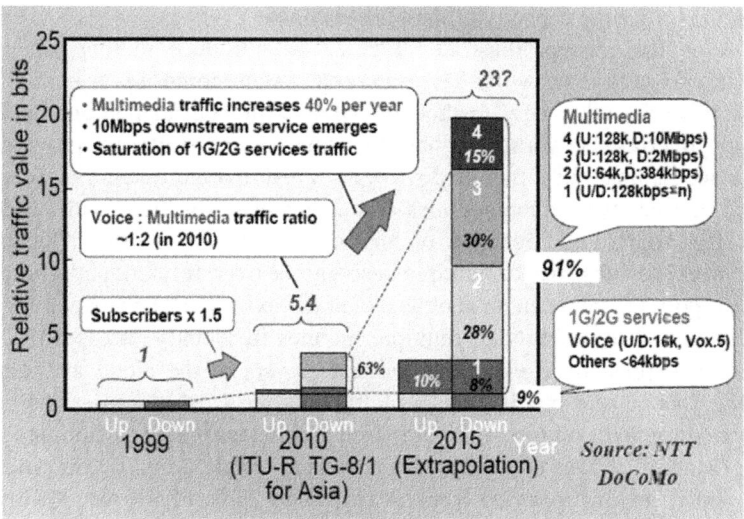

Fig. 1. Forecasted traffic vs. cellular capacity increase

To increase the access link capacity, a mobile device may contract a neighboring cooperative device and utilize the access link of the cooperative neighbor to support its traffic. Subsequently, multiple paths can be established between a pair of source and destination devices to increase the capacity and the reliability of the transport for the multimedia application. This concept is shown in Fig. 2.

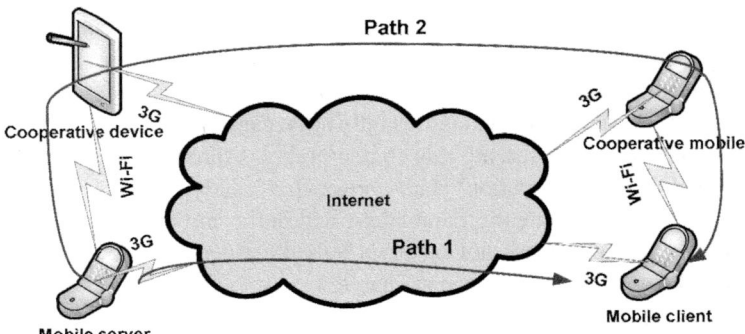

Fig. 2. Multiple access links are used by a multimedia application to increase the capacity and the reliability of its network access

In Fig. 2, a mobile server can stream the data simultaneously on two disjoint paths to the mobile client; one is a direct path while the other traverses across cooperative devices and utilize the capacity of their access links.

Based on the concept illustrated Fig. 2, we introduce in this paper a mobile multipath cooperative network. The proposed system comprises a mobile multipath overlay network and a new method of MDC with rate adaptation. In the mobile multipath overlay network, neighboring devices cooperate to provide multiple wireless access links for a mobile device. As a result, a multipath transport facility is provided between two remote mobiles. The new method of MDC with rate adaptation adjusts the source coding rate of the content upon capacity fluctuation and synchronizes the deliveries of multiple descriptions over different paths.

The proposed system can be deployed over the existing IP network. It leverages the proliferation and availability of multiple mobiles in vicinity. It eases the impact of information explosion on enterprise content servers in the cloud and subsequently opens the door to new mobile driven multimedia services. Many service models can be conveniently built on top of the proposed architecture. Using Multiple Description Coding (MDC) [5] over two different paths can provide a quality improvement of a QVGA video stream from 300 Kbps at 15fps and 28dB PSNR to 600Kbps at 30fps and 30dB PSNR. High quality multimedia content sharing can also be enabled when cooperate devices distribute individually obtained descriptions among each other.

In Section 2, we introduce the mobile multipath overlay network architecture and its protocols. In Section 3, we present the MDC module that utilizes the mobile multipath overlay network for real-time streaming. In Section 4, we introduce the multipath rate adaptation system. In Section 5, we present our experiment results that demonstrate the performance gain provided by our proposed system. In section 6, we give conclusions.

2   Mobile Multipath Overlay Network Architecture

In this section, we introduce the mobile multipath overlay network for transporting multiple descriptions of a multimedia content. Several algorithms have been proposed to transport multiple descriptions of a multimedia content across the network [2][3][4]. These algorithms focus on selecting multiple paths in an infrastructure based content distribution network and utilizing multiple access links directly available at a mobile handset. Different from those approaches, in our proposed overlay network, neighboring devices communicate among one another to provide multiple wireless access links for a mobile device.

We call a traffic source as a source and a traffic destination as an aggregator. The cooperative device contracted by a source is called a source helper. The cooperative device contracted by an aggregator is called an aggregator helper. A device may recruit multiple helpers. A multimedia content is encoded into multiple descriptions, each traversing one path in the overlay network to reach the aggregator. Each description can be rendered at an aggregator independently. And the rendered quality increases incrementally based on number of descriptions received.

The data plane of the multipath overlay network introduces an overlay switching layer between the application and the underlying transport protocol. The function of the overlay switching layer is to route the traffic from the source to the aggregator over overlay network helpers. An illustration of the multipath overlay network data plane is shown in Fig. 3. We use RTP as an example for the application layer.

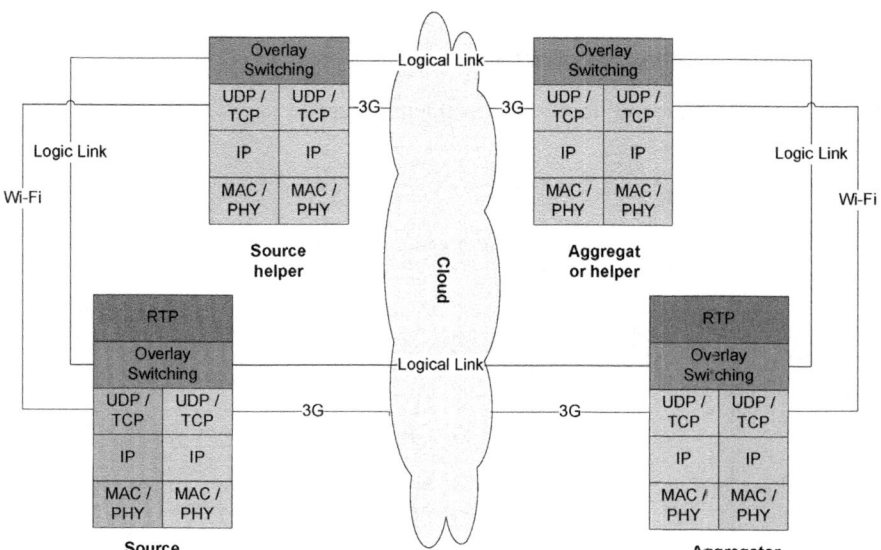

Fig. 3. The data plane of the multipath overlay network

The overlay switching layer employs a label switching architecture similar to that in MPLS [6]. The label switching architecture reduces the routing processing complexity and enables differentiating per hop behaviors for different streaming sessions. Each packet is tagged with a label when travelling from one overlay device to another. The logical link between two overlay devices on an established overlay network path is called a segment. The label tagged on a packet is uniquely assigned by the downstream overlay device of that segment during the session establishment phase (whose details are in Section 2.1). An overlay device determines the QoS treatment it should give to an incoming packet based on the label. When sending the packet downstream, an overlay network device should swap the label to another label assigned by the downstream overlay device on the successive segment.

A device may function as source/aggregator/helper and may support one or more sessions for a service and one or more services.

The control plane of the multipath overlay network is used to establish, replace and release a data plane path between a source and an aggregator. The multipath overlay network uses TCP for transporting overlay network control plane messages. The control plane does not include the logic link between a source helper and an aggregator helper. During the signaling phase, the source and the aggregator can contract its neighboring cooperative devices to function as helpers. The selection of helpers is based on local P2P device and service discovery procedures that take into consideration physical channel capacities and processing capabilities.

In the following, we briefly describe the protocols to establish and maintain a multipath transport for a streaming session in the mobile overlay network.

2.1   Multipath Management Protocol

In Fig. 4, we show the multipath establishment procedure.

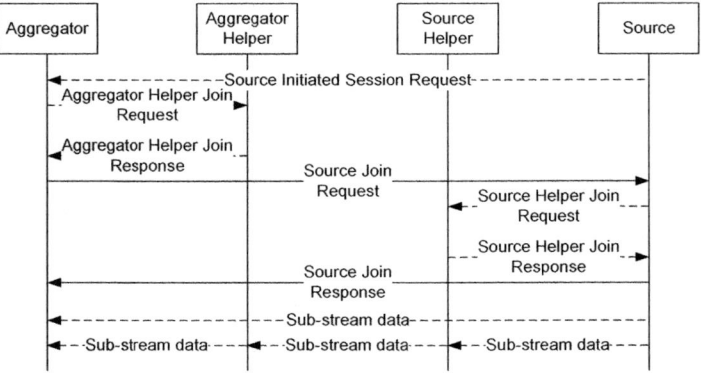

Fig. 4. Multipath establishment for a streaming session. Helper Join Request/Response may be optional.

When a mobile device intends to download data, it acts as an aggregator. When a mobile device intends to upload data, it acts as a source. If a mobile device intends to upload data, it needs to first send a session request to an aggregator to bootstrap the multipath establishment procedure.

The multipath establishment procedure is as follows. An aggregator, if needed, sends a request to a cooperative device to contract it as an aggregator helper. The request contains a designated label that should be tagged to the packet of this streaming session sent towards the aggregator on this path segment. An aggregator helper sends a response back to the aggregator with a designated label that should be tagged to the packet of this streaming session sent from the source side to the aggregator helper on upstream path segment. Subsequently, the aggregator requests the designed source device to join the session and provides the label to be used on the direct path and the label to be used on the alternative path to reach the aggregator helper. The source device determines whether a source helper is needed to send the alternative description of the same streaming session based on its access capacity. If a source helper is contracted to cooperate, the source should give the label to be used for forwarding the traffic towards the aggregator helper to the source helper. And the source helper should give the label to the source for sending packets of this streaming session to the source helper. At the end, the source sends a response to the aggregator. The source can send a description without completing the path establishment for another description.

One significant advantage can be provided by the multipath overlay network is that, when one path is down, the streaming can still be supported by another path. To help ensuring the QoE of streaming over multipath, a seamless path replacement procedure is introduced. In Fig. 5, we show the multipath replacement procedure.

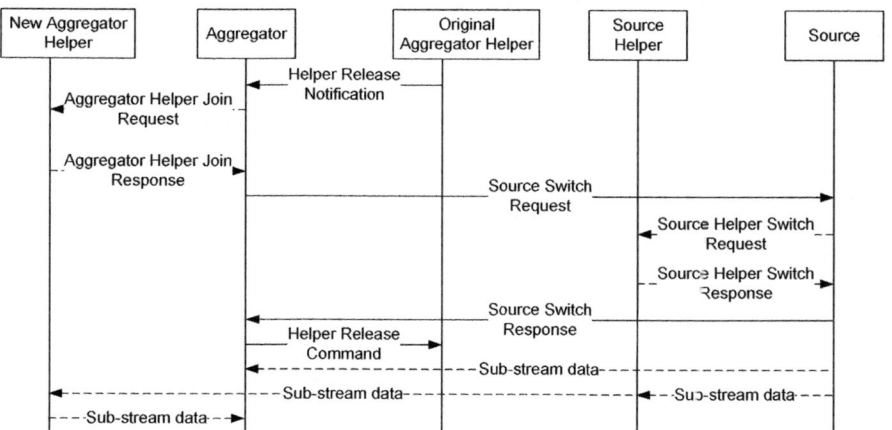

Fig. 5. Multipath replacement for a streaming session

A beneficiary device, the aggregator or the source, should contract another cooperative device to replace the original helper if the original helper is determined to be unsuitable for the delivery of a description of the multimedia stream.

3   Multiple Description Coding Module

Multiple Description Coding (MDC) [5][7] is proposed as an alternative to layered coding for streaming over unreliable channels. The goal of the MDC is to create several independent descriptions that can contribute to one or more characteristics of video, such as spatial or temporal resolution, signal-to-noise ratio, frequency content etc. These independent descriptions can be sent over different paths and aggregated at the receiver.

MDC is different from Scalable Video Coding (SVC) [8] [9] in that SVC generates a base layer and one or more enhancement layer bit streams. The base layer is required for the stream to be decoded. Enhancement layers provide improved quality. In MDC, all descriptions are independent and hence the more descriptions the decoder receives, the better the quality. For scenarios where the packet transmission schedules can be optimized in a rate-distortion sense, layered coding provides a better performance. The converse is true for scenarios where the packet schedules are not rate-distortion optimized [10] [11].

Besides increased fault tolerance, MDC allows content providers to send descriptions of a stream without paying attention to the client. Receivers that can't sustain the data rate only subscribe to a subset of these descriptions, thus freeing the content provider from sending streams at lower bit rates. The robustness of MDC comes from the fact that it is highly unlikely that the same portion of the same picture is corrupted in all the descriptions. Various experiments have shown that the MDC is very robust and the quality is acceptable even at high loss rates.

We utilize MDC in H.264 in a number of ways as those described in [12] to achieve better QoE, including Group Of Pictures (GOP) level MDC, Frame Level MDC, Slice Level MDC, Macro Block (MB) Level MDC and Region Of Interest (ROI) and/or Persistence based MDC. Each one of the options has its own advantages and disadvantages.

MDC in combination with Rate Adaptation is used to improve the performance of the proposed MDC system. This is due to the fact that the rate adaptation makes a more efficient use of the instantaneous network capacity. MDC enables real-time rate adaptation through Selective Combining at frame or slice level using the above methods to achieve an average 3dB improvement in PSNR with perceivable visual quality enhancement. Multiple descriptions also decrease the peak-variations in a given sequence and improve the peak to average ratios.

4   Multipath Rate Adaptation Algorithm

4.1   Rate Adaptation Algorithm

Many algorithms [13][14][15][16] have been introduced to optimize MDC data transmission over multiple paths. These algorithms either focused on optimizing the distribution the coded data among different paths or adjust the data transmission rate of the coded content. Instead, our proposed rate adaptation algorithm adjusts the source coding rate of an individual description so that the highest feasible quality of a description can be delivered without jeopardizing delay and delay jitter experience of the real-time streaming. The proposed rate adaptation system also performs throttling and seeking forward operations on each path to synchronize multiple descriptions of the same multimedia content.

The rate adaptation system operates on feedback information, including aggregator's buffer occupancy status, buffer difference status among multiple paths, packet loss ratio, network delay, delay jitter and throughput on individual paths.

The buffer occupancy report of each individual path sent to the source identifies the buffer state including overflow, underflow, or normal. If buffer overflow occurs, the source throttles the transmission of the corresponding description data so that the delivery of the description can be in synchronization with the overall playback of the stream. Buffer underflow signals potential network congestion. If underflow occurs, the source reduces the source coding rate in an attempt to restore buffer normal buffer status. If underflow persists, the source performs seeking forward to synchronize delivery of the description with the overall playback of the stream.

It is critical to maintain synchronization among multiple descriptions. In doing so, the aggregator calculates a reference buffer occupancy level which is the average buffer size of healthy receiving buffers. The buffer difference of each individual buffer from the reference level is then calculated. The buffer difference status is reported back to source and identified as early, late, and normal. If description early occurs, the source throttles the transmission of the corresponding description. If description late occurs, the source first reduces the source coding rate to restore the description normal status. If description late persists, the source performs seeking forward.

The source also monitors traffic performance metrics, including packet loss ratio, throughput, delay and jitter. Desirable ranges of these traffic performance metrics are calculated. The desirable range of throughput is set based on the transmission data rate. The desirable ranges of delay and delay jitter are set to be vicinity of the current calibrated network delay. The calibrated network delay needs to be calculated carefully, as the delay may vary based on the adjusted source coding rate. To avoid oscillation, the calibrated delay is initially set based on the result obtained by burst transmissions at the session initialization. The calibrated delay is then calculated as a low pass filtered version of the instantaneous delay. The source reduces the source coding rate when any of the performance metrics deteriorates with respect to their desirable ranges. The source increases the source coding rate when the throughput and the delay outperform with respect to the desirable ranges, since serves to restore the high source coding rate after temporary network congestion events.

The proposed rate adaptation system also takes into account the scene change. When the multimedia streaming operates at a scene that generates a very low data rate, the system cannot effectively obtain the capacity information based on feedback. The data rate instructed by the rate adaptation shall remain at the previous level, when the intensity of the scene goes down. When the multimedia content is switched from a lower data rate generating scene to a higher data rate generating scene, the source coding rate should be allowed to go beyond the rate instructed the rate adaptation system, since the instructed data rate does not necessarily represents the capacity of the path. This manner, a scene change opportunistically explores the capacity of the network.

4.2   Overlay Feedback System

A feedback system is designed to report the buffer status and traffic performance to the source. The data flows of the feedback system are shown in Fig. 6.

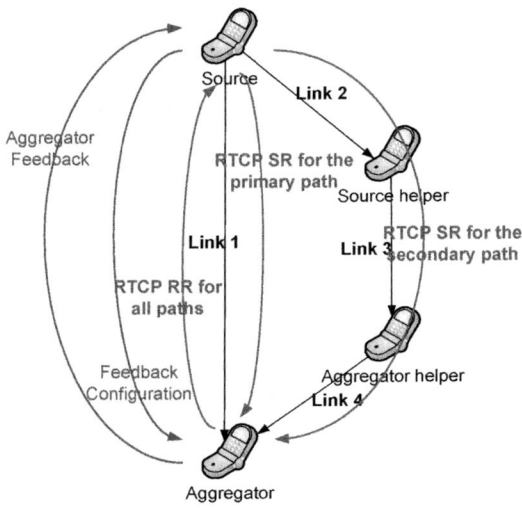

Fig. 6. Data flows of the feedback system

An overlay network control plane message, called Feedback Configuration is sent whenever necessary from the source to the aggregator to configure the aggregator to perform feedback, including feedback time interval and the buffer difference window.

RTCP is used to feedback delay, jitter, and packet loss ratio. To measure the delay, RTCP Sender Reports (SR) need to be sent to the aggregator on each individual path tagged with transmission timing information. The RTCP Sender Report (SR) for the secondary path utilizes the overlay network path. To remove the necessity of establishing a reverse overlay network path, all RTCP Receiver Reports (RR) are sent from the aggregator to the source directly. RTCP only provides round trip time information of a path. The delay of the primary path can be estimated as half of its round trip time. However, the delay of an alternative path has to be calculated based on the delay of the primary path. Denote the round trip time for the ith path by $RTT(i)$. The primary path is the 0th path. The delay of the ith path $D(i)$ is estimated as follows:

$$D(0) = RTT(0)/2,$$
$$D(i) = RTT(i) - D(0), \forall i > 0.$$

An overlay network control plane message, called Aggregator Feedback, is sent periodically to the source to report the buffer occupancy status, the buffer difference status, and the receiving throughput.

4.3   Performance of the Rate Adaptation System

We demonstrate the effectiveness of our rate adaptation system in Fig. 7. The study is done by writing a rate adaptation system simulator.

In the simulator, a movie trailer's data are streamed down from a source device to an aggregator device over two paths. The frame rate is 30fps. The movie trailer roughly goes through a scene change every 4 seconds and has 6 source coding levels of each description. The source data rate varies between 200Kbps to 3Mbps.

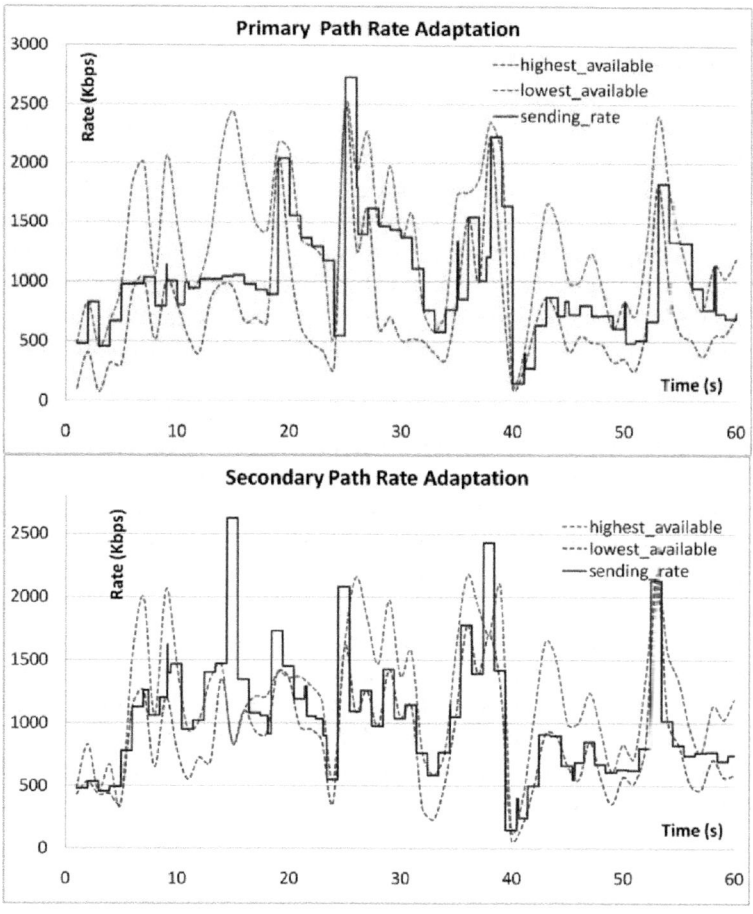

Fig. 7. Capacity utilization of individual paths under source coding rate adaptation

The primary path has a latency 100ms. The secondary path has a latency 500ms. The network capacity of each path varies between 2Mbps and 500Kbps. The feedback system signals the statistics described in Section 4.2 to the source every 0.5 second.

In Fig. 7, the source coding rate over time is plotted against a capacity region identified by the highest available rate and lowest available rate. The highest available rate is equal to the minimum of the network capacity and the highest available source coding rate of the scene. The lowest available rate is equal to the minimum of the network capacity and the lowest available source coding rate of the scene. As the network capacity varies and the source rate of the scene varies, a desirable source coding rate should be close to the highest available rate curve, yet not degrading traffic performance seen at the receiver. The rate adaptation maintains the two descriptions to be synchronized and provides 80% capacity utilization with less than 30ms delay jitter on average.

5   Experiment Results

We investigated the end-to-end distortion performance of the proposed system and compare the results with a single description coded bitstream.

The clips used for testing are movie trailers, including Golden Eye (GE), Despicable Me (DM) and Sorcerer's Apprentice (SA). The DM and SA clips are of WVGA (800x480) resolution while the GE clip is of VGA (640x480) resolution. The GE clip is a fast moving scene which has High Texture and High Motion. The SA clip is a Medium Motion and High Texture clip with frequent scene changes. The DM is a cartoon trailer that has fast scene cuts, High Texture and a combination of Low Motion and High Motion scenes.

First, the sequences are encoded using an un-modified Joint Model 17 (JM) encoder using the constant Quantization Parameter (QP) mode. Then the JM 17 code is modified to include the necessary changes required for the MDC system, including but not limited to inserting the SEI messages and having the QP values read from a file. All the clips are again encoded using this modified encoder and the results are investigated. All other settings for both the encoders are kept as close to the same as possible.

The lab system setup is as shown in Fig. 8. In the set up, we have a laptop acting as an MDC aggregator in the cloud, a mobile acting as a MDC source and a mobile server, and a cooperative mobile acting as a source helper. The source and its helper are connected using a Wi-Fi connection, while both of them access the cloud via their 3G access links. The source intends to upload its data, hence streams a description directly to the aggregator and streams another description via the helper to the aggregator.

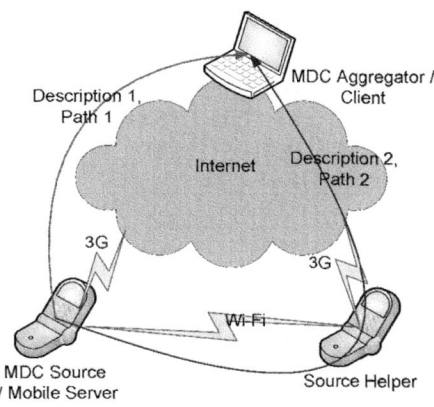

Fig. 8. Lab System Set Up

In a simple experiment, a video sequence is encoded using two slices, the first slice with a Quantization Parameter (QP) value of 25 and the second slice with a QP value of 35. This sequence is called Description 1 (D1). The same sequence is encoded again using two slices, but now the first slice is encoded using a QP value of 35 and the second slice is encoded using a QP value of 25 (Description 2, D2). Fig. 9 shows the example. When the client receives both the descriptions D1 and D2, highest quality slices among available descriptions are selected for decoding and hence obtains a better video quality equivalent to QP 25 coded frames.

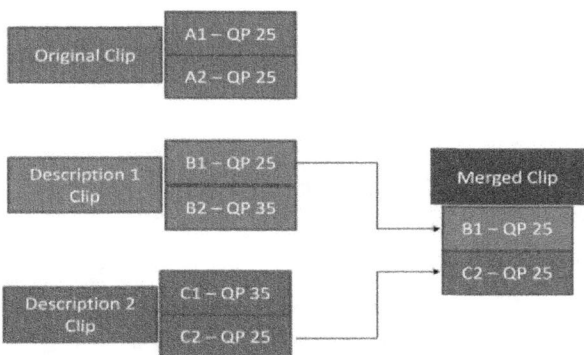

Fig. 9. MDC Encoder – Slice Level Example

Various experiments have been conducted in our labs. For each experiment, we calculated the bit rate and the PSNR of the luma (Y) component and averaged across the total number of frames in each sequence. The first set of experiments shows a typical use case scenario in which each description is of acceptable quality and the MDC selective combined clip shows a marked quality improvement. The second set of experiments shows the capability of the system to adapt to the varying capacity of individual paths. Table 1 shows the results from both configurations.

Table 1. Performance of the Slice Level MDC System

Content	Description 1, D1		Description 2, D2		Combined (D1 U D2)		
	Bit Rate (Kbps)	PSNR (dB)	Bit Rate (Kbps)	PSNR (dB)	PSNR (dB)	ΔPSNR (dB)	ΔRate (%)
Typical Configuration : QP 25 / QP 35							
GE	732	35.9	702	35.4	38.6	2.7	22.1
DM	754	38.3	858	39.3	42.4	4.1	26.7
Edge Configuration : QP 25 / QP 50							
GE	618	27.9	577	26.8	39.6	11.7	4.6
SA	729	28.9	750	27.6	45.8	16.9	6.8

Another advantage of using the proposed MDC system is the reduction in the peak frame size. This helps reduce traffic congestion in the network since peaks are not only reduced but staggered in time when descriptions are transmitted over two paths. Fig. 10 shows the distribution of frame sizes for the SA sequence at different QP settings. As it shows, if the sequence is encoded using QP 25 (blue curve) across the frame, the individual frame sizes have peaks that are quite large compared to the average frame size. This is due to the fact that the scene changes cause I-frames to have much larger size.

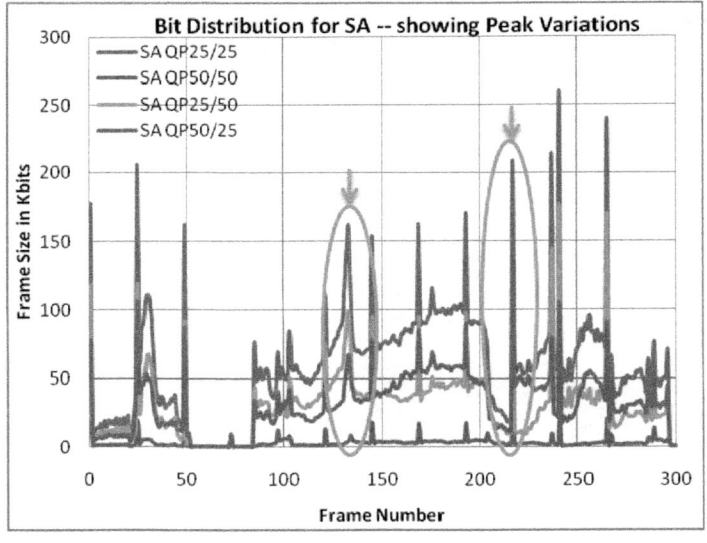

Fig. 10. Bit Distribution for Sorcerer's Apprentice, showing peak reduction

In the case of the proposed MDC scheme, by encoding half of the frames at a QP value of 25 and the other half at QP value of 50, the peaks are distributed across both descriptions. Fig. 11 and Fig. 12 show the quality of the decoded video at the client side. The image on the right shows the quality when both descriptions are received and the left side shows the quality of the video when only one description is received and decoded. Fig. 11 shows the I-frame quality and Fig. 12 shows a succeeding P-frame quality. The figures show that the quality is acceptable even if one of the descriptions is received and the quality improves when both the descriptions are received and decoded.

Fig. 11. Visual Video Quality for an I-frame (L) D1 only (R) D1 U D2

Fig. 12. Visual Video Quality for a Succeeding P-frame (L) D1 only (R) D1 U D2

As shown in Table 1, when both descriptions are received and decoded, the typical increase in Luma (Y) PSNR as compared to the single description case is about 3dB for the typical configuration with a 24% increase in the total bit rate and is about 13dB for the edge configuration case with a 5% increase in the total bit rate. We have also computed the VQM (voice quality metric) scores and the SSIM (structural similarity) values for the merged sequences and for the individual sequences and noticed that the increase is consistent with the visual observations.

From experiments conducted in our lab and from the data presented here, it could be observed that the proposed system consisting of just two descriptions can greatly increase the quality of the video in the case of High Motion and High Texture sequences. In general, it is known that these types of sequences usually require high coding rates and also have high peak-to-average ratios since the frames changes much more frequently. Hence by using the proposed system, we could not only reduce the peaks but also increase the overall quality of the bitstream.

6   Conclusions

In this paper, we introduce a mobile multipath cooperative network for real-time streaming. The system can be deployed over the existing IP network. The architecture enables a mobile device to access the cloud using multiple access links including those provided by its neighboring mobile devices. Different streaming sessions served by the same mobile device can be given distinct per hop QoS and routing treatments. As a result, the network access capacity, the reliability and subsequently the quality of experience for a multimedia streaming originated from or destined to a mobile device are significantly enhanced. We also introduced a new method of MDC with rate adaptation that performs source coding rate adjustment and synchronization for multiple descriptions across different paths. Our source coding rate adaptation system is shown to well utilize the network capacity and maintains synchronization of multiple descriptions over different paths without inducing undesirable delay and delay jitter.

References

[1]   Gantz, J., Reinsel, D.: The Digital Universe Decade – Are You Ready. In: Annual Report of IDC iView (2010)
[2]   Mao, S., Cheng, X., Hou, Y.T., Sherali, H.D.: Multiple description video multicast in wireless ad hoc networks. In: Proc. of the First International Conference on BroadNets, pp. 671–680 (2004)
[3]   Ghareeb, M., Viho, C.: A Multiple Description Coding Approach for Overlay Multipath Video Streaming Based on QoE Evaluations. In: Proc. of the Int'l Conference on MINES, pp. 39–43 (2010)
[4]   Gogate, N., Chung, D.-M., Panwar, S.S., Wang, Y.: Supporting image and video applications in a multihop radio environment using path diversity and multiple description coding. IEEE Trans. on Circuits and Systems for Video Technology 12(9), 777–792 (2002)
[5]   Goyal, V.K.: Multiple Description Coding: Compression Meets the Network. IEEE Signal Processing Magazine 18(5), 74–94 (2001)
[6]   Rosen, E., Viswanathan, A., Callon, R.: Multiprotocol Label Switching Architecture. IETF RFC 6178 (2001)
[7]   Vitali, A.: Multiple Description Coding - a new technology for video streaming over the Internet. EBU Technical Review, Bergamo (2007)
[8]   Schwarz, H., Marpe, D., Wiegand, T.: Overview of the scalable video coding extension of the H.264/AVC standard. IEEE Trans. on Circuits System Video Technology 17, 1103 (2007)
[9]   Radha, H., Chen, Y., Parthasarathy, K., Cohen, R.: Scalable Internet Video using MPEG-4. Signal Processing Image Communication 15(1-2), 95–126 (1999)

[10] Chakareski, J., Han, S., Girod, B.: Layered Coding vs. Multiple Descriptions for Video Streaming over Multiple Paths. In: Multimedia Systems. Springer Online Journal Publication (2005)

[11] Lee, Y.-C., Kim, J., Altunbasak, Y., Mersereau, R.M.: Performance Comparisons of Layered and Multiple Description Coded Video Streaming over Error-Prone Networks. In: Proc. of IEEE Int'l Conf. Communications, vol. 1, pp. 35–39 (2003)

[12] Suehring, K.: H.264/AVC JM Reference Software Download, http://iphome.hhi.de/suehring/tml/download/

[13] Kim, J., Mersereau, R.M., Altunbasak, Y.: Network-adaptive video streaming using multiple description coding and path diversity. In: Proc. of IEEE ICME 2003, vol. 1, pp. 653–656 (2003)

[14] Han, H., Shakkottai, S., Hollot, C.V., Srikant, R., Towsley, D.: Overlay TCP for multi-path routing and congestion control. IEEE/ACM Trans. Networking (2006)

[15] Key, P., Massoulié, L., Towsley, D.: Path selection and multipath congestion control. In: Proc. IEEE Infocom (2007)

[16] Wang, W.H., Palaniswami, M., Low, S.H.: Optimal flow control and routing in multi-path networks. Performance Evaluation 52, 119–132 (2003)

Multi-stream Rate Adaptation Using Scalable Video Coding with Medium Grain Scalability

Sergio Cicalò, Abdul Haseeb, and Velio Tralli

Engineering Department of University of Ferrara (ENDIF), Italy
{sergio.cicalo,abdul.haseeb,velio.tralli}@unife.it

Abstract. Multiple video streaming in a shared channel with constant bandwidth requires rate adaptation in order to optimize the overall quality. In this paper we propose a multi-stream rate adaptation framework with reference to the scalable video coding (SVC) extension of the H.264/AVC standard with medium grain scalability (MGS) and quality layer (QL). We first provide a general discrete multi-objective problem formulation with the aim to maximize the sum of assigned rates while minimizing the differences among distortions under a total bitrate constraint. A single-objective problem formulation is then derived by applying a continuous relaxation to the problem. We also propose a simplified continuous semi-analytical model that accurately estimates the rate-distortion relationship and allows us to derive an optimal and low-complexity procedure to solve the relaxed problem. The numerical results show the goodness of our framework in terms of error gap between the relaxed and its related discrete solutions, the significant performance improvement with respect to an equal-rate adaptation scheme, and the lower complexity with respect to a sub-optimal algorithm proposed in the literature.

Keywords: SVC, MGS, rate-distortion modeling, rate adaptation, quality fairness.

1 Introduction

H.264 Advanced Video coding (AVC) standard with scalable extension, also called Scalable Video Coding (SVC) [1], provides flexibility in rate adaptation by coding an original video sequence into a scalable stream. Three scalability methods are possible in SVC, named temporal, spatial and SNR scalability, that allow to extract a sub-stream in order to meet a particular frame rate, resolution and quality, respectively.

Due to the different complexities of the scenes composing a video sequence, the relationships between the rate and the quality of a set of videos can be really different among them. If individual video streams are transmitted to different users in a broadcast dedicated channel, as for instance in the case of on-demand IPTV services [2], an equal rate allocation can lead to unacceptable distortion

L. Atzori, J. Delgado, and D. Giusto (Eds.): MOBIMEDIA 2011, LNICST 79, pp. 152–167, 2012.
© Institute for Computer Sciences, Social Informatics and Telecommunications Engineering 2012

of high-complexity videos with respect to low-complexity ones. Adaptive transmission strategies must be investigated to dynamically optimize the quality of experience (QoE) of each end-user.

In this paper, we focus on rate adaptation, also called in literature statistical multiplexing, of SNR-scalable video streams, with a fixed temporal and spatial resolution. Many contributions exist in the literature that provide rate adaptation exploiting the Fine Granularity Scalability (FGS) tool, e.g. [3],[4] and [5]. FGS coding allows to extract an arbitrary rate-distortion (R-D) point while maintaining the monotonic non-decreasing behavior of the R-D curves. Nevertheless FGS mode has been removed from SVC, due to its complexity.

Two different possibilities for the SNR scalability tool are now available in SVC standard and implemented in the reference software [6], namely Coarse Grain Scalability (CGS) and Medium Grain Scalability (MGS). CGS can be achieved by coding quality refinements of a layer using a spatial ratio equal to 1 and inter-layer prediction. However, CGS scalability can only provide a small discrete set of extractable points equal to the number of coded layers. MGS provides a finer granularity of quality scalability by dividing a CGS layer into up to 16 MGS layers. The granularity can be also improved if a post-processing quality layer (QL) insertion and a consequent quality-based extraction is performed with the aim to optimize the R-D performance [7]. With this tool MGS can be seen as alternative to the FGS coding.

The first aim of this work is to analyze the performance of the MGS with QL and to provide a general R-D model. Other contributions exist in literature that estimate the R-D model for SNR-based scalable stream, with CGS and MGS, e.g. [8],[9], either analytical and semi-analytical. The analytical models are dependent on the probability distribution of discrete cosine transform (DCT) coefficients and often incur in a loss of accuracy. To achieve higher accuracy, semi-analytical R-D models are preferable. The semi-analytical models are based on parametrized functions that follow the shape of analytically derived functions, but are evaluated through curve-fitting from a subset of the rate-distortion empirical data points. In [9], the authors proposed an accurate semi-analytical square-root model for MGS coding and compared it with linear and semi-linear model. They concluded that the best performance is obtained by changing the model according to a parameter that estimates the temporal complexity, evaluated before encoding the entire sequence. However, a general model, that is able to estimate the R-D relationship of a large range of video sequences, is necessary to perform analytical optimization of the rate-adaptation problem. Besides, they did not consider the post-processing QL insertion that produces a variation of the R-D performance.

In [10] the authors proposed a general semi-analytical rate-distortion model for video compression, also verified in [11] for SVC FGS layer, where the rate and the distortion have an inverse relationship. Three sequence-dependent parameters must be estimated through the knowledge of six empirical R-D points. We have also verified this model with reference to SNR scalability with MGS and QL. The high accuracy of the results led us to investigate a simplified model with lower complexity, where the number of R-D points can be reduced by eliminating one of the parameters to estimate. Thus, we propose and compare a simplified two-parameters semi-analytical rate-distortion model. This simplification has two main advantages: (i) only four empirical points are needed by the curve fitting algorithm to achieve good performance, (ii) it allows the derivation of a low-complexity optimal procedure to solve the multi-stream rate-adaptation problem, with a maximum number of iterations equal to the number of streams involved in the optimization.

In summary, this paper collects the following main contributions: in section 2, a general optimization problem is formulated with the aim to provide the maximum quality to each video while minimizing their distortion difference, and by fulfilling the available bandwidth. In section 3 we analyze and verify two similar semi-analytical models for MGS with QL by comparing them with respect to complexity and the normally used goodness parameters: the root mean square error (RMSE) and the coefficient of determination R^2 [12]. An optimum and computationally efficient procedure to solve the relaxed general problem is derived in section 4, with a discussion about complexity and optimality. Finally the numerical results, discussed in section 5, show (i) the goodness of our framework by looking at the error between the relaxed and discrete solutions, (ii) the performance improvement with respect to a blind adaptation, and (iii) the complexity of the proposed algorithm with respect to a sub-optimal golden search algorithm proposed in literature.

2 General Problem Formulation for Multi-stream Rate Adaptation

In general, the aim of multi-stream rate adaptation is to optimize a certain number of utility functions U_i with respect to a quality metric and according to rate constraints [13]. Before or after the encoding process the original high quality video must be adapted, to meet a particular QoE metric depending on spatial, temporal and SNR resolutions.

In this section we provide a general problem formulation for multi-stream rate adaptation. Let K be the number of streams involved in the optimization. Given a set of lossy compression techniques $\{1, \ldots, N_k\}$, we can define in general $\mathcal{D}_k = \{d_{1,k}, \ldots, d_{N_k,k}\}$, $k = 1, \ldots, K$ as the set of distortion values for the k-th stream. Let us note that its cardinality $|\mathcal{D}_k| = N_k$ is generally not equal for each video source, as in the case of high-flexibility SNR-based compression techniques.

The rate-distortion theory evaluates the minimum bit-rate R_k required to transmit the k-th stream with a given distortion $d_{n,k}$, by defining a function $\mathcal{F}_k$ that maps the distortion to the rate, i.e.

$$\begin{aligned} \mathcal{F}_k : \quad & \mathcal{D}_k \to \mathbb{R}^+ \\ & d_{n,k} \to R_k = \mathcal{F}_k(d_{n,k}) \end{aligned} \tag{1}$$

One of the desirable properties of $\mathcal{F}_k$ is the strictly decreasing monotony, i.e.

$$\mathcal{F}_k(d_{i,k}) > \mathcal{F}_k(d_{j,k}), \quad \forall d_{i,k}, d_{j,k} : d_{i,k} < d_{j,k}. \tag{2}$$

When multiple streams have to be transmitted in a shared channel the rate adaptation algorithm must choose at each time slot and according to one optimization strategy, the best vector $\boldsymbol{D}^* = [D_1^*, ..., D_K^*] \in \boldsymbol{D} = \mathcal{D}_1 \times ... \times \mathcal{D}_K$. $\boldsymbol{D}$ contains all the possible combinations of the elements of $\mathcal{D}_k$, $k = 1, ..., K$ and has cardinality $N = \prod_{k=1}^K N_k$.

The main purpose of multi-stream rate adaptation is to provide the minimum distortion, or equivalently the maximum rate according to assumption (2), to each video under a total bit-rate constraints R_c. However, the solution of such problem can generally lead to large distortion variations among different streams, due to the different complexity of video sources. Quality fairness is an important issue that must be addressed when multiple videos from different sources are transmitted in a shared channel. In [4] the authors have shown that, given a continuous decreasing exponential R-D relationship with a constant exponent equal for each source, the solution to the problem of minimizing the distortion variations is also the solution to the problem of minimizing the total average distortion. However, an exponential R-D relationship is not an accurate model for all the different video compression techniques, particularly for the SVC SNR scalable stream [4]. Thus, a general multi-objective problem has to be formulated and a continuous relaxation of the problem leads to some particular simplification under certain assumptions. The general objective of our proposed framework is to minimize the differences among the distortions provided to each video stream while maximizing the sum of the rates until a maximum bit-rate is met. As mentioned above, these two objectives alone can generally lead to different solutions.

Thus, we formulate the general problem as a multi-objective problem:

$$\min_{\boldsymbol{D} \in \boldsymbol{D}} \sum_i \sum_{j<i} \Delta(D_i, D_j) \tag{3}$$

$$\max_{\boldsymbol{D} \in \boldsymbol{D}} \sum_{k=1}^K \mathcal{F}_k(D_k) \tag{4}$$

$$s.t. \ \sum_{k=1}^K \mathcal{F}_k(D_k) \leq R_c \tag{5}$$

where

$$\Delta(D_i, D_j) = \begin{cases} 0 & \text{if } (i,j) \in \mathbb{X}_D \lor (j,i) \in \mathbb{X}_D \\ |D_i - D_j| & \text{otherwise} \end{cases} \tag{6}$$

with

$$\mathbb{X}_D = \{(i,j) \in \mathbb{Z}^2 : (D_i = D_{max,i} \land D_j > D_i) \lor (D_i = D_{min,i} \land D_j < D_i)\} \tag{7}$$

and $D_{min,i} = \min_n d_{n,i}$, $D_{max,i} = \max_n d_{n,i}$. The operators $\land$ and $\lor$ are the logic "AND" and "OR", respectively.

Ideal fairness among the distortion values assigned to the multiple video streams, i.e. $D_i = D_j$, $\forall i \neq j$, is hard to be achieved. This fact is due to

(i) the discretization of the R-D relationship and (ii) the presence of the minimum and the maximum distortion values for each source that are related to the complexity of each video and which can be very different. The definition of the fairness metric takes this fact into account. In fact, the difference among video distortions $\Delta(D_i, D_j)$ is slightly modified to take into account the minimum and the maximum constraints. It is worth noting that, under the assumption (2), this problem admits a feasible solution only if at least the sum of the minimum rates of the video sequences is supported by the transmission bandwidth R_c, i.e

$$\sum_{k=1}^{K} \mathcal{F}_k(D_{max,k}) \leq R_c \tag{8}$$

otherwise a certain number of videos are not admitted in the transmission until this constraint is not satisfied. The solution of the problem in (3)-(5) requires in general an exhaustive search in the space $\mathcal{D}$ of all possible vectors. If N becomes large the required complexity can be not suitable for real-time adaptation. On the other hand if N is small, i.e there are few video sources as well as few related R-D points, the problem solution can lead to a waste of the available bandwidth and a large distortion differences among multiple videos.

In the next section we will propose a semi-analytical R-D model with reference to the SNR scalability tool of SVC with MGS and QL layers [7]. This continuous model will allow us to apply a continuous relaxation to the optimization problem leading to a simplification in a single-objective problem formulation.

3   Rate Distortion Model for MGS with Quality Layer

We consider here SNR scalability obtained through the MGS coding and QL post-processing insertion, with a fixed temporal and spatial resolution. In this case the components of $\mathcal{D}_k$ are the distortion values of the extractable substreams from the high quality original encoded stream.

MGS coding allows to distribute the transform coefficients obtained from a macro-block by dividing them into multiple sets. The number of sets identifies the number of weights, often named MGS layers, in the MGS vector. Thus, the elements of the MGS vector correspond to the cardinality of each set.

The R-D relationship and its granularity depend on the number of MGS layers and the coefficient distribution [14], [15]. In [15] the authors analyzed the impact on performance of different numbers of MGS layers with different configurations used to distribute the transform coefficients. We also verified their results, by noting that more than five MGS layers reduce the R-D performance without giving a substantial increase in granularity. This is mainly due to the fragmentation overhead that increases with the number of MGS layers.

While extracting an MGS stream two possibilities are available in the reference software: a flat-quality extraction scheme, and a QL-based extraction scheme. The second scheme requires a post-encoding process that computes a priority index for each NAL unit, but achieves higher granularity, as well as better R-D-performance [7]. However, differently to flat-quality extraction scheme, the quality-based extraction process does not give substantial variations in granularity and R-D performance when varying the distribution of the coefficients, as also shown in [15]. In our extensive simulation campaign the best results in terms of granularity and R-D performance are obtained with a MGS vector equal to [3 2 4 2 5].

When the SVC video has to be adaptively transmitted it is common practice to analyze the R-D model with respect to a fixed set of frames identified by one group of pictures (GOP). In this way, the adaptation module can follow the complexity variations of the different scenes. Therefore, throughout this paper we assume that the reference time interval used to analyze the R-D relationship as well as to optimize the distortion of multiple streams is the GOP interval.

In [10] the authors propose a general continuous semi-analytical R-D model for video compression, also verified in [11] for SVC FGS layers, with the following relationship :

$$\mathcal{R}_k(D) = \frac{\eta_k}{D + \theta_k} + \phi_k. \tag{9}$$

The distortion D is evaluated as the average mean square error (MSE) of the decoded video. The drawback of this approach is the need to estimate the three sequence/encoder dependents parameters, η_k, θ_k and ϕ_k, by using curve-fitting from a subset of the rate-distortion data points. The curve-fitting algorithm requires a relevant number of iterations and function evaluations and six empirical R-D points. To reduce the complexity, we can simplify this parametrized model by eliminating one parameter, i.e.

$$\mathcal{R}_k(D) = \frac{\alpha_k}{D} + \beta_k \tag{10}$$

In this case, only four R-D points need to be evaluated to estimate the two sequence-dependent parameters α_k and β_k, and as a result the number of iterations and function evaluations decreases. Beside the complexity reduction, this model allows a simple derivation of the solution of the problem (3)-(5), as we will show later.

Table 1 compares the goodness of the two models with respect the coefficient of determination R^2, the RMSE, the number of iterations and function evaluations required by a non-linear Least Square Trust-Region (LSTR) algorithm to converge. It can be noted how the number of function evaluations as well as the number of iterations decrease while a minimum loss occurs in the goodness parameter. In Figure 1, we plot the empirical R-D relationship for the five sequences, used to obtain numerical results, as well as their related R-D curves based on model (10). All of them are referred to the GOP with the worst RMSE value (the minimum in Table 1). We can also appreciate in this figure the achievable granularity of the quality-based extraction.

In the next section we will apply a continuous relaxation to the problem (3)-(5) by exploiting the model (10) and we will provide a low-complexity optimal procedure to solve it.

Table 1. Comparison between the two semi-analytical model in (9) and (10) with respect to the minimum and maximum $RMSE$ and the coefficient of determination R^2 evaluated for each GOP (GOP size equal to 16) of five video sequence with CIF resolution and frame rate of 30 fps. The video are encoded with one base layer (QP equal to 38) and two enhancement layers (QP equal to 32 and 26), both with 5 MGS layers and a weights vector equal to [3 2 4 2 5].

Video	Model	R^2 [min,max]	$RMSE$ [min,max]	Av. No. iteration	Av. No. Function Evaluation
Coastguard	Model (10)	[0.9842 , 0.9934]	[37.895 , 79.992]	30.2	89.6
	Model (9)	[0.9956 , 0.9982]	[22.261 , 36.724]	34.7	155.9
Crew	Model (10)	[0.9752 , 0.9944]	[23.038 , 89.130]	30.9	94.2
	Model (9)	[0.9914 , 0.9972]	[20.019 , 52.489]	35.6	159.9
Football	Model (10)	[0.9662 , 0.9891]	[53.403 , 205.572]	29.0	89.5
	Model (9)	[0.9809 , 0.9993]	[12.940 , 99.810]	38.0	169.3
Foreman	Model (10)	[0.9669 , 0.9955]	[19.710 , 53.371]	25.7	73.2
	Model (9)	[0.9906 , 0.9980]	[13.516 , 33.745]	34.1	154.3
Harbour	Model (10)	[0.9854 , 0.9907]	[51.860 , 73.344]	37.5	129.8
	Model (9)	[0.9952 , 0.9991]	[18.883 , 44.822]	45.3	164.3

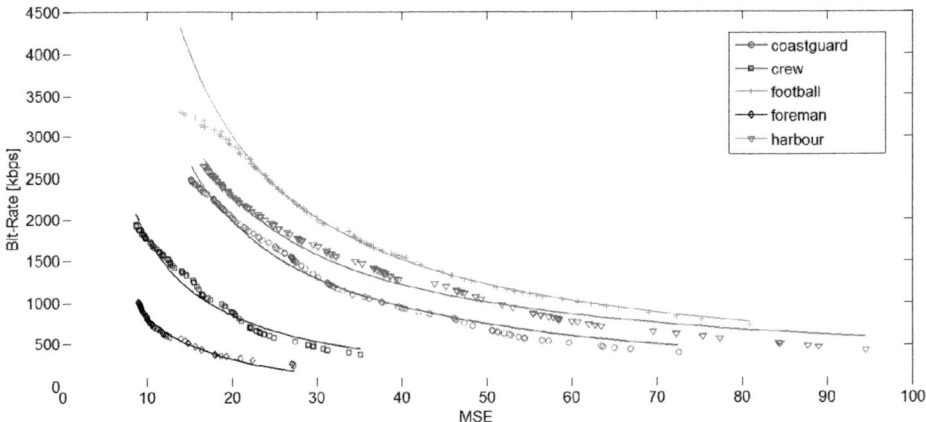

Fig. 1. R-D Model (straight line), according to eq. (10) fitting the empirical R-D relationship for the GOP with the worst $RMSE$ with reference to Table 1

4   GOP-Based Multi-stream Rate Adaptation Framework

Without loosing generality we assume that each video is coded with the same GOP size and the rate allocation is performed at the GOP boundaries. Thus, from now on we focus on one GOP interval. Considering all the discussions in the previous sections, we apply a continuous relaxation to the optimization problem based on the model (10). Therefore we assume that the discrete variable D_k becomes continuous (denoted by $\tilde{D}_k$), but limited by the minimum and maximum distortion, i.e.

$$\tilde{D}_k \in [D_{min,k}, D_{max,k}]. \tag{11}$$

With reference to the SNR scalability, the points $\{D_{max,k}, \mathcal{F}_k(D_{k,max})\}$ and $\{D_{min,k}, \mathcal{F}_k(D_{k,min})\}$ are the base layer and the highest enhancement layer points, respectively. Those values are two of the four R-D points required by the curve-fitting algorithm.

It is worth noting that a trivial solution can be derived if the sum of the full quality encoded stream rates is less then or equal to the available bandwidth, that corresponds to transmit the entire encoded streams without adaptation. Thus, we analyze the non-trivial case where the following constraint holds :

$$\sum_{k=1}^{K} \mathcal{F}_k(D_{k,min}) > R_c \tag{12}$$

According to the continuous relaxation (11) and the assumptions (8) and (12), a feasible solution is obtained when the constraint on the overall channel bandwidth is active with equality. A single-objective problem, where the second objective, i.e (4) in the problem formulation, is eliminated and replaced by an equality constraints can be then formulated. Nevertheless, as a result of the relaxation of the problem, the two constraints referred to the maximum and minimum available rates of each stream must be added. They imply that each video sequence has to obtain at least the base layer and not more than the maximum available bit-rate must be allocated to each video source to save bandwidth.

Thus, the relaxed problem can be formulated as

$$\min_{\tilde{D} \in \mathbb{R}^K} \sum_{i} \sum_{j<i} \Delta(\tilde{D}_i, \tilde{D}_j) \tag{13}$$

$$s.t. \sum_{k=1}^{K} \mathcal{R}_k(\tilde{D}_k) = R_c \tag{14}$$

$$\mathcal{R}_k(\tilde{D}_k) \geq \mathcal{F}_k(D_{k,max}) \quad \forall k \tag{15}$$

$$\mathcal{R}_k(\tilde{D}_k) \leq \mathcal{F}_k(D_{k,min}) \quad \forall k \tag{16}$$

Note that the model $\mathcal{R}_k(\tilde{D}_k)$ replaces the actual R-D relationship $\mathcal{F}_k(D_k)$. In the next subsection we will derive an optimal procedure to solve this relaxed problem using methods that are computationally efficient and without the use of heuristics or brute-force search.

4.1   Problem Solution

A solution to the relaxed problem (13)-(16) can be derived by using sub-optimal procedures as the golden search algorithm proposed in [3] for a piecewise linear model. Nevertheless, the continuous formulation of model (10) allows us to derive a low-complexity optimal procedure, by noting that the solutions to the problem without the constraints (15) and (16) can be easily derived as follows:

$$\tilde{D}^* = \tilde{D}_k^* = \frac{\sum_{k=1}^{K} \alpha_k}{R_c - \sum_{k=1}^{K} \beta_k}, \quad \forall k. \tag{17}$$

Since those constraints imply that a minimum (maximum) or a maximum (minimum) rate (distortion) has to be allocated to each video stream, these solutions can be improved successively through a simple iterative procedure.

Let $x_k, y_k \in \{0,1\}$, $k = 1, \ldots, K$, be binary variables that indicate whether or not the two constraints are active for the video stream k and will be updated during the procedure. We can then define:

$$A_{\boldsymbol{x},\boldsymbol{y}} = \sum_{k=1}^{K} x_k y_k \alpha_k \tag{18}$$

$$B_{\boldsymbol{x},\boldsymbol{y}} = \sum_{k=1}^{K} x_k y_k \beta_k \tag{19}$$

$$R_{\boldsymbol{x},\boldsymbol{y}}^{av} = R_c - \sum_{k=1}^{K} (1 - x_k) \mathcal{F}_k(D_{k,max}) - \sum_{k=1}^{K} (1 - y_k) \mathcal{F}_k(D_{k,min}) \tag{20}$$

where $R_{\boldsymbol{x},\boldsymbol{y}}^{av}$ is the available rate for the videos which have not active constraints. The iterative procedure works as follows:

1. Initialize: $x_k = 1$ and $y_k = 1 \quad \forall k = 1, \ldots, K$
2. For each $k : x_k \cdot y_k = 1$ Compute:
   $\tilde{D}_k^* = \frac{A_{\boldsymbol{x},\boldsymbol{y}}}{R_{\boldsymbol{x},\boldsymbol{y}}^{av} - B_{\boldsymbol{x},\boldsymbol{y}}}$
   $\tilde{R}_k^* = \mathcal{R}_k(\tilde{D}_k^*)$, based on model (10)
   $condition = 0$
   2a. If $\tilde{R}_k^* > \mathcal{F}_k(D_{k,min})$ then
      $\tilde{R}_k^* = \mathcal{F}_k(D_{k,min})$
      $\tilde{D}_k^* = D_{k,min}$
      $y_k = 0$
      $condition = 1$
   2b. elseif $\tilde{R}_k^* < \mathcal{F}_k(D_{k,max})$ then
      $\tilde{R}_k^* = \mathcal{R}_k(D_{k,max})$
      $\tilde{D}_k^* = D_{k,max}$
      $x_k = 0$
      $condition = 1$
3. If $condition = 1$
   Go to step 2.
4. else break

The final relaxed solutions, given x_k and y_k, $k = 1, \ldots, K$, are then given by:

$$\tilde{R}_k^* = \begin{cases} \frac{\alpha_k}{\tilde{D}_k^*} + \beta_k & \text{if } x_k \cdot y_k = 1 \\ \mathcal{F}_k(D_{k,max}), & \text{if } x_k = 0 \\ \mathcal{F}_k(D_{k,min}), & \text{if } y_k = 0 \end{cases} \quad (21)$$

with

$$\tilde{D}_k^* = \begin{cases} \frac{A_{x,y}}{R_{x,y}^{av} - B_{x,y}} & \text{if } x_k \cdot y_k = 1 \\ D_{k,max}, & \text{if } x_k = 0 \\ D_{k,min}, & \text{if } y_k = 0 \end{cases} \quad (22)$$

The algorithm requires in the worst case, a maximum of K iterations with $(K - 1)/2$ rate and distortion evaluations. At the first iteration, due to the initialization, $\tilde{D}_k^*$ is computed as in (17). At each iteration the algorithm checks if the related rate solutions violate one of the constraints (15), (16). If it happens for one video, the algorithm assigns the relative minimum or maximum rate to this particular video and re-evaluates the distortion for the other video streams.

The optimality of the solutions (21) and (22) can be easily proved, by noting that the sum of the difference functions in (13) is always kept to zero, i.e. $\sum_i \sum_{j<i} \Delta(\tilde{D}_i^*, \tilde{D}_j^*) = 0$ and the sum of the rates is always equal to the available bandwidth. In fact, if at the n-th iteration a maximum rate constraint (condition of step 2a) is violated for the i-th video, the distortion of the other videos at the next iteration, $\tilde{D}_k^*[n + 1]$, will decrease, i.e.

$$\tilde{D}_k^*[n + 1] < \tilde{D}_k^*[n] < D_{i,min}, \quad \forall k \neq i : x_k[n+1] \cdot y_k[n+1] = 1, y_i[n] = 0. \quad (23)$$

Vice versa, when the second constraint (condition of step 2b) is violated for the j-th video the distortion $\tilde{D}_k^*[n + 1]$ of the other video will increase, i.e.

$$\tilde{D}_k^*[n + 1] > \tilde{D}_k^*[n] > D_{j,max}, \quad \forall k \neq j : x_k[n+1] \cdot y_k[n+1] = 1, x_j[n] = 0. \quad (24)$$

For all other videos with $x_k \cdot y_k = 1$ the solutions are left untouched, as shown in (22). The inequalities (23) and (24) follow from the monotony property of the R-D function.

Let us finally note that the conditions of steps 2a and 2b are auto-exclusive for each video source if

$$D_{s,max} > D_{p,min}, \quad \forall s \neq p, \quad s, p = 1, \ldots, K \quad (25)$$

When two or more video streams have a very different scene complexity in the same GOP, the inequality (25) may not be verified and the evaluated distortion $\tilde{D}_k^*$ may fall inside the interval $[D_{s,max}, D_{p,min}]$. In this particular case, to assure the best fairness, the algorithm would require some temporary additional steps to evaluate which constraints has to be applied first, which leads to a small increase in the complexity. In order to keep the complexity low we propose for this case to prioritize the distortion minimization. Thus, we first apply the constraints on the maximum rate (step 2a) by assigning the minimum distortion

$D_{p,min}$ to the p-th video. At the next iteration, the distortion will decrease, due to the convexity of the R-D functions. If the distortion decreases in such way that the evaluated rate of the s-th video do not violate its maximum distortion constraint, the algorithm will be able to assign a lower distortion to it. Let us note that this choice does not compromise the optimality of the solution of the problem according to eq. (6).

From a mathematical perspective the optimal discrete solution $\boldsymbol{D}^*$, starting from the relaxed one $\tilde{\boldsymbol{D}}^*$, should be derived by applying optimization techniques, e.g. branch & bound search. Nevertheless, such techniques require the knowledge of all the empirical discrete R-D points or a subset of R-D points close to the relaxed optimum solutions, with an increase in complexity. To keep the complexity low, it is common practice to extract the higher discrete bit-rate under the optimal relaxed solution, by paying a minimum waste of bandwidth due to the granularity of the empirical R-D relationship.

5   Numerical Results

In this section we evaluate the performance of the proposed rate adaptation framework by using the JSVM reference software [6]. We encode five video sequences with different scene complexity, i.e. *coastguard, crew, football, foreman, harbour* in CIF resolution with a frame rate of 30 fps. The SNR-scalability is obtained through 2 enhancement layers, each one split in 5 MGS layers with vector distribution [3 2 4 2 5]. The quantization parameter (QP) of the base and enhancement layers are equally spaced and set to 38, 32 and 26, respectively. Each sequence is coded GOP-by-GOP with a GOP size equal to 16, and the post-processing quality-based process is then applied, as mentioned throughout the paper. We first provide the performance metrics for a particular case of bandwidth, i.e. $R_c = 3000$ kbps, then we study the impact of different R_c values. The fairness is evaluated through two metrics: the average MSE difference $\delta_{av} = (1/S) \sum_i \sum_{j<i} |D_i^* - D_j^*|$, where the average is computed with respect to the number $S = K(K-1)/2$ of terms in the sum, and the most used MSE variance for each GOP. We first compare the solution of our algorithm (OPT) with an equal-rate (ER) scheme where no adaptation is performed, i.e. the same proportion of the available bandwidth is assigned to each video. To have a fair comparison we apply to ER scheme the constraints (15) and (16) in order to guarantee the base-layer to each video and to fulfill the available bandwidth. Therefore, after sorting the streams in two vectors into decreasing order according to base-layer bit-rate and into increasing order according to highest layer bit-rate, respectively, we iteratively check if the bit-rate $R_k = R_c/K$ required by each ordered stream violates one of those constraints. If it happens, we assign the corresponding bit-rate and equally re-distribute the remaining bandwidth to the other streams.

Table 2 shows the average MSE resulting from the rate assigned to each video sequences for the first 15 GOPs. As expected, the ER scheme is able to provide less distortion to the low-complexity video, i.e. *crew, foreman*, by compromising

Table 2. Average MSE of each video sequence with equal-rate (ER) assignment and rate adaptation with the proposed algorithm (OPT). Total bandwidth is equal to 3000 kbps.

Gop index	Guard ER	Guard OPT	Crew ER	Crew OPT	Football ER	Football OPT	Foreman ER	Foreman OPT	Harbour ER	Harbour OPT
1	53.71	53.71	18.59	34.64	80.86	55.87	18.40	31.66	74.28	55.52
2	57.35	54.57	19.79	37.85	74.65	59.56	18.24	29.96	81.23	56.98
3	69.45	54.63	23.52	38.67	64.02	54.06	24.63	29.99	94.54	58.27
4	81.35	59.02	39.87	39.87	63.69	56.29	17.75	33.34	75.92	57.75
5	53.71	47.36	24.89	41.67	49.53	43.55	17.73	31.58	71.93	50.97
6	55.16	41.70	28.22	38.26	16.85	24.55	19.51	34.00	73.82	46.48
7	49.11	42.22	39.87	44.31	20.23	31.36	12.40	27.35	82.14	49.31
8	49.38	42.64	33.87	38.57	31.49	39.12	14.21	28.35	73.47	48.10
9	45.79	44.11	37.47	41.71	43.89	44.20	19.20	36.12	73.51	50.37
10	42.02	46.06	42.85	43.02	47.94	45.19	19.51	32.24	69.64	52.51
11	44.49	49.17	34.40	45.68	59.81	48.88	17.77	31.33	67.82	53.78
12	42.07	40.36	25.56	39.42	41.44	41.17	18.73	30.32	71.87	46.47
13	40.17	43.18	27.09	41.48	50.24	43.84	16.55	27.87	72.23	50.91
14	42.11	56.76	23.86	35.08	82.50	56.45	25.39	45.48	68.08	57.95
15	38.29	60.28	24.81	38.76	84.63	56.84	25.92	57.12	69.48	55.82
Av.	50.95	49.05	29.64	39.93	54.12	46.73	19.06	33.78	74.66	52.74

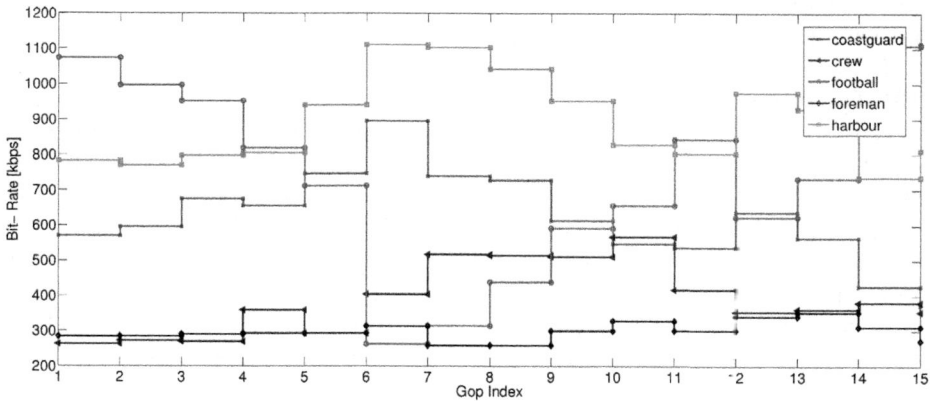

Fig. 2. Rate assigned by our adaptation algorithm in each GOP, with bandwidth equal to 3000 kbps

the distortion of the video sequences with more complexity. Our algorithm, while providing fairness, is able to improve the performance of the complex videos, by allocating more bits to video with more complex scenes. This is more clear in figure 2 where we plot the rate assigned to each video sequence GOP-by-GOP.

Table 3. Average modified MSE difference Δ_{av}, average MSE difference δ_{av} and MSE variance in each GOP interval. Comparison between the proposed algorithm (OPT) and equal-rate (ER) assignment with bandwidth equal to 3000 kbps.

GOP index	Δ_{av}		δ_{av}		Variance	
	ER	OPT	ER	OPT	ER	OPT
1	36.12	0.43	36.12	13.86	884.40	145.41
2	35.51	1.00	36.17	15.67	889.50	171.43
3	33.78	0.84	37.37	14.50	941.76	148.35
4	19.53	0.55	32.65	13.85	705.43	139.62
5	24.79	1.48	27.44	8.89	489.84	53.75
6	29.92	1.64	29.92	10.31	614.97	69.38
7	33.67	1.42	33.67	11.37	752.18	84.72
8	27.28	2.21	27.28	8.72	495.50	52.27
9	23.39	2.01	23.39	6.20	382.93	26.39
10	21.24	1.93	21.24	8.72	319.33	54.28
11	24.30	1.46	25.10	9.68	398.50	73.46
12	24.56	1.22	24.56	6.81	420.64	34.09
13	26.90	1.54	26.90	9.69	463.11	70.69
14	32.00	0.30	32.00	11.40	680.44	98.23
15	32.64	1.05	32.64	8.87	730.21	73.16
Av.	28.37	1.21	29.76	10.57	611.25	86.35

More bit-rate is assigned to *coastguard, football* and *harbour* video sequences, allowing them to achieve more quality.

In Table 3, we show the improvements of our proposed schemes with respect to ER. The average MSE difference is significantly reduced and equivalently the variance is decreased up to ten times. However, in this particular case of bandwidth, the MSE difference (variance) is still quite high, due to the minimum rate constraints. The average modified MSE difference $\Delta_{av} = (1/S) \sum_i \sum_{j<i} \Delta(D_i^*, D_j^*)$ according to definition in (6), is also evaluated in Table 3. Let us note that this metric also give us the information of the error generated when the discrete solution replaces the continuous solution of the relaxed problem, whose Δ_{av} is zero. This error includes two contributions: the estimation error of the model and the integrality gap. As expected the average error is not small due to mainly the low granularity of the low-rate points.

In figure 3, the MSE variance averaged over 15 GOPs is evaluated for different bandwidths. In the bandwidth interval considered, the assumptions (8) and (12) hold for each GOP. When the bandwidth is very low the two schemes provide approximately the same MSE because the optimization range is limited by the minimum rate constraints. When the bandwidth increases, our procedure improves the fairness leading the variance close to 0. A slight variance increase occurs at large bandwidths when the maximum rate constraints limit the achievable distortion. On the other hand the ER scheme generally increases the MSE variance until the base-layer constraints are active for most of the streams.

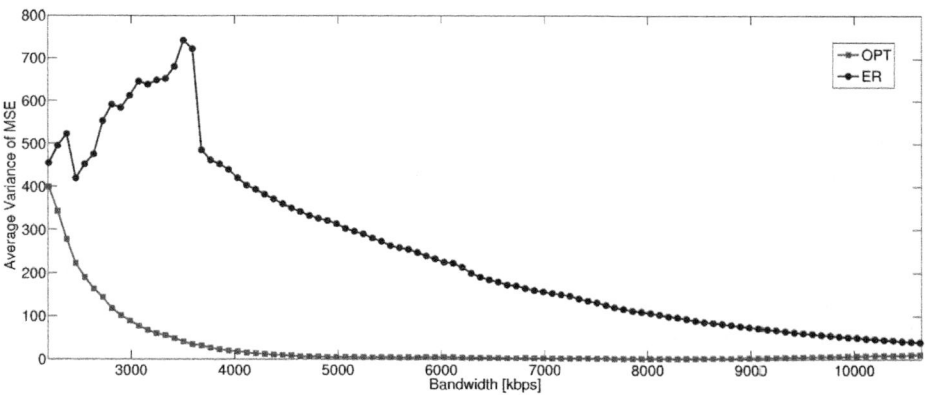

Fig. 3. Variance of the MSE averaged over 15 GOPs, with different bandwidth values. Comparison between the proposed algorithm (OPT) and equal-rate (ER) assignment.

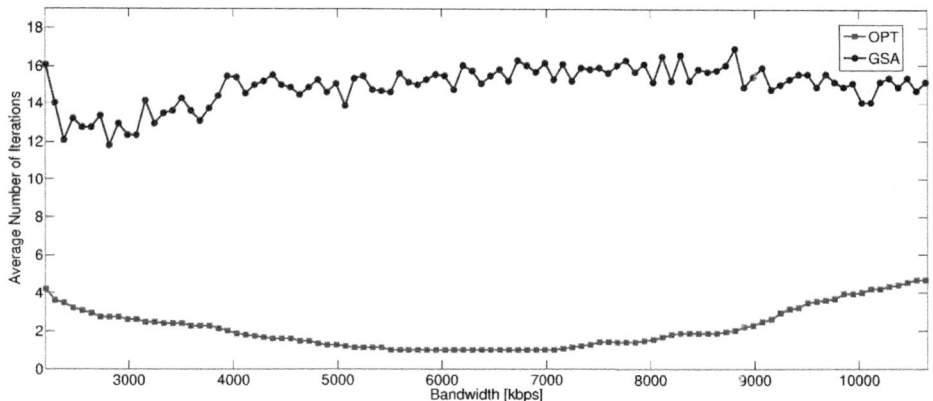

Fig. 4. Average number of iterations required by our adaptation algorithm (OPT) and golden search algorithm (GSA) to converge

This behavior can be partially reduced by controlling the base-layer bit-rate [16] to each video according to their complexity as performed for instance in [3].

To further assess our proposed scheme, we compared it to the golden search algorithm (GSA) proposed in [3], to solve the problem (13)-(16). This algorithm can be seen as a suboptimal version of our procedure. The initial solution is computed as function of the golden-section value and the difference between the lower and higher bounds, i.e. $a = \min_k D_{k,min}$ and $b = \max_k D_{k,max}$, identified by the minimum and the maximum distortion among the videos. At each iteration the solution is updated by applying the per-video constraints and by compressing the search interval consequently. The GSA terminates when the difference between the sum of the assigned rates and the available bandwidth is less of a chosen value ϵ. Nevertheless, an additional termination condition must be introduced to assure the convergence of the algorithm, that is usually indicated by the tolerance τ, i.e. $|a - b| \leq \tau$. In order to provide a fair comparison we set $\epsilon = 0.0002R_c$, and $\tau = 0.01$, leading to a sub-optimality error under 0.5% over all the investigated cases. In figure 4 we plot the average number of iterations required by the two algorithms for different bandwidths. The number of iterations of our algorithm is limited by the number of video sequences, as mentioned in sub-section 4.1, and decreases away from the minimum and the maximum bandwidths obtained as the sum of minimum and maximum rates of each video. The GSA algorithm requires generally more iterations due to the sub-optimal choice of the starting-point. This result does not change by increasing the number of videos involved in the optimization, as we also verified.

6   Conclusions

In this work we proposed a multi-stream rate adaptation framework with reference to SNR-scalability of SVC with MGS and QL. We formulated a general discrete problem with the aim to minimize the average distortion while providing fairness to different video sources. Two similar semi-analytical model that estimate the R-D relationship of each video source GOP-by-GOP are evaluated and compared with respect to goodness parameters and complexity. The general discrete problem was then relaxed and an optimal procedure was derived based on a low-complexity model. In the numerical results we showed the feasibility of our framework by analyzing the gap between the relaxed and discrete solution according to fairness metrics, the improvements with respect to an equal-rate scheme and the lower complexity of the proposed procedure with respect to an existing algorithm in the literature.

Acknowledgments. The work has been supported in part by the Italian Ministero dello Sviluppo Economico within the project WEBS, and in part by University of Ferrara, Italy. The work has been inspired by research and discussions within the cooperation framework of project e-GAP[2] of the Royal Society, UK.

References

1. Schwarz, H., Marpe, D., Wiegand, T.: Overview of the Scalable Video Coding Extension of the H.264/AVC Standard. IEEE Transactions on Circuits and Systems for Video Technology 17(9) (September 2007)
2. Wiegand, T., Noblet, L., Rovat, F.: Scalable Video Coding for IPTV Services. IEEE Transactions on Broadcasting 55(2) (June 2009)
3. Wang, Y., Chau, L., Yap, K.: Joint Rate Allocation for Multiprogram Video Coding Using FGS. IEEE Transactions on Circuits and Systems for Video Technology 20(6) (June 2010)
4. Zhang, X., Vetro, A., Shi, Y., Sun, H.: Constant Quality Constrained Rate Allocation for FGS Coded Video. IEEE Transactions on Circuits and Systems for Video Technology (February 2003)

5. Jacobs, M., Tondeur, S., Paridaens, T., Barbarien, J., Van de Walle, R., Schelkens, P.: Statistical Multiplexing Using SVC. In: Proc. IEEE Int. Symp. Broadband Multimedia Syst. Broadcast, pp. 1–6 (2008)
6. JSVM 9.19.11, Reference Software (February 2011)
7. Amonou, I., Cammas, N., Kervadec, S., Pateux, S.: Optimized Rate-Distortion Extraction with Quality Layers in the Scalable Extension of H.264/AVC. IEEE Transactions on Circuits and Systems for Video Technology 17(9) (September 2007)
8. Mansour, H., Krishnamurthy, V., Nasiopoulos, P.: Rate and Distortion Modeling of Medium Grain Scalable Video Coding. In: Proc. of 2008 IEEE 15th International Conference on Image Processing, San Diego, October 12-15 (2008)
9. Cesari, M., Favalli, L., Folli, M.: Quality Modeling for the Medium Grain Scalability Option of H.264/SVC. In: Mobimedia 2009, London, UK, September 7-9 (2009)
10. Stuhlmuller, K., Fraber, N., Link, M., Girod, B.: Analysis of Video Transmission over Lossy Channels. IEEE Journal On Selected Areas In Communications 18(6) (June 2000)
11. Mansour, H., Krishnamurthy, V., Nasiopoulos, P.: Channel Adaptive Multi-User Scalable Video Streaming with Unequal Erasure Protection. In: Eight International Workshop on Image Analysis for Multimedia Interactive Services, WIAMIS 2007 (2007)
12. Kwon, D., Shen, M., Kuo, C.: Rate Control for H.264 Video with Enhanced Rate and Distortion Models. IEEE Transactions on Circuits and Systems for Video Technology 17(5), 517–529 (2007)
13. Thang, T., Kim, J., Kang, J., Yoo, J.: SVC Adaptation: Standard Tools and Supporting Methods. Journal, Image Communication Archive 24(3) (March 2009)
14. Gupta, R., Pulipaka, A., Seeling, P., Karam, L.J., Reisslein, M.: H.264 Coarse Grain Scalable (CGS) and Medium Grain Scalable (MGS) Encoded Video: A Trace Based Traffic and Quality Evaluation (2011) (submitted)
15. Gorkemli, B., Sadi, Y., Tekalp, M.: Effects Of MGS Fragmentation, Slice Mode And Extraction Strategies on the Performance of SVC With Medium-Grained Scalability. In: Proc. of 2010 IEEE 17th International Conference on Image Processing, Hong Kong, September 26-29 (2010)
16. Liu, Y., Guo Li, Z., Chai Soh, Y.: Rate Control of H.264/AVC Scalable Extension. IEEE Transactions on Circuits and Systems for Video Technology 18(1), 111–116 (2008)

The Role of Log Entries in the Quality Control
of Video Distribution

Ismo Hakala, Sanna Laine, Mikko Myllymäki, and Jari Penttilä

University of Jyväskylä, Kokkola University Consortium Chydenius
P.O. Box 567, FIN-67701, Kokkola, Finland
{ismo.hakala,sanna.laine,mikko.myllymaki,
jari.penttila}@chydenius.fi

Abstract. Diversification of university teaching with the help of video lectures has become much more common during the past few years. Once videos have become an essential part of teaching arrangements, whoever organizes the teaching must also pay attention to factors related to videos in quality system work for teaching. In the video production process it is the factors related to video transmission that exert influence on the usability of videos and set limitations for their production. A lot of information about those factors can be obtained from the media server log files. The particular focus of this paper is on the functionality of the connection between a server and a client and its effect on users. The paper deals with information obtained from a media server's log file, describes the activities around collection and handling of log data, and introduces a preliminary classification for monitoring video transmissions. The results obtained from the media server's log files are presented in accordance with that classification system at the end of the paper.

Keywords: Quality of Service, log file, streaming video lecture.

1 Introduction

During the last few years there has been a strong increase in the use of video lectures in many universities. The motivation behind this increased use of videos has typically been the desire to make studying more flexible as regards time and place. The idea is to provide students who live in places that are geographically distant better chances for studying and, at the same time, ensure that the lectures are available also for temporarily absent students. These arrangements also allow students to revise any difficult or complicated lecture items and topics. Diversification of face-to-face teaching, by delivering it as real time video and by providing recorded (on-demand) videos alongside it, makes it possible for a student to study solely with the help of videos without a need to participate in any traditional type of contact teaching. In teaching arrangements, the organizer of the education must then, in addition to video production, pay attention to the possibilities that the student has for viewing videos and to the factors related to video transmission.

L. Atzori, J. Delgado, and D. Giusto (Eds.): MOBIMEDIA 2011, LNICST 79, pp. 168–179, 2012.

Internet connections in homes in recent years have speeded up considerably and the problems related to video transmissions have become less common. However, portable computers and new mobile devices, that are suitable for study, have increased the number of mobile broadband connections which are slower and more unstable in congested networks than traditional broadband connections. Evaluation of the factors related video distribution and habits and activities formed around them is an important part of a quality system for teaching, especially in blended teaching arrangements, where the role of videos alongside face-to-face teaching is important.

The education organizer must, among other things, know the problems associated with video transmission and use, how to detect those problems, how to react on them, the reasons for them and how to influence them. Regardless of the quality of a produced video, the student becomes aware of the problems in video transmission, for example, if the video keeps on stopping during its viewing or in the worst of cases if there is an abnormal break during it [1],[5],[6]. In a quality system framework, these related factors can be taken into account to some extent when soliciting course feedback from the students. Information received through course feedback, however, is subjective and general. The media server's log files provide, for the organizer, an information source that is adequate and accurate but also fairly difficult and little used. The log files provide information, among other things, about server work load, the system that the students use and bandwidth employed as well as about possible problems in transmission or abnormal breaks that are independent of viewers. Partial answers are obtained as to the causes of the problems found, which makes it possible to take anticipatory measures to those problems.

Log data can be used for example to solve problem situations and to monitor the amount of usage and availability of service. Statistical summaries based on log entries can be helpful when deciding about development directions related to the use of videos in teaching and to their production process. For example, information obtained from the log data about the bandwidth used in video transmissions can be useful when considering questions related to image quality and size. The log data can also be used when drawing up guidelines for video use.

This paper deals with information obtained from a server's log file, describes the activities around collection and handling of log data, and introduces a preliminary classification for monitoring the development of video transmissions. The results obtained from analyses of Windows Media Server's (WMS) log files of 2008-2010 for Master Studies in Mathematical Information Technology at the Kokkola University Consortium of the University of Jyväskylä are presented in accordance with a preliminary classification at the end of the paper.

2   Log File Handling

Windows Media Server's log file comprises information that is obtained from the server as well as from its clients. The server collects some of the information to be entered in the log immediately when a client contacts the server. The client collects information throughout the duration of the connection, and only sends small amounts of information to the server while a video is being viewed. During the viewing, the client sends information related to the user's activities on the media player (forward, rewind, pause). On the other hand, the server constantly collects data needed for the log. At the end of a successful connection, the client sends the data it has collected to the server. The server combines this data with the data it itself has collected during the connection. In situations where the client cannot enter data in the log, the server simulates the log entry and enters there the information that it has collected dynamically, in addition to the information it collected when forming the connection. In cases where streaming does not start at all (for example, when there is a wrong user name or incorrect file name), streaming related data naturally is missing and only connection time data is entered in the log. [4]

2.1   WMS Log File

WMS records data on 52 fields in its log (Table 1). With the help of these, it is possible to monitor, in multiple ways, the use of video files distributed from the media server [4].The fields give information about the client, server, data network, video transmission, video to be transmitted, and the use of the transmitted video.

Table 1. Windows Media Server's log file fields

c-ip	c-cpu	c-pkts-recovered-resent
date	filelength	c-buffercount
time	filesize	c-totalbuffertime
c-dns	avgbandwidth	c-quality
cs-uri-stem	protocol	s-ip
c-starttime	transport	s-dns
x-duration	audiocodec	s-totalclients
c-rate	videocodec	s-cpu-util
c-status	channelURL	cs-user-name
c-playerid	sc-bytes	s-session-id
c-playerversion	c-bytes	s-content-path
c-playerlanguage	s-pkts-sent	cs-url
cs(User-Agent)	c-pkts-received	cs-media-name
cs(Referer)	c-pkts-lost-client	c-max-bandwidth
c-hostexe	c-pkts-lost-net	cs-media-role
c-hostexever	c-pkts-lost-cont-net	s-proxied
c-os	c-resendreqs	
c-osversion	c-pkts-recovered-ECC	

When the factors related to video transmission are being examined, the information obtained from the media server's log entries that is usually made use of in these circumstances includes the duration of connection, status, bandwidth used for video transmission, number of rebufferings and the time used for buffering, as well as the number of packets sent and received ([1], [5], [6]).

The X-duration field contains information about the client's viewing time. Information about network connection used for the viewing is given by the *avgbandwidth* field, which indicate the average bandwidth used by the client during the connection. Success of the connection is indicated by the value of the *c-status* field. The values in this field reveal whether the connection was a complete success or whether there were breaks and reconnections during the connection, or whether the viewing was terminated as a result of a connection breakdown or a server internal error. The field also tells about unauthorized attempts to view videos and of attempts to view videos not on the server at all.

The *transport* field tells about the transport protocol used. The log has several fields related to data transmission. Most of these fields are given a value only when the UDP protocol is in use. This means that some of the functions related to transmission reliability are moved to the application layer. For example, the *c-pkts-lost-client* field, which tells about the packets lost by the client, is given a value only when the UDP protocol is used. The same applies to the *c-pkts-lost-net* and *c-pkts-recovered-resent* fields. The *s-pkts-sent* field contains the number of packets the server has sent to the client and the *c-pkts-received* field, in turn, the number of packets correctly received at first attempt by the client. Both of these fields are given a value whether the UDP or TCP protocol is used.

Fields *c-buffercount* and *c-totalbuffertime* give information about the number of bufferings and the duration of time used for buffering. The media player normally uses buffering as its default to reduce the effect of lags in the data network between it and the media server. Changes along the network, such as server overload or bandwidth changes, may regardless cause emptying of the buffer, the effect of which is to stop playback on the viewer's computer for the time it takes to rebuffer.

2.2   Handling and Parsing of Log Files

There exist suitable commercial programs to analyze log files. The programs enable, in a limited manner, filtering of log files, but they do not fully meet needs for joining log entries or removing or combining certain data. For this reason preparatory joining of the log files and their filtering was done with a separate parser. The transmission of videos was then examined with the help of parsed log files by analyzing the data with software dedicated for statistical analysis.

The problem with Windows Media Server's logs is that they contain a lots of lines about client activities and every line has 52 fields. Thus a log file may contain tens of thousands of lines, leaving its structure quite unorganized. As each movement (forward, rewind) along the timeline leaves its mark on the log, the use of log data as such with an analysis software is often not very practical. For this reason, it seemed a reasonable idea to assemble a file on the basis of the log file. In that new file, all the user operations during a session were combined into a single entry. This joining of data was realized with a simple PHP parser, which automatically combines log files that are input as parameters into a single file in a manner desired.

During parsing, only operations related to successful viewings, wrong logins, and unsuccessful viewing requests were combined into one log entry. In case of an unsuccessful viewing request, of the attempts directed to the same file only one attempt per hour was left on the parsed file. In case of a forgotten password or missing user rights, it was also removed all consecutive unsuccessful login attempts except one. Viewing sessions that ended abnormally were kept as such also in the parsed log. During parsing, any entries by our own personnel were removed from the log data, too.

When combining successful operations related to the same session, the fields were subjected to various operations in accordance with the data contained in them. For example, the fields describing transmitted data (among them s-pkts-sent, c-pkts-received, c-pkts-lost-client, c-pkts-lost-net and c-pkts-recovered-resent) were summed up, as was the field defining the duration of viewing time (x-duration). When parsing, the weighted average of the avgbandwidth field was calculated in relation to the values of the x-duration field, to get the average bandwidth truthfully into the parsed file. Depending on the contents, the maximum (e.g., s-totalclients) and minimum (e.g., c-quality) values of some of the fields were also taken.

3   Analysis of Log Data

3.1   Classification of Log Data

When assessing the success of video transmissions we have first roughly classified the log data, from the users' viewpoint, into successful and unsuccessful viewing sessions as well as unsuccessful connection attempts. In [5], successful connections have been classified by their quality into three different groups: good, average or low. In line with the criteria in [5], we have classified successful connections as follows:

A. During a connection, at most 10% packets are retransmitted and there is no rebuffering at all.
B. During a connection, more than 10% packets are retransmitted and there is no rebuffering at all.
C. During a connection, rebuffering takes place.

Unsuccessful connections, in their turn, belong to class

D. The connection terminates abnormally with status 408 or 500.

and unsuccessful connection attempts belong to class

E. No connection is created and status 401 or 404 is entered in the log.

In a class A type connection the log file status is 200 and the portion of retransmitted packets of all packets (packet loss rate) transmitted remains below 10%. The number of packets that are retransmitted is calculated from the log files as the difference between the s-pkts-sent and c-pkts-received field values. This means that the information can be obtained regardless of whether the protocol used is UDP or TCP. Thus the new field gives only mediated information about the number of the packets lost by the client. We define a packet as lost by the client if it cannot arrive at the client by the playback time. According to [6], only when the packet loss rate is above 10% it is shown clearly as packet loss by the client. Therefore, on the basis of log data, in the class A type connection video is transmitted to the viewer without problems.

In a class B type connection the log file status 200 tells that the connection is successful and without problems, but that more than 10% of all the packets are retransmitted. If the number of retransmitted packets is great, it indicates problems with the connection as the number of packets lost by the client starts increasing. For the user, loss of packets may appear as malfunctioning while watching videos, for example, as dropping of video frames [5], [6].

In a class C type connection the status of the log file is either 200 or 210 and there is buffering during the connection. Buffering presents itself as jitteriness in the data transfer connection, and when watching videos these problems become manifest as a halting video image [5], [6]. If the data transfer connection is jittery, the size of the buffer is not necessary big enough to smooth down the variations in the connection. As a result the buffer might become empty. The time taken for rebuffering is usually very short, but the user's video stops for the duration of the operation. In some cases the player itself can create a connection after a short connection break and continue playback from the same point (status 210). For the user, however, it appears that the video stops once the buffer is emptied.

The quality of a class D type connection before the connection terminated abnormally might have been anywhere between A and C. If the log ends up with status 408 or 500, much less information about the connection quality remains than would remain were the session to end with status 200 or 210. If the log file ends with status 408, it means that the client's connection to the server has been broken or slowed down to the extent that it makes watching videos impossible. In that case the client cannot create a log entry. Instead the server will create a 408 entry in the log on behalf of the client. There are many reasons for a connection to slow down or break, among them congestion in data communications, problems related to network devices (switch, router, firewall, proxy server, etc.) or a network cable becoming disconnected. Also, a media server can become overloaded, which may lead to a situation where data communications slow down to such an extent that the media server may even have to drop some clients. In that case the log will end up with status 500.

In a class E type connection there has been an error when trying to create a connection and the connection has not been created. If the viewer tries to watch a non-existent recording the log file will have 404 entered, and if the user has not been authorized to watch a video, 401 will be entered.

3.2   Statistics That Can Be Derived from Log Entries

All lecture teaching for the students in Mathematical Information Technology at the Kokkola University Consortium is recorded on video, which is offered to the students both as real time video and on-demand. During 2008-2010 44 courses recorded on video were offered. Of these 11 were recorded in 2008, 15 in 2009, 13 in 2010 and the rest were from 2007. The bandwidth allocated for the users was at least 512 Kbps, also mobile users. The minimum bandwidth required by the videos was 340-380 Kbps. Altogether 187 students watched real time or on-demand videos. Of them, 86% watched in real time and 61% watched on-demand recordings. The average length of on-demand session was 42 min and the average length of real time session was 81min.

Fig. 1 shows the number of sessions viewed in 2008-2010. The videos were viewed altogether 12 673 times (9959 hrs) in this period. The on-demand recordings were viewed altogether 10 850 times (7507 hrs) and real time video 1823 times (2452 hrs). The number of all viewed sessions almost doubled from 2008 to 2010.

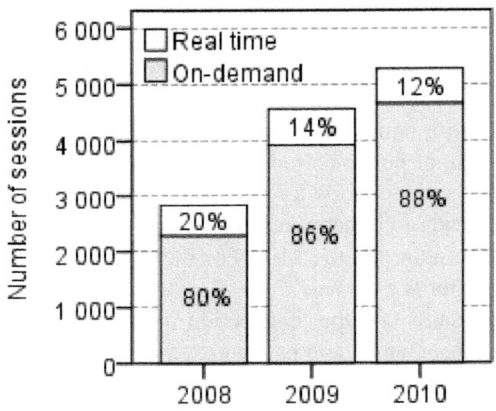

Fig. 1. Number of sessions viewed in 2008-2010

In 2008-2010, there were a total of 1555 class E type unsuccessful connection attempts entered in the parsed log. Status 401 came up in 890 of these and status 404 occurred 665 times. These unsuccessful connection attempts consist of human errors committed by the service provider or by the user. Typical errors made by the service provider include writing mistakes (a link on the learning environment or the name of a video file on the server misspelled) and links written beforehand (a link that is added on the learning environment before the video is moved to the server). Also the user errors consist typically of writing mistakes, such as a misspelled user name or password. In the following statistical evaluation, class E type unsuccessful connection attempts have been left out.

Fig. 2 shows both real time and on-demand connections in the master studies in 2008-2010. The figure is based on the classification discussed earlier. Fig. 2A shows all the connections formed. Fig. 2B shows only the connections that were created successfully.

Of all these connections, approximately 23% (2956 times) terminate abnormally, and belong to class D. One exception notwithstanding, all the terminated connections ended with status 408. Of the on-demand recordings 24% (2581 times) and of the real time videos 21% terminate abnormally. In these cases the average session time was 47 minutes for the on-demand videos and 29 minutes for the real time videos. Most of the students (74%) that used on-demand videos have experienced an abnormal termination of an on-demand recording at least once. The corresponding figure for the real time videos is 47%.

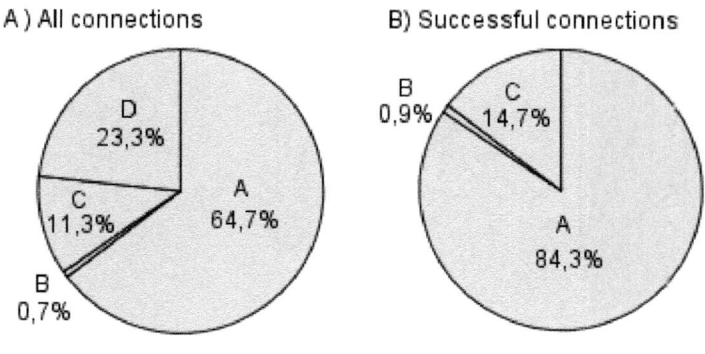

Fig. 2. Real time and on-demand video connections in 2008-2010

The data from 2008-2010 shows malfunctioning due to buffering having occurred in 11% of all connections and in 15% of successful connections. According to Fig. 3 class C can further be subdivided into three subclasses based on how common the malfunction is. A malfunction was regarded as occasional and almost unperceivable for the user if rebuffering took place at most twice. The malfunction was regarded as occasionally repeated if rebuffering took place 3-5 times and often repeated if it happened at least 6 times. In the successful connections most (8.5%) of the malfunctions were occasional, 2.9% occasionally repeated and 3.4% often repeated. During successful viewing sessions, malfunctioning due to buffering occurred slightly less in real time videos (13.6%) than in on-demand videos (14.9%).

Numerous retransmittings of packets may appear to the user as different kinds of faults in the videos. Based on the data from 2008-2010, there were only a few cases of this kind belonging to class B, a mere 0.7% of all connections and 0.9% of successful connections. Of the successful connections, the real time videos belonging to class B formed 0.1% while the corresponding figure among the on-demand recordings was 1.1%.

The portion of all connections that belonged to class A was 65%, and the portion of successful connections belonging to that same class was 84%. According to the log data, these connections appeared faultless to the users. Of the successful connections, on-demand recordings belonging to class A formed 84% and real time videos about 86%.

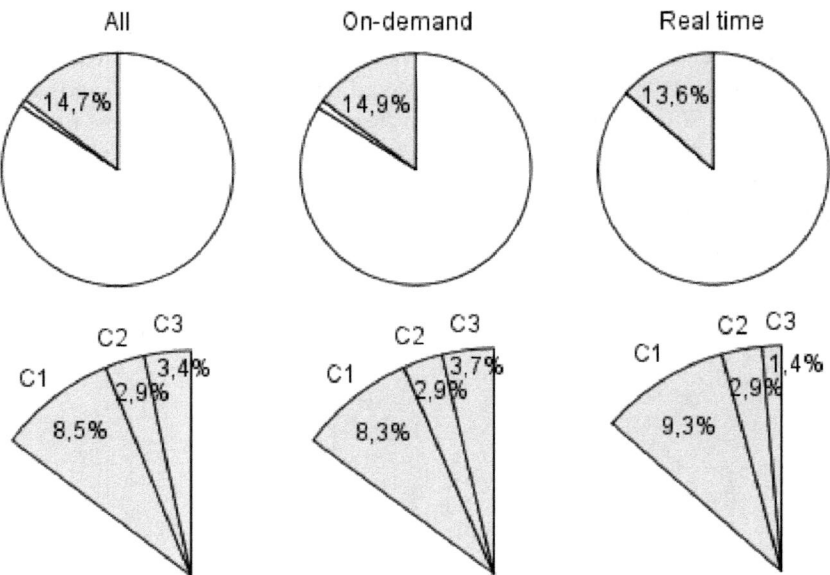

Fig. 3. Class C type connections of all successfully created real time and on-demand video connections in 2008-2010. *(C1=occasional, C2=occasionally repeated, C3=often repeated).*

3.3   Discussion

The classification above enables us to examine video transmission on the basis of the media server's log file more accurately and reliably than before. Judging from the results of the current log data analysis, information about connection quality obtained with the help of student questionnaires remained clearly inadequate. We were surprised, for example, of the relatively large portion of abnormally terminated and faulty connections, which had not been brought to our attention by earlier student questionnaires or other feedback. Based on our earlier research work, only about 3% of our students regarded problems with video image as disturbing [2].

The relative share (23%) of the abnormally terminated connections is fairly large. One cannot obtain any explicit reason for these connection breakdowns because very little information related to data connections is entered to the log. A partial explanation can be found in the transport protocols used for data communications. When the UDP protocol is used, 21% of on-demand video connections terminate abnormally. For the TCP protocol the same figure is 24%. Moreover, according to Fig. 4 the relative percentages for connections lasting less than 5mins are 23% (TCP) and 45% (UDP), and for connections lasting more than 40mins 47% (TCP) and 16% (UDP).

Another possible explanation regarding the on-demand recordings is that the errors might have accumulated for a small number of students. In case of those recordings, about 40% of all abnormal terminations of the recordings were caused by 10 students.

On-demand connections

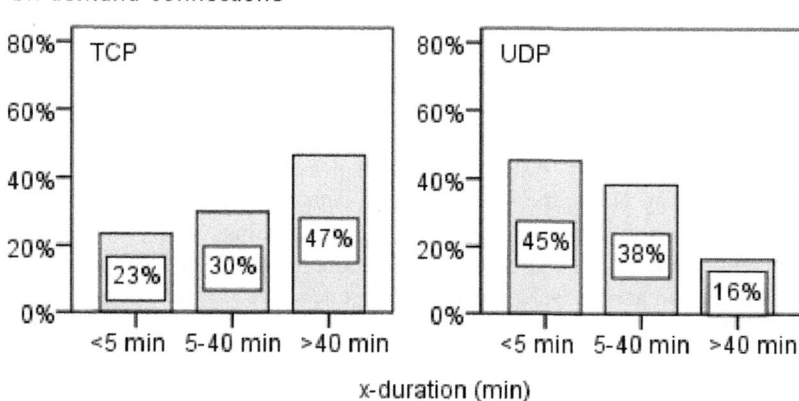

Fig. 4. Abnormally terminated connections for the on-demand recordings

The total share of faulty connections is 12% overall. Depending upon video material used, faulty connections may appear as malfunctions to the user also. In video lectures, the relative portion of static images is predominant. It is therefore likely that, to a large extent, faults in these go unnoticed by the user. The biggest source of the malfunctions is rebuffering. Nevertheless, it is quite rare to find that it happens repeatedly. Increasing the size of the buffer reduces the number of malfunctions, but on the other hand it also increases the delay in playback. Here the latter has no great importance, however.

Some of the factors that affect connection quality and which the user can influence are buffer size and transport protocol. The increase of the buffer size can be done in the media player. In Windows Media Player, the size of the buffer can be selected between 0 and 60 secs, the default value being 5 secs. The user also has a say in what transport protocol to use. Depending on the server settings, the server offers either the UDP or TCP protocol as a default, but the user can force the player or computer to use any of the two. By instructing the students in how to increase the adjustable buffer size in the media player and by making TCP the obligatory transport protocol, some of the problems mentioned above can be eliminated to some extent though not entirely.

The log file also provides information which has more to do with the user or the organizer of the education than with the network connection. In some cases, entries of this type convey useful information that can be used in the development of teaching arrangements. For example, if the user is not authorized to watch a video, a separate entry will be left in the log file about the matter. Usually this is due to that the user has misspelled his/her user name or password or that the user profile has expired. Because these types of entries are quite common in the log file, a facility has been created for the students to allow them to check the functionality of their user ID and also to remotely change their password if needed.

4   Conclusions

The use of lecture videos is already an essential part of educational arrangements in many universities. For this reason, the role of quality control of video transmissions has become more important. Regular monitoring of video distribution quality should form a part of the quality system work. When assessing the quality of operations related to streaming video, the analysis of log data collected by the media server has proved a very useful tool. Analyzing log data helps, above all, in directing development work associated with video technologies. It makes possible also to instruct students in making selections that help one succeed in video viewing.

One of the aims of the quality monitoring is to create indicators for measuring video transmission quality. With the help of the kind of classification that measures quality it is possible to compare the quality of video transmission at different times. This makes it possible to follow the productivity of the development work in video transmission. In addition, these indicators provide us with a tool that enables us to react whenever the quality of transmission for one reason or another may have deteriorated.

If the videos have been made available through Windows Media Server, it is possible to analyze the log data in many different ways. The log data alone, however, is not enough for analyzing how the students have experienced the quality of the videos. To complete the picture, we need to resort to questionnaires directed to the students. More useful material could be collected by mapping students' experiences of watching videos immediately after watching the video. This, however, requires a development of suitable video sharing application.

To ensure that maximum benefit is derived from monitoring log data, that monitoring must be regular. Further automation in future should make it easier to arrange regular monitoring of large log files and would help in the detection of problem situations. With the help of automated monitoring, the education organizer could, for example, be notified daily about problems occurred.

References

1. Chang, C.: Constructing a Streaming Video-Based Learning Forum for Colloborative Learning. Journal of Educational Multimedia and Hypermedia 13(3), 245–263 (2004)
2. Hakala, I., Myllymäki, M.: Video lectures alongside with contact teaching. In: Proceedings of the 18th EAEEIE Annual Conference on Innovation in Education for Electrical and Information Engineering, Praha, Czech Rebublic (2007)
3. Hakala, I., Myllymäki, M.: Quality Management for Mathematical Information Technology Teaching at the Kokkola University Consortium Chydenius. In: Proceedings of the 19th EAEEIE Annual Conference on Innovation in Education for Electrical and Information Engineering, Tallinn, Estonia (2008)
4. Koyun, H.: Logging Model for Windows Media Services (2007),
   http://www.microsoft.com/windows/windowsmedia/howto/
   articles/LoggingModel.aspx

5. Terada, N., Kawai, E., Sunahara, H.: Extracting Client-Side Streaming QoS Information from Server Logs. In: Proceedings of IEEE Pacific Rim Conference on Communications, Computers and Signal Processing, Victoria, B.C., Canada (2005)
6. Wang, Z., Banerjee, S., Jamin, S.: Studying Streaming Video Quality: From An Application Point of View. In: Proceedings of ACM MultiMedia Conferenc 2003, Berkeley, CA (2003)

Standards for Query Formalization
in Mobile Visual Search

Ruben Tous and Jaime Delgado

Universitat Politècnica de Catalunya (UPC BarcelonaTech), Barcelona, Spain
{rtous,jaime.delgado}@ac.upc.edu

Abstract. The research around visual search is gaining relevance due to the evolution in the generation and usage of digital images. A significant push comes from the mobile visual search topic, due to the widespread proliferation of camera enabled mobile devices. The new scenarios are increasing the urgency of novel solutions for challenging problems such as the efficient coding of compact visual descriptors and the interoperability of distributed visual search query interfaces. Currently, almost every visual search service offers a different retrieval interface and image metadata description format, preventing unified and efficient access. In this context, standardization groups such as ISO/IEC SC29/WG11 (MPEG) and ISO/IEC SC29/WG1 (JPEG) have been working to create unified interfaces for image repositories. In one hand, MPEG provides the ISO/IEC 15938-12 (MPEG Query Format, MPQF), which standardizes a query language for multimedia repositories and has also started an activity for standardizing compact descriptors for visual search (CDVS). On the other hand, JPEG is now finishing the ISO/IEC 24800 (JPSearch), which provides solutions to the image metadata interoperability problem. This paper analyzes how these standardization activities can be combined to satisfy the requirements posed by the mobile visual search scenario, which are their limitations and which are the necessary actions to be taken by the standardization committees in order to overcome them.

Keywords: visual search, cbir, metadata, interoperability, image, information retrieval, standards, jpsearch, mpqf.

1 Introduction

Visual search allows overcoming the inherent limitations of text-based information retrieval systems. Relying on structured or unstructured text-based annotations is impractical for very large image databases, or for images that are generated continuously or ephemerally (e.g. from surveillance cameras, medical devices or smartphones). The increasing amount of digitally stored images and the widespread proliferation of camera enabled mobile devices, such as mobile phones, PDAs or tablets, poses new challenges to researchers and practitioners in this area. One of the problems to be addressed is the one related to query interface interoperability. There are already multiple existing mobile visual search

L. Atzori, J. Delgado, and D. Giusto (Eds.): MOBIMEDIA 2011, LNICST 79, pp. 180–193, 2012.

systems but every one provides a different search interface and multimedia meta-data description format. This fact prevents users from experiencing a unified access to the repositories. Systems aiming to provide a unified query interface to search images hosted in different systems without degrading query expressive-ness need to address several questions which include but are not limited to the following:

- Which mechanism is going to be used for query formalization? How is this mechanism going to cope with visual search queries?
- Which metadata schema or schemas are going to be exposed to user queries? Is the system going to expose a mediated/pivot schema?
- How the mappings among the underlying target metadata schemas are going to be generated?
- Which formalism is going to be used to describe the mappings?
- How is the system going to use the mappings during querying?

Currently many standardization efforts are trying to provide answers to some of these questions. Two of the most relevant initiatives are the ISO/IEC 15938-12:2008 (MPEG Query Format or simply MPQF) [1,2,3] and ISO/IEC 24800 (JPEG's JPSearch framework) [4,5]. While MPQF offers a solution for the query interface interoperability, JPSearch (whose Part 3 makes use of MPQF) faces the difficult challenge to provide an interoperable architecture for images' metadata management. On the other hand, MPEG has also started an activity for stan-dardizing compact descriptors for visual search (CDVS) [6]. This standard will unify the way to exchange image summaries in visual search queries. Develop-ers of mobile visual search applications willing to provide a standard interface will face the problem of selecting the proper components of each one of these standards and try to figure out how to combine them. This paper analyzes how these three standardization activities can be combined to satisfy the require-ments posed by the mobile visual search scenario, which are their limitations and which are the necessary actions to be taken by the standardization commit-tees in order to overcome them.

2   ISO/IEC 15938-12:2008 Standard (MPEG Query Format or MPQF)

A key element in all the different approaches to distributed image search& retrieval is the interchange of queries and API calls among all the involved par-ties. The usage of different proprietary interfaces for this task makes extremely difficult the deployment of distributed image search services without degrading the query expressiveness. The progressive adoption of a unified query interface would greatly alleviate this problem. In december 2008 the MPEG standardiza-tion committee (ISO/IEC JTC1/SC29/WG11) released a new standard which provides a standardized interface to search functionalities in distributed multi-media databases (including not only still images but also video and audio), the

MPEG Query Format (MPQF). To achieve this goal, the MPQF standard specifies precise input and output parameters to express multimedia requests and uniform client side processing of result sets, respectively. Essentially, MPQF is an XML-based query language that defines the format of the queries and replies exchanged between clients and servers in a distributed multimedia search-and-retrieval system. In one hand, standardizing such kind of language fosters *interoperability* between parties in the multimedia value chain (e.g. content providers, aggregators and user agents). On the other hand, MPQF favors also *platform independence*; developers can write their applications involving multimedia queries independently of the system used, which fosters software reusability and maintainability. Figure 1 shows a diagram outlining the basic MPQF scenario. In the simplest case, the requester may be a users agent and the responder might be a retrieval system. However, MPQF has been specially designed for more complex scenarios, in which users interact, for instance, with a content aggregator. The content aggregator acts at the same time as responder (from the point-of-view of the user) and as a requester to a number of underlying content providers to which the user query is forwarded.

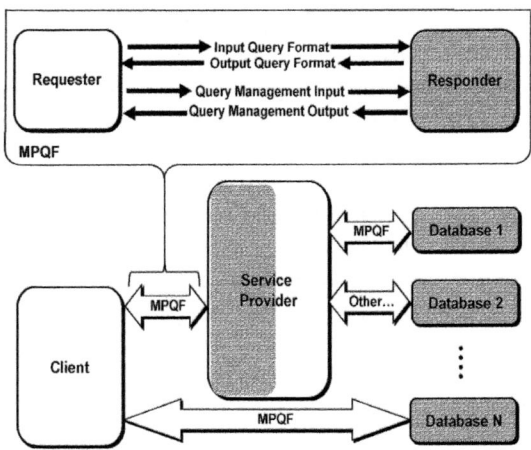

Fig. 1. MPEG Query Format diagram

2.1 MPEG Query Format Syntax and Terminology

MPQF queries (requests and responses) are XML documents that can be validated against the MPQF XML schema (see Figure 2). MPQF instances include always the *MpegQuery* element as the root element. Below the root element, an MPQF instance includes the *Query* element or the *Management* element. MPQF instances with the *Query* element are the usual requests and responses of a digital content search process. The *Query* element can include the *Input* element or the *Output* element, depending if the document is a request or a response. The part of the language describing the contents of the *Input* element (requests) is named the Input Query Format (IQF) in the standard. The part of the language describing the *Output* element (responses) is named the Output Query Format (OQF) in the standard. IQF and OQF are just used to facilitate understanding, but do not have representation in the schema. Alternatively, below the root element, an MPQF document can include the *Management* element. Management messages (which in turn can be requests and responses) provide means for requesting service-level functionalities like interrogating the capabilities of a MPQF processor.

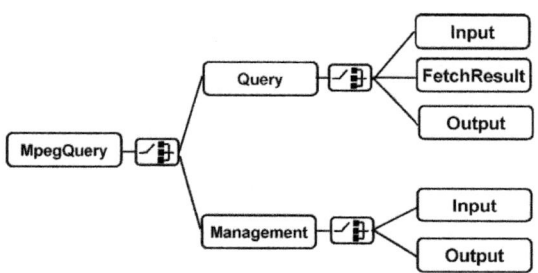

Fig. 2. MPQF language parts

Example in Code 1 shows an input MPQF query asking for JPEG images related to the keyword *"Barcelona"*.

2.2   MPEG Query Format Database Model

As happens with any other query language, an MPQF query is expressed in terms of a specific database model. The MPQF database model formally defines the information representation space which constitutes the evaluation basis of an MPQF query processor. MPQF queries are evaluated against one or more multimedia databases which, from the point-of-view of MPQF, are unordered sets of *Multimedia Contents*. The concept of *Multimedia Content* (MC) refers to the combination of multimedia data (resource) and its associated metadata. MPQF allows retrieving complete or partial MC's data and metadata by specification of a condition tree. So, MPQF deals with a dual database model (see Figure 3) constituted by content and metadata.

Example in Figure 4 shows a graphical representation of an MPQF condition tree carrying two different conditions which reflect the duality of the database model used by the language. In one hand, there's an CBIR-like condition using the *QueryByExample* query type and including the Base64 encoding of the binary contents of an example JPEG image. On the other hand, there is a simple XML-like condition specifying that the metadata field *FileSize* must be inferior to 1000 bytes. In order to deal with this model duality, MPQF operates over sequences of what the standard calls *evaluation-items*. By default, an *evaluation-item* (EI) is a multimedia content in the multimedia database, but other types

Code 1. Example MPQF input query

```
<MpegQuery>
  <Query>
    <Input>
      <OutputDescription outputNameSpace="//purl.org/dc/elements/1.1/">
        <ReqField>title</ReqField>
        <ReqField>date</ReqField>
      </OutputDescription>
      <QueryCondition>
        <TargetMediaType>image/jpg</TargetMediaType>
        <Condition xsi:type="AND" preferenceValue="10">
          <Condition xsi:type="QueryByFreeText">
            <FreeText>Barcelona</FreeText>
          </Condition>
          <Condition xsi:type="GreaterThanEqual">
            <DateTimeField>date</DateTimeField>
            <DateValue>2008-01-15</DateValue>
          </Condition>
        </Condition>
      </QueryCondition>
    </Input>
  </Query>
</MpegQuery>
```

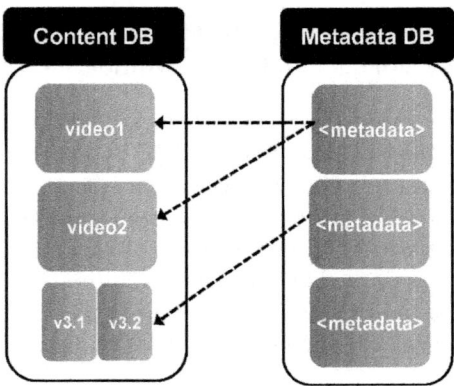

Fig. 3. Dual database model

of EIs are also possible. For instance, an EI can be a segment of a multimedia resource, or an XPath-item related to a metadata XML tree. The scope of query evaluation and the granularity of the result set (the granularity of EIs) can be determined by a *EvaluationPath* element specified within the query. If this *EvaluationPath* element is not specified, the output result is provided as a collection of multimedia contents, as stored in the repository, all satisfying the query condition.

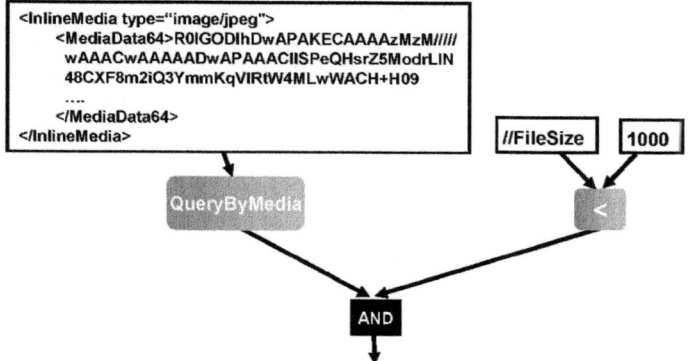

Fig. 4. Example condition tree

The *EvaluationPath* element determines the query scope on basis to a hypothetical XML metadata tree covering the entire database. So, we can redrawn our visual representation of the model in order to show this hierarchical nature, as shown in Figure 5.

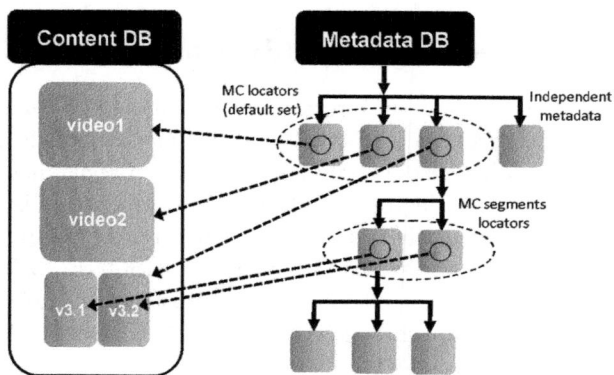

Fig. 5. Hierarchical MPQF model

2.3 MPQF Evaluation Model

The condition tree of an MPQF query is constructed combining filtering elements (conditions) from the *BooleanExpressionType* and interconnecting them with Boolean operators (*AND, OR, NOT* and *XOR*). Each condition acts over a sequence of evaluation-items and, for each one, return a value in the range of [0..1]. In the case of XML-like conditions (e.g. it "the size of the file must be smaller than 1000 bytes") the condition can return just 1 or 0, which mean *true* and *false* respectively. In the case of IR-like conditions (keywords or CBIR), they can evaluate to any value in the range of [0..1]. A *threshold* value within a condition is used to indicate the minimum value the score of an evaluation-item is required to have. Otherwise the evaluation-item is not considered further during evaluation.

So, with respect to XML-like conditions, MPQF acts as a conventional Boolean-based filtering language, while with respect to IR-like conditions MPQF acts preserving scores as a fuzzy-logic system. The standard specifies the behaviour of the provided Boolean operators in presence of non-Boolean values.

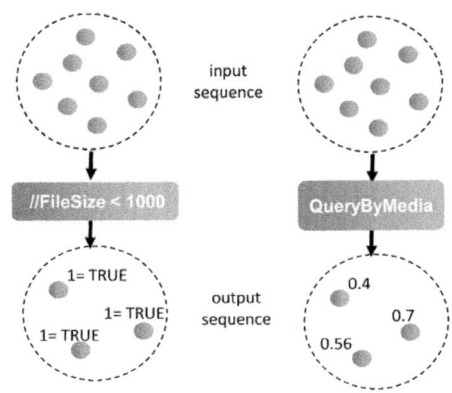

Fig. 6. Two evaluation styles in MPQF: Boolean and fuzzy-logic

3   ISO/IEC 24800 Standard (JPSearch)

The selection of a unified query interface is not enough to guarantee interoperability if it is not accompanied with a proper mechanism to manage metadata heterogeneity. The need of dealing with the management of metadata translations is a common factor in all the approaches to distributed image search&retrieval. Currently there is a standard solution to this problem proposed by the JPEG Committee, named JPSearch (ISO/IEC 24800). JPSearch provides a set of standardized interfaces of an abstract image retrieval framework. On one hand, JPSearch specifies the pivot JPSearch's Core Metadata Schema as the main component of the metadata interoperability strategy in ISO/IEC 24800. The core schema contains a set of minimal core terms which serve as metadata basis supporting interoperability during search among multiple image retrieval systems. The core schema is used by clients to formulate, in combination with the MPEG Query Format, search requests to JPSearch compliant search systems. In addition to the definition of JPSearch Core Metadata Schema, ISO/IEC 24800 provides a mechanism which allows a JPSearch compliant system taking profit from proprietary or community-specific metadata schemas. A translation rules language, JPSearch Translation Rules Declaration Language (JPTRDL), is defined, allowing the definition of translations between proprietary metadata schemas and the JPSearch Core Metadata Schema.

3.1   JPSearch Core Schema

The JPSearch Core schema defines the metadata structures which must be understood (directly or by means of applying translations to a native metadata format) by any JPSearch compliant system. The core schema has been developed by investigating available image metadata formats and selecting those elements or its semantic concepts that show a broad overlap (e.g. have an equivalent representation) among the individual metadata formats. A similar approach of choosing the core properties was also chosen be W3C group[1]. For instance, the element *Title* of our core schema has similar tag names in Dublin Core (*dc:title*), DIG35 (*ipr_title*) or MPEG-7 (\ *CreationInformation* \ *Creation* \ *Title*). Based on this analysis, the current version of the core schema supports 20 main elements including spatial annotation. The main elements represent basic information such as *Title, Description, Keyword, CreationDate* and so forth. The broader set concentrates on semantic descriptions of the content such as events, persons, etc. Furthermore, a mechanism for the integration of parts of other formats (e.g., low level descriptors of MPEG-7) during search is provided.

Code 2 shows an example XML image description compliant with the JPSearch Core Schema:

Code 2. Example JPSearch image description

```
<ImageDescription>
  <Identifier>urn:unique:identifier:1:2:3</Identifier>
  <Creators>
    <GivenName>John</GivenName>
    <FamilyName>Smith</FamilyName>
  </Creators>
  <CreationDate>2011-12-17T09:30:47.0Z</CreationDate>
  <ModifiedDate>2011-12-17T09:30:47.0Z</ModifiedDate>
  <Description>Sample description</Description>
  <Keyword>Sardinia</Keyword>
  <Keyword>Italy</Keyword>
  <Keyword>50th JPEG meeting</Keyword>
  <Title>Example Instance document of the JPSearch core schema</Title>
  <GPSPositioning latitude="34" longitude="34" altitude="10"/>
  <RegionOfInterest>
    <RegionLocator>
      <Region dim="2"> 0 0 100 100</Region>
    </RegionLocator>
    <Description>A short description about the selected region
    </Description>
    <Keyword>plenary meeting</Keyword>
  </RegionOfInterest>
  <Width>640</Width>
  <Height>480</Height>
</ImageDescription>
```

---

[1] http://www.w3.org/2008/WebVideo/Annotations/drafts/ontology10/WD/ mapping_table.html

3.2   JPSearch Translation Rules Declaration Language (JPTRDL)

The JPSearch Translation Rules Declaration Language (JPTRDL) allows the publication of machine-readable translations between metadata terms belonging to proprietary metadata schemas and metadata terms in the JPSearch Core Metadata Schema. Users can choose which metadata language to use in a JPSearch-based interaction if the proper translations are available. JPTRDL is a key component of the metadata interoperability strategy in ISO/IEC 24800, and it aims to allow a JPSearch compliant system benefit from proprietary or community-specific metadata schemas.

Code 3 shows a one-to-many translation rule which maps the JPSearch Core Schema date element into three fields.

Code 3. Example JPSearch translation rule

```
<?xml version="1.0" encoding="iso-8859-1"?>
<TranslationRules>
  <TranslationRule xsi:type="OneToManyFieldTranslationType">
    <FromField xsi:type="FilteredSourceFieldType">
      <XPathExpression>date</XPathExpression>
      <FilterWithRegExpr>(\d\d)/(\d\d)/(\d\d\d\d)</FilterWithRegExpr>
    </FromField>
    <ToField xsi:type="FormattedTargetFieldType">
      <XPathExpression>day</XPathExpression>
      <ReplaceWithRegExpr>£1</ReplaceWithRegExpr>
    </ToField>
    <ToField xsi:type="FormattedTargetFieldType">
        <XPathExpression>month</XPathExpression>
        <ReplaceWithRegExpr>£2</ReplaceWithRegExpr>
    </ToField>
    <ToField xsi:type="FormattedTargetFieldType">
        <XPathExpression>year</XPathExpression>
        <ReplaceWithRegExpr>£3</ReplaceWithRegExpr>
    </ToField>
  </TranslationRule>
</TranslationRules>
```

3.3   JPSearch Registration Authority

According to the JPSearch specification, ISO/IEC 24800 compliant systems can manage multiple proprietary or community-specific metadata schemas, besides the JPSearch Core Metadata Schema. The multiplicity of schemas is solved by allowing the publication of machine-readable translations between metadata terms belonging to proprietary metadata schemas and metadata terms in the JPSearch Core Metadata Schema. In order to rationalize the usage of schemas and translation rules across different JPSearch systems, Subclause 3.3.3 of Part 2 of ISO/IEC

24800-2 specifies that a global authority for schemas and their translation rules will be established where all JPSearch compliant retrieval applications can obtain the information needed.

The establishment of a JPSearch Registration Authority (JPSearch RA) was formally approved during the 54th JPEG meeting in Tokyo, Japan, in February 2011, and will be operative in July 2011. The JPSearch RA will maintain a list of Metadata Schemas together with their related Translation Rules, if any. Those schemas and rules will be directly stored in the JPSearch RA web site or the JPSearch RA web site will provide a link to an external organization in charge of keeping that information updated. Registration forms will be available from the Registration Authority. Any person or organization will be eligible to apply. More information about the JPSearch RA can be obtained at www.iso.org/iso/maintenance_agencies/or directly at the JPEG home page (www.jpeg.org).

4   Unifying MPEG Query Format and JPSearch, a Proposal

IT professionals aiming to take profit from the MPQF and JPSearch standards will encounter several obstacles. In principle, the straightway approach would be to just apply the JPSearch specification, as it defines the complete framework and its Part 3 relies in MPQF as the query language. However, the way JPSearch adapts MPQF to the still images domain suffers from several flaws, which results in queries which are not complaint with MPQF. These defects can be summarized as:

- JPSearch Part 3 [4] redefines the XML root hierarchy of the MPQF queries (called JPQF queries in that context). Figure 7 shows graphically the XML root hierarchy of JPQF, which is different to the one in Figure 2. These differences make JPQF queries not compliant with the MPEG Query Format standard. This is inconvenient, especially considering that JPQF is just the subset of MPQF which does not deal with audio or video.
- Because JPSearch Part 3 replicates several syntax elements from MPQF instead of referencing them, it does not consider the latest corrigenda and amendments from MPEG. For instance, the 2nd MPQF corrigenda [7] fixes an error in the MPQF's *QueryConditionType* which disallowed the expression of empty queries (queries without conditions). Because *QueryConditionType* is redefined in JPSearch Part 3 the error has not been fixed there.

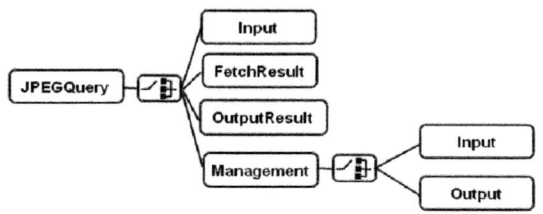

Fig. 7. JPSearch Part 3 (query format) root elements

In order to solve these problems we propose:

1. Withdraw JPSearch Part 3 and, optionally, simply refer to the MPEG Query Format as the query language for JPSearch compliant systems.

2. Within the MPEG Query Format standard, establish the JPSearch metadata strategy as the normative way to handle metadata in MPQF. MPQF is metadata neutral and does not specify neither predefined metadata structures nor visual search descriptor formats to which address the queries. This poses an interoperability problem cause queries referring to different metadata schemas or CBIR descriptors would be incompatible. Taking profit from the JPSearch metadata approach would provide MPQF with the metadata infrastructure it currently lacks, but keeping it open to other metadata formats as the JPSearch metadata strategy also consider this possibility. This approach also would allow MPQF to take benefit from JPSearch Registration Authority (JPSearch RA).

Example in Code 4 shows an MPQF query addressing metadata fields as defined in the JPSearch Core Schema.

Code 4. Example MPQF input query addressing metadata fields as defined in the JPSearch Core Schema

```
<MpegQuery>
  <Query>
    <Input>
      <OutputDescription outputNameSpace="//purl.org/dc/elements/1.1/">
        <ReqField>JPCore:Identifier</ReqField>
        <ReqField>JPCore:Creators/JPCore:GivenName</ReqField>
        <ReqField>JPCore:Creators/JPCore:FamilyName</ReqField>
        <ReqField>JPCore:CreationDate</ReqField>
        <ReqField>JPCore:Width</mpqf:ReqField>
        <ReqField>JPCore:Height</mpqf:ReqField>
        <ReqField>JPCore:GPSPositioning/@latitude</ReqField>
        <ReqField>JPCore:GPSPositioning/@longitude</ReqField>
      </OutputDescription>
      <QueryCondition>
        <TargetMediaType>image/jpg</TargetMediaType>
          <Condition xsi:type="Equal">
            <DateTimeField>JPCore:CreationDate</DateTimeField>
            <DateTimeValue>2009-10-07T08:46:45</DateTimeValue>
          </Condition>
      </QueryCondition>
    </Input>
  </Query>
</MpegQuery>
```

5 MPEG Query Format Visual Search Operations and Limitations

Visual search Examples in Code 5 and Code 6 show two possible ways to specify a query-by-example condition in MPQF. *QueryByMedia* and *QueryByDescription* query types are the operations of MPQF which allow specifying information about an example multimedia content. The individual difference lies in the used sample data. The *QueryByMedia* query type uses a media sample such as image or video as a key for search, whereas *QueryByDescription* allows querying on the basis of an XML-based description key. The example query in Code 5 embeds an example JPEG image in Base64 encoding. It is also possible to just specify the URI of the example image.

Code 5. Example MPQF input visual search query inlining the query image in Base64 encoding

```
<MpegQuery>
  <Query>
    <Input>
      <OutputDescription outputNameSpace="//purl.org/dc/elements/1.1/">
        <ReqField>title</ReqField>
        <ReqField>date</ReqField>
      </OutputDescription>
      <QueryCondition>
        <TargetMediaType>image/jpg</TargetMediaType>
        <Condition xsi:type="QueryByMedia">
          <MediaResource xsi:type="MediaResourceType" resourceID="id1">
            <MediaResource>
              <InlineMedia type="image/jpeg">
                <MediaData64>ROlGODlhDwAPAKECAAAAzMzM/////wAAACwAAAAADwA
                PAAACIISPeQHsrZ5ModrLlN48CXF8m2iQ3YmmKqVlRtW4MLwWACH+HO9
                wdGltaXplZCBieSBVbGVhZCBTbWFydFNhdmVyIQAAOw==</MediaData64>
              </InlineMedia>
            </MediaResource>
          </MediaResource>
        </Condition>
      </QueryCondition>
    </Input>
  </Query>
</MpegQuery>
```

The example query in Code 6 shows how an example description can be included in a query to form a condition. The query requests JPEG images which expose similar descriptors to the attached MPEG-7 metadata descriptor. Descriptors containing low-level features (e.g. the MPEG-7 *Color Layout* descriptor) can be used, becoming a way to express CBIR queries.

Code 6. Example MPQF input visual search query referring to the query image as a URI

```xml
<?xml version="1.0" encoding="UTF-8"?>
<MpegQuery>
 <Query>
  <Input>
   <QueryCondition>
    <TargetMediaType>image/jpeg</TargetMediaType>
    <Condition xsi:type="QueryByDescription" matchType="exact">
     <DescriptionResource resourceID="desc07">
      <AnyDescription xmlns:mp7="urn:mpeg:mpeg7:schema:2004">
       <mp7:Mpeg7>
        <mp7:DescriptionUnit xsi:type="mp7:CreationInformationType">
         <mp7:Creation>
          <mp7:Title>World Cup 2006</mp7:Title>
         </mp7:Creation>
        </mp7:DescriptionUnit>
       </mp7:Mpeg7>
      </AnyDescription>
     </DescriptionResource>
    </Condition>
   </QueryCondition>
  </Input>
 </Query>
</MpegQuery>
```

The main limitation of the MPQF's operations for expressing CBIR queries is the lack of reference visual descriptors and search parameters. MPQF's only provides placeholders where any information about an example image can be introduced, without constraining which are the valid descriptor formats and which kind of visual query has to be performed. It's up to the server to decide how to process the example information. This feature clearly constraints the interoperability of visual search operations.

5.1   Enriching the MPEG Query Format with Predefined Compact Descriptors for Visual Search

Recently MPEG has issued a Call for Proposals for compact descriptors for visual search (CDVS) [6]. This initiative aims to standardizing technologies for visual content matching in images or video, including the format of visual descriptors and descriptor extraction process. The future standard will ensure interoperability of visual search applications and databases as well as reduce load on wireless networks carrying visual search-related information. Without still knowing which would be the format of the visual descriptors in the future standard, it is clear that the MPEG Query Format should include them at least as the default way of expressing visual search conditions. The flexibility of the *QueryByDescription* query type will allow to directly inline the descriptor within the query or to just refer to the descriptor URI.

6 Conclusions

In this paper we have evaluated the standards ISO/IEC 15938-12 (MPEG Query Format, MPQF) and ISO/IEC 24800 (JPSearch) as candidates to provide interoperability to distributed visual search query interfaces. MPQF has proven to be a powerful multimedia query language, while JPSearch provides the necessary mechanisms to address the problem of metadata interoperability. However, we have identified some problems related to the integration of both standards and to the way they cover the visual search scenario. On one hand, our proposed solution consists on including the JPSearch metadata interoperability model within the MPQF standard, while fixing the way the MPQF language is adapted within JPSearch's Part 3. On the other hand, we suggest to include the future MPEG standard for compact descriptors for visual search (CDVS) as the default way of expressing visual search conditions within MPQF. The MPEG Query Format standard was published in December 2008. After almost three years, two amendments and two corrigenda, we envisage that the time has come for a second edition of ISO/IEC 15938-12, which could include the improvements proposed in this paper.

Acknowledgments. This work has been partly supported by the Spanish government (TEC2008-06692-C02-01).

References

1. Döller, M., Tous, R., Gruhne, M., Yoon, K., Sano, M., Burnett, I.S.: The MPEG Query Format: On the Way to Unify the Access to Multimedia Retrieval Systems. IEEE Multimedia 15(4) (2008) ISSN: 1070-986X
2. Gruhne, M., Tous, R., Döller, M., Delgado, J., Kosch, H.: MP7QF: An MPEG-7 Query Format. In: Proceedings of the 3rd International Conference on Automated Production of Cross Media Content for Multi-channel Distribution (AXMEDIS 2007), Barcelona, Spain, pp. 15–18 (November 2007)
3. ISO/IEC 15938-12:2008 Information Technology - Multimedia Content Description Interface - Part 12: Query Format (2008)
4. ISO/IEC 24800-3:2010 Information technology - JPSearch - Part 3: JPSearch Query format (2010)
5. ISO/IEC 24800-2:2011 Information technology - JPSearch - Part 2: Registration, identification and management of schema and ontology (2011)
6. MPEG's Requirements Subgroup. ISO/IEC JTC1/SC29/WG11/N12038 Call for Proposals for Compact Descriptors for Visual Search. In: Output Document of the 96th MPEG Meeting, Geneva, CH (March 2011)
7. ISO/IEC 15938-12:2008/Cor.2:2010, Information technology – Multimedia content description interface – Part 12: Query format. Technical Corrigendum 2 (2010)

3D Wide Baseline Correspondences Using Depth-Maps

Marco Marcon, Eliana Frigerio*, Augusto Sarti, and Stefano Tubaro

Politecnico di Milano - Dipartimento di Elettronica e Informazione,
P.zza Leonardo Da Vinci, 32, 20133 Milano, Italy
marco.marcon@polimi.it, efrigerio@elet.polimi.it
http://home.dei.polimi.it/marcon

Abstract. Points matching between two or more images of a scene shot from different viewpoints is the crucial step to defining epipolar geometry between views, recover the camera's egomotion or build a 3D model of the framed scene. Unfortunately in most of the common cases robust correspondences between points in different images can be defined only when small variations in viewpoint position, focal length or lighting are present between images. While in all the other conditions ad-hoc assumptions on the 3D scene or just weak correspondences can be used. In this paper, we present a novel matching method where depth-maps, nowadays available from cheap and off the shelf devices, are integrated with 2D images to provide robust descriptors even when wide baseline or strong lighting variations are present.

Keywords: Machine vision, feature extraction, 3D descriptors.

1 Introduction

Feature points matching between two shots of a scene from different viewpoints is one of the basic and most tackled computer vision problems. In many common applications, like objects tracking in video sequences, the baseline is relatively small and features matching can be easily obtained using well known feature descriptors [14,4]. However many other applications require feature matching in much more challenging contexts, where wide baselines, lighting variations and non-lambertian surfaces reflectance are considered. Many interesting approaches based on two single images have been proposed in the literature, starting from the pioneering work of Schmid and Mohr [12] many other interesting approaches followed: Matas et al. [8] introduced the maximally stable extremal regions (MSER) where affinely-invariant stable subset of extremal regions are used to find corresponding *Distinguished Regions* between images, or moment descriptors for uniform regions [10] while other approaches are based on clearly distinguishable points (like corners) and affine-invariant descriptors of their neighborhood. One of the most popular approaches in the last few years becomes the Scale Invariant Feature Transform (SIFT) proposed by Lowe [3] thanks to its outperforming capabilities, as shown by Mikolajczyk and Schmid [7]. The SIFT algorithm is

---

* Corresponding author.

L. Atzori, J. Delgado, and D. Giusto (Eds.): MOBIMEDIA 2011, LNICST 79, pp. 194–203, 2012.
© Institute for Computer Sciences, Social Informatics and Telecommunications Engineering 2012

based on a local histogram of oriented gradient around an interest point and its success is mainly due to a good compromise between accuracy and speed (is as also been integrated in a Virtex II Xilinx Field Programmable Gate Array, FPGA [13]). Actually some other approaches, always based on affine invariant descriptors, got growing interest like the Gradient Location and Orientation Histogram (GLOH) [7] which is quite close to the SIFT approach but requires a Principal Component Analysis (PCA) for data compression, or the Speeded-Up Robust Features (SURF) [1] a powerful descriptor derived from an accurate integration and simplification of previous descriptors. All of the aforementioned approaches assume that, even if nothing is known of the underlying geometry of the scene, the defined features, since are describing a very small portion of the object, will undergo a simple planar transformation that can be approximated with an affine homography. This simplification has two main drawbacks, first of all the extracted features are very general and weak since wide affine transformations must provide very similar results, moreover, whenever the framed object present abrupt geometrical discontinuities (e.g. geometrical edges or corners) the affine approximation is not valid anymore. A possible solution to such problems could be a rough description of the underlying 3D geometry. In particular, within the Astute Artemis project, we are investigating the opportunity to use scene depth-maps to have a rough estimation of 3D underlying geometry: We use depth-maps to estimate the orientation of the plane, where the considered feature is laying, with respect to the observing camera and then we apply an homography to make this plane parallel to the camera image plane. Accordingly to this, our descriptors can be just *similarity invariant* with 2 Degrees of Freedom, scale and rotation, with respect to the 4 Degrees of Freedom present in an affine transformation (disregarding in both cases the translation on 2 axes). The proposed descriptors can then be less generic becoming more robust and discriminative. Another important aspect which we have been dealing with is geometric discontinuities in objects surface, in particular, when detected corners or edges are not due to texture of a locally planar surface but to the abrupt folding of the surface itself, affine approximation between two wide baseline views is not valid any more. Projection on the average tangent plane or the unfolding of the discontinuity (edge or corner) can significantly improve matching capabilities. In the following we will show how low-cost depth-map acquisition devices (like Microsoft Kinect) can be fruitfully adopted to prove effectiveness of the aforementioned approach.

2 Surface vs. Texture Relevant Points

Actually the, by far, most used algorithm to define significant points in a picture that can be used to be matched with corresponding points in another image, is the corner Harris detector. This pioneering algorithm from Harris and Stephens [5] is still the basic element for localization of feature descriptors: [9]. Applying this algorithm to depth-maps provides us with surface discontinuities like geometrical corners or edges. In most of cases this features are a sub-set of corner and edges imputable to texture variations, so, once we have the depth-map registered with its corresponding image and we perform the Harris detector we are able to distinguish between:

- edges and corners due to textural variation but belonging to a flat surface.
- edges in the depth-maps corresponding to a folded or truncated surface.
- corners in the depth-maps (that are usually corners in the image too) corresponding to abrupt variations in the surface: e.g. spikes, corners or holes.

The capability to characterize different Harris features as geometrical or not (i.e. if they are also present or not in the depth-maps) is particularly important for definition of robust invariant descriptors. In particular the knowledge of the underlying geometry allows us to apply geometrical transformations to the textures on each slice in the neighborhood of the identified point in such a way to make their representation invariant from the view point. The opportunity to recover univocally a plane where the features in the neighborhood of the significant point lay is particularly important since it allows us, applying e.g. the proper homography, to obtain a frontal view of the neighborhood of a considered point independently from the viewpoint. The direct effect of this transformation is that the comparison between significant points for images acquired from different viewpoints can be simply performed comparing two frontal views of the regions around the points themselves: these regions can undergo only rotation and scaling: i.e. *similarity transform* where translation is disregarded since comparing neighborhood of two points implies the assumption that we are examining regions spatially already aligned).

3   Fusion of Geometric and Texture Descriptors

Many techniques have been developed to find flat planes in depth-maps, a significant example can be found in [15], and also surface curvature from cloud of points has been deeply investigated [16].

In our case we followed a simplified approach to define tangent plane to the surface around the interest point: it can be adopted even in case of discontinuities like corners, edges or generic surface folds. In fig. 1 there is a sample image where a Rubik's cube presents textural corners and edges on faces and abrupt geometrical corners and edges due to surface folds.

To find the tangent plane we followed a Principal Component Analysis for the spatial dispersion of depth-map points surrounding the interest point, in particular, accordingly to [6], we evaluated the covariance matrix (3×3) of the depth-map around the point (we used a 15×15 neighborhood window centered at the considered point but it can be adapted accordingly to the surface roughness or curvature) and then we performed the eigenvector decomposition. The resulting eigenvector associated to the lower eigenvalue represents the direction cosines for the "tangent" plane where we project the texture from the color image: this plane represents the locus where the points surrounding the interest point are maximally dispersed and it will always be the same independently

Fig. 1. A synthetic representation of a Rubik cube

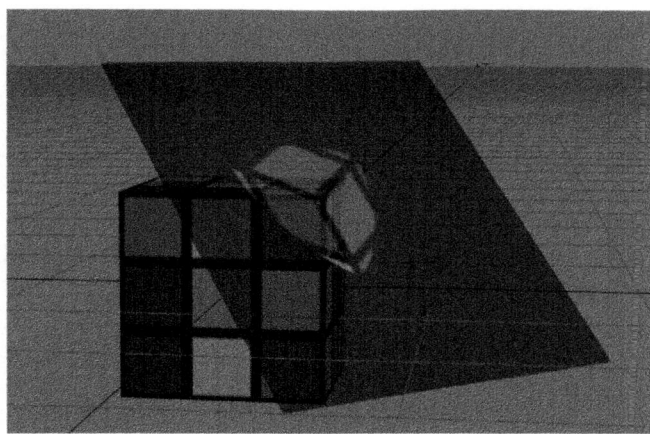

Fig. 2. The definition of the "tangent" plane and the reprojection on it of the neighborhood of the interest point

from the viewpoint (if all the sides are still visible). The image pixels are the projected onto this plane accordingly to their 3D position (recovered from the depth-map). Fig. 2 shows the reprojection of the texture on the "tangent" plane.

Then, through the homography that transforms the tangent plane into a frontal plane (a plane parallel to the image plane of the camera) we can recover a frontal view which is independent from the viewpoint apart for rotation and scaling (in fig. 3).

Fig. 3. The neighborhood of the interest point after the homography that provides a frontal view

4   Similarity Invariant Transform

Accordingly to the aforementioned steps we are able to obtain a 2D representation of the same 3D object part whose misalignment can be modeled by a four-parameter geometric transformation that maps each point (x_f, y_f) in F to a corresponding point (x_g, y_g) in G according to the matrix equation (in homogeneous coordinates):

$$\begin{bmatrix} x_g \\ y_g \\ 1 \end{bmatrix} = \begin{bmatrix} \rho\cos\vartheta & \rho\sin\vartheta & -\varDelta x \\ \rho\sin\vartheta & \rho\cos\vartheta & -\varDelta y \\ 0 & 0 & 1 \end{bmatrix} \begin{bmatrix} x_f \\ y_f \\ 1 \end{bmatrix}$$

Equivalently, defining the two images as two functions denoted by f and g, representing a gray-level image defined over a compact set of R^2, for any pixel (x, y) is true that:

$$f(x, y) = g\left(\rho(x\cos\vartheta + y\sin\vartheta) - \varDelta x, \rho(-x\sin\vartheta + y\cos\vartheta) - \varDelta y\right)$$

where $\varDelta x$ and $\varDelta y$ are translations, ρ is the uniform scale factor, and θ is the rotation angle. In other words, when we speak about similarity transformation we refer to the operations in this order:

$$RST = RS_{\rho,\theta} \cdot T_{\varDelta x, \varDelta y}$$

Since we are comparing image regions centered around interest points the translation invariance has no relevance in our case and the similarity invariance can be limited to rotation and scaling. Many approaches are present in the literature to tackle this problem [11], anyway most of them are incomplete like geometric moments and complex moments, while we oriented our research toward complete descriptors, that means that only representations retaining all the information of an image, except for orientation and scale, are considered. In particular we used the Fourier-Mellin transform (FMT) that is the Fourier Transform of the image $f(x, y)$ mapped in its corresponding Log-Polar coordinates $\hat{f}_{LP}(\mu, \xi)$:

$$f_{LP}(\mu, \xi) = \begin{cases} f\left(e^{\mu}\cos\xi, e^{\mu}\sin\xi\right) & \xi \in [0, 2\pi) \\ 0 & otherwise \end{cases}$$

The FMT is defined as:

$$F_m(w, k) = \int\limits_{0}^{\infty}\int\limits_{0}^{2\pi} f_{LP}(\mu, \xi)\, e^{-j(w\mu + k\xi)} d\xi d\mu$$

Then we explored two possible invariant for orientation and scale: the Taylor Invariant and the Hessian Invariant, which are described in the following sections. In particular we recall that after a Log-polar transformation a rotation corresponds to a circular shift along the axis representing the angles while a scaling corresponds to a shift along the logarithmic radial axis. Applying the 2D Fourier transform to the Log-polar transform the aforementioned shifts are reflected in phase shifts while the amplitude will remain unchanged.

5   Taylor and Hessian Invariant Descriptors

In this section we depict the two orientation-scale invariant descriptors that we used, both of them are based on the FMT described in the previous section. The Taylor invariant descriptor [2] is focused on eliminating the linear part of the phase spectrum by subtracting the linear phase from the phase spectrum. Let $F(u, v)$ be the Fourier transform of an image $f(x, y)$, and $\phi(u, v)$ be its phase spectrum. The following complex function is called the Taylor invariant:

$$F_{Tl}(u, v) = e^{-j(au + bv)} F(u, v)$$

where a and b are respectively the derivatives with respect to u and v of $\phi(u, v)$ at the origin $(0, 0)$, i.e.:

$$a = \varphi_u(0, 0),$$
$$b = \varphi_v(0, 0)$$

The Taylor invariant is rotationally symmetric, but not reciprocally scaled. It can be modified accordingly to the Laplacian invariant:

$$F_L(u, v) = \left(u^2 + v^2\right) F_{Tl} = \left(u^2 + v^2\right) e^{-j(au + bv)} F(u, v)$$

The effect is then the registration of the input features in such a way that the phase spectrum is flat in the origin, i.e. if we should take the inverse transforms, all of them will be rotated and scaled to accomplish to this constrain.

The idea behind the Hessian Invariant Descriptor [2] is to differentiate the phase spectrum twice to eliminate the linear phase, the invariant parts are then the modulus of the spectrum and the three, second order, partial derivatives of the phase spectrum:

$$F_H\left(u,v\right) = \left[\left|F\left(u,v\right)\right|, \varphi_{uu}\left(u,v\right), \varphi_{uv}\left(u,v\right), \varphi_{vv}\left(u,v\right)\right]$$

As described in the following sections, we evaluated both descriptors obtaining very similar results.

6   Results

We applied the previous descriptors to real images together with their depth-maps. The proposed algorithm can be summarized as follow:

- for each shot of the scene, significant points are extracted using Harris corner detector applied on the picture;
- the PCA was applied on the neighborhood 15×15 of the corresponding point of each detected point on the depth map and the eigenvector associated to the lower eigenvalue is used to determine the homography that transform the tangent plane into a frontal plane;
- the Fourie Mellin Transform is applied to the reoriented neighborhood;
- at last the Laplacian invariant is applied to $F_m\left(w, k\right)$ (only Laplacian translation invariant is used for these test).
- The resulted vector is used as feature descriptor of the significant point and correct match from different images are selected as those for which the Euclidean distance is minimized.

For completeness we summarize also the main step of the SIFT algorithm implemented for comparing the performances:

- Maximally Stable Extremal Regions (MSER) [8] are found for each shot of the scene;
- all the MSER are approximated as elliptical and oriented so that each major axis is horizontal;
- the ellipsis are deformed in circles and the intensity gradient for each pixel is computed;
- each circular region is divided in rectangular subregions and the histogram of the gradient's direction is computed for each subregion;
- the feature vector is made linking all the histograms computed on the circular neighborhood and, as for the proposed algorithm, correct match from different images are selected as those for which the Euclidean distance is minimized.

We performed some experiments using snapshots similar to those visible in fig. 4. No databases of pictures and depthmaps associated are yet available nowadays, so we decided to test our algorithm taking 20 pictures of the box illustrated in fig. 4 from different viewpoints. We used a Kinect device for the acquisition in an indoor environment and without any restriction except avoid that sun light directly on the IR device's camera. In fig. 5 we show how the planes, where the interest points lay, are reprojected in frontal views; the homographies have been defined accordingly to the PCA analysis of the underlying depth-map.

Fig. 4. A box acquired from different viewpoints and its depth-maps

Fig. 5. Images of interesting points after the homography to obtain a frontal view of framed surface by the depth-map

We checked the discriminative power of the proposed descriptors, in particular we compared the correct match rate and the euclidean distance from the closest match and from the second candidate. With the SIFT descriptor applied to the images, we obtained a correct match rate of 73%. For correct matches the mean ratio of the euclidean distances between the correct one and the second one is around 0.8. Using the proposed approach we obtained a correct match rate of 85% with an average ratio of distances for the first match and the second one of 0.65.

7   Conclusion

In this paper we propose a novel approach to define putative correspondences between images where the information from corresponding depth-maps are fruitfully integrated to reduce variability in the neighborhood around interest points, in particular projective or affine distortions are reduced to similarity transforms making available more robust and complete descriptors like Taylor or Hessian invariants applied to the Fourier-Mellin Transform.

The resulting approach demonstrates the profitable integration of depth-maps with acquired images to strengthen matching capabilities. Examples have been obtained by a low cost Kinect device.

Acknowledgement. This work was supported by the ASTUTE project: a 7 Framework Programme European project within the Joint Technology Initiative ARTEMIS.

References

1. Bay, H., Ess, A., Tuytelaars, T., Gool, L.V.: Surf: Speeded up robust features. Computer Vision and Image Understanding 110(3), 346–359 (2008)
2. Brandt, R., Lin, F.: Representation that uniquely characterize image modulo translation, rotation and scaling. Pattern Recognition Letters 17, 1001–1015 (1996)
3. Lowe, D.: Distinctive image features from scale-infariant keypoints. Int'l J. Computer Vision 2(60), 91–110 (2004)
4. Fusiello, A., Trucco, E., Tommasini, T., Roberto, V.: Improving features tracking with robust statistics. Pattern Analysis and Applications 2, 312–320 (1999)
5. Harris, C., Stephens, M.: A combined corner and edge detector. In: Proc. Alvey Vision Conf., pp. 147–151 (1988)
6. Jolliffe, I.: Principal Component Analysis, 2nd edn., vol. XXIX - 487. Springer, NY (2002)
7. Mikolajczyk, K., Schmid, C.: A performance evaluation of local descriptors. IEEE Trans. on Pattern Analysis and Machine Intelligence 27(10) (2005)
8. Matas, J., Chum, O., Urban, M., Pajdla, T.: Robust wide-baseline stereo from maximally stable extremal regions. Image and Vision Computing 22(10), 761–767 (2004)
9. Mikolajczyk, K., Schmid, C.: Scale and affine invariant interest point detectors. International Journal of Computer Vision 60(1), 63–86 (2004)
10. Mindru, F., Tuytelaars, T., Gool, L.V., Moons, T.: Moment invariants for recognition under changing viewpoint and illumination. Computer Vision and Image Understanding 94(1-3), 3–27 (2004)
11. Mukundan, R., Ramakrishnan, K.: Moment Functions in Image Analysis: Theory and Applications. World Scientific Publishing Co. Pte. Ltd., Singapore (1998)

12. Schmid, C., Mohr, R.: Local grayvalue invariants for image retrieval. Pattern Analysis and Machine Intelligence 19(5), 530–535 (1997)
13. Se, S., Ng, H., Jasiobedzki, P., Moyung, T.: Vision based modeling and localization for planetary exploration rovers. In: Proceedings of International Astronautical Congress (2004)
14. Shi, J., Tomasi, C.: Good features to track. In: Computer Vision and Pattern Recognition (1994)
15. Yang, M., Foerstner, W.: Plane detection in point cloud data. Technical Report TR-IGG-P-2010-01, Department of Photogrammetry Institute of Geodesy and Geoinformation University of Bonn (2010)
16. Yang, P., Qian, X.: Direct computing of surface curvatures for point-set surfaces. In: Proceedings of 2007 IEEE/Eurographics Symposium on Point-based Graphics, PBG (2007)

Automatic Object Classification and Image Retrieval by Sobel Edge Detection and Latent Semantic Methods

Vesna Zeljkovic[1] and Pavel Praks[2]

[1] School of Engineering & Computing Sciences, New York Institute of Technology (NYIT),
Nanjing Campus
[2] Dept. of Applied Mathematics, VSB-Technical University of Ostrava, Czech Republic
dr.zeljkovic@gmail.com, pavel.praks@vsb.cz

Abstract. We perform in this paper a comparative study of ability of the proposed novel image retrieval algorithms to provide automated object classification invariant of rotation, translation and scaling. We analyze simple cosine similarity coefficient methods and the SVD-free Latent Semantic method with an alternative sparse representation of color images. Considering applied cosine similarity coefficient methods, the two following approaches were tested and compared: i) the processing of the whole image and ii) the processing of the image that contains edges extracted by the application of the Sobel edge detector. Numerical experiments on a real database sets indicate feasibility of the presented approach as automated object classification tool without special image pre-processing.

Keywords: Object Classification, Image Retrieval, Sparse Image Representation, SVD-free Latent Semantic Method, Cosine Similarity Coefficient, Sobel Edge Detector.

1 Introduction

Automatic object recognition and classification is very important and has numerous applications, such as image retrieval and robot navigation.

Rapid development of information technologies provides users an easy access to a large amount of multimedia data, for instance images and videos. Unfortunately, wide popular text retrieval techniques, which are based on keyword matching, are not efficient for describing rich multimedia context. Recently, wavelets and various methods of numerical linear algebra are successfully used for automated information retrieval and identification tasks [10-15]. Moreover, genetic programming is used as a tool for image feature synthesis and recognition [16, 17]. In this paper, a comparison of modified Sobel edge detection and Latent Semantic methods with an alternative sparse representation of color images for automatic object classification and retrieval is presented.

L. Atzori, J. Delgado, and D. Giusto (Eds.): MOBIMEDIA 2011, LNICST 79, pp. 204–216, 2012.
© Institute for Computer Sciences, Social Informatics and Telecommunications Engineering 2012

2   Proposed Methods for Automatic Object Classification

We propose in this paper three different methods invariant of rotation, translation or scaling of the classified objects that successfully perform object classification on set of three different groups of objects Dinosaurs, Mummies and Sculls represented by images taken under various rotational, scaling and zooming conditions.

2.1   Sobel Edge Detector and Similarity Coefficient Methods

We applied two techniques for automatic object classification. We used Sobel edge filtered images for similarity computation in the first method and in the second method we applied simple cosine similarity coefficient on plain gray images with the goal to classify them.

The first technique implies procedure with an image converted to gray image with the extracted edges using Sobel edge detection method [1-7]. The idea behind this method is to significantly reduce the amount of data and filter out useless information, while preserving the important structural properties of an image and the targeted object.

Every image is processed as a two-dimensional $m \times n$ matrix image. We apply the two-dimensional Sobel masks to gray images. The Sobel operator performs a 2-D spatial gradient measurement on an image. It is used to find the approximate absolute gradient magnitude at each point in an input grayscale image. The Sobel edge detector uses a pair of 3×3 convolution masks, one estimating the gradient in the x-direction (columns) and the other estimating the gradient in the y-direction (rows). After that the magnitude of the gradient is calculated. In the next step we applied cosine similarity coefficient [8]-[11] in order to extract the image containing the most similar object in the database.

In the second approach, we convert color images to gray scale images and process them. Then we apply simple cosine similarity coefficient [8]-[11] as in the first method.

In our initial study, we applied for both techniques image de-noising and pre-processing by wavelet filter application. Our numerical results pointed out, that the application of de-noising methods does not have any influence of the proposed algorithms to perform more successful object recognition. This additionally slowed down the algorithm so we concluded to omit that pre-processing stage.

We applied different edge detection functions and we have concluded, based on the obtained results that the Sobel edge detector gave the most clear and emphasized edge extracted results for the first proposed method.

In the current computer implementation of the proposed object recognition procedures, no pre-processing of images is assumed. The presented numerical experiments indicate optimistic application of the proposed techniques for object recognition and classification.

The colors of images are coded in Matlab (tm) as non-negative integral numbers and we did not use any scaling. The application of the proposed procedures can be written in Matlab as follows.

```
% Input:
% A ... the m × n document matrix
% Output:
% sim ... the vector of similarity coefficients
[m,n] = size(Image);
```

1. Calculate the gray image presentation for both proposed techniques:

```
Gray = rgb2gray(Image);
```

2. Apply Sobel edge detector on gray scale Image

```
ImageSobel = edge(Gray,'sobel');
```

3. Compute the similarity coefficients between two inspected images

```
xx = ImageSobel '*ImageSobel0;
%for the first method
%or
xx = Gray '*Gray0;
%for the second method
xx= xx/(norm(ImageSobel0)*norm(Sobel));
sim(i) = 1-acos(xx);
```

The proposed two algorithms give at the output the similarity coefficients *sim*. The absolute value of *i*-th element of *sim* coefficient is a measure of the similarity between two compared images.

Both algorithms give acceptable and competitive results. They are efficient, easy for implementation and fast enough for real application.

2.2   Latent Semantic Indexing Method

The Latent Semantic Indexing method (LSI) [12] was originally developed for automated text retrieval because of efficient matching of polysemy and synonymy. Moreover, LSI can be extended image retrieval [13-15]. A raster $m \times n$ image can be represented as a sequence of $m \times n$ pixels. Elements of this sequence represent colors of the original image. In order to achieve sparsity character of LSI-based image descriptor, FFT or similar technique with quantization can be applied [14]. The image preprocessing and retrieval can be done by the following steps:

```
Procedure IP [Image Preprocessing]
Input:
N images with the same resolution m × n,
Output: mn x N document matrix A
for j=1:N {for j-th image}
```

• Step A: Represent j-th image as a sequence of one-dimensional signal [14]. Let symbol A denote a $mn \times N$ term-document matrix related to mn keywords (pixels) in N images. The (i, j)-element of the term-document matrix A represents the color of i-th position in the j-th image document:

```
A(:,j)= reshape(j-th image,m*n,1)
```

• Step B: Sparse representation of images by DST transformation leaving unchanged top 1 % coefficient. The remaining 99 % unsignificant coefficients are set to zero (a quantization).

```
A(:,j)= dst(A(:,j)); A(1,j)= 0;
A(:,j)= quantize(A(:,j),0.01);
end;
```

After Step B, image database is represented by the sparse $mn \times N$ document matrix A.

• Step C: Latent Semantic Indexing.
Following [12, 14] the Latent Semantic Indexing method can be written as:
```
Procedure LSI [Latent Semantic Indexing]
function sim = lsi(A,q,k)
Input:
A . . . the mn × N matrix
q . . . the query vector
k  .  .  . Compute  k  largest  singular  values  and
vectors; k ≤ N
Output: sim  .  .  .  the  vector  of  similarity
coefficients
[m,n] = size(A);
```
1. Compute the co-ordinates of all images in the k-dim space by the partial SVD of a document matrix A.
```
[U,S,V] = svds(A,k);
```
{Compute the k largest singular values of A; The rows of V contain the co-ordinates of images.}

2. Compute the co-ordinate of a query vector q
```
qc = q' * U * pinv(S);
```
{The vector q_c includes the co-ordinate of the query vector q; The matrix pinv(S) contains reciprocals of nonzeros singular values (a pseudoinverse). For more details please see Fig. 5 of [12].}

3. Compute the similarity coefficients between the query vector and images.
```
for j = 1:N  Loop over all images
sim(j)=(qc*V(j,:)')/(norm(qc)*norm(V(j,:)));
end;
```

{Compute the similarity coefficient for i-th image; $V(j, :)$ denotes the j-th row of V.}

The procedure LSI returns to a user the vector of similarity coefficients *sim*. The j-th element of the vector *sim* contains a value which indicates a measure of a semantic similarity between the j-th document and the query document.

3   Numerical Results

The collection of three different groups, each containing 24 images, in total 24×3=72 color images was analyzed. The sample images representing versatility in scale, rotation and distance from the camera for all three groups is presented in Figure 1. The dimensions of images varied so all of them were set to the same width of 2000 pixels and the height of 2000 pixels. So the each picture is characterized by 4,000,000 attributes. For example, the name "D_3.jpg" implies that the third image from Dinosaurs group of images is considered. The analyzed image database is available for research purposes under an e-mail request.

a) Dinosaur   b) Mummy

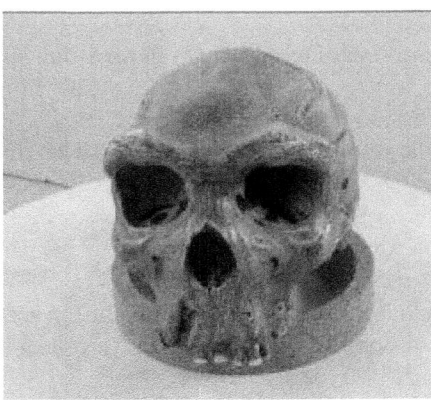

c) Scull

Fig. 1. Image database examples

The queries were represented as images from the collection.

3.1   Results for Sobel Edge Detector and Similarity Coefficient Methods

Figure 2 represents: a) Gray scale converted and rescaled Dinosaur image No.1 and the same image after Sobel edge detector, b) Gray scale converted and rescaled Mummy image No.1, and the same image after Sobel edge detector and c) Gray scale converted and rescaled Scull image No.1, and the same image after Sobel edge detector. It can be observed that the filtered images contain an emphasized unique structure and edge elements present in the classified object. The Sobel edge extraction application enables us to extract useful information necessary for further object comparison and identification. It is obvious that all three filtered images also contain the background edges that could be drawback in object classification due to introduction of emphasized non useful information.

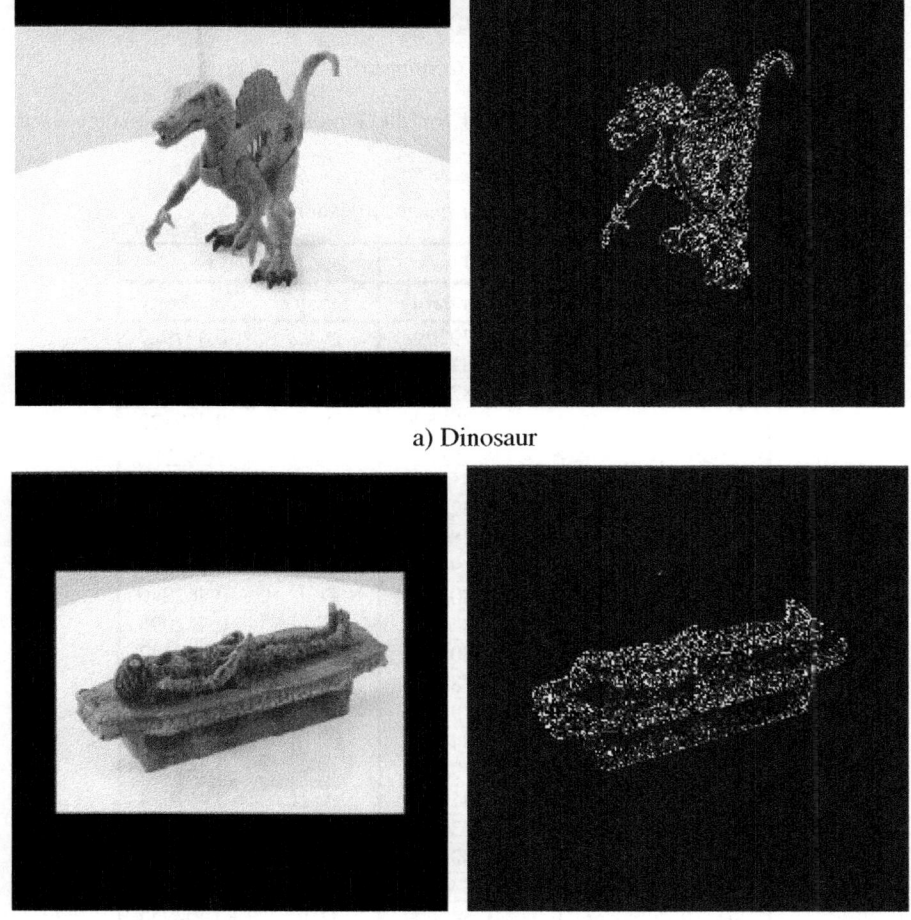

a) Dinosaur

b) Mummy

Fig. 2. Gray scale images and Sobel edge detector applied on database images

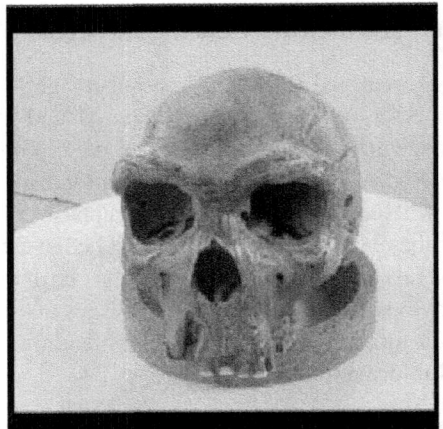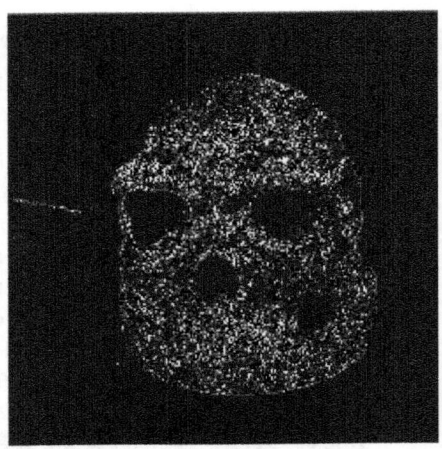

c) Scull

Fig. 2. (*continued*)

Table 1 contains the results obtained for the Dinosaur group of images used as query with both proposed algorithms.

Table 1. Dinosaur image retrieval result

	Method II		Method I	
Image	Similar	Similarity	Similar	Similarity
D_0	D_1	0.792098	D_22	-0.477922
D_1	D_2	0.825600	D_9	-0.462840
D_2	D_6	0.827910	M_11	-0.452429
D_3	D_2	0.793506	D_11	-0.467122
D_4	D_3	0.781580	D_13	-0.467995
D_5	D_2	0.818380	M_16	-0.468481
D_6	D_2	0.827910	D_20	-0.462955
D_7	D_0	0.781627	D_15	-0.480580
D_8	D_9	0.794382	D_7	-0.483005
D_9	D_10	0.801764	D_1	-0.462840
D_10	D_14	0.844410	M_11	-0.462095
D_11	D_10	0.809539	D_3	-0.467122
D_12	D_11	0.783882	D_3	-0.478209
D_13	D_11	0.791611	D_4	-0.467995
D_14	D_10	0.844410	D_10	-0.467348
D_15	D_14	0.795961	D_6	-0.463719
D_16	D_20	0.846660	M_21	-0.467024
D_17	D_16	0.839282	M_3	-0.479152
D_18	D_17	0.835905	M_16	-0.481522
D_19	D_16	0.834931	D_5	-0.480134
D_20	D_16	0.846660	D_6	-0.462955
D_21	D_20	0.840781	M_21	-0.488171
D_22	D_21	0.829590	D_0	-0.477922
D_23	D_16	0.838982	D_1	-0.473084

The first column in Table 1 represents the query image, second column represents the most similar image retrieved by each method and the third column is the maximum similarity coefficient determined by the applied technique.

Analyzing Table 1 we can conclude that we have 100% correct classification results obtained with the second method. The compared objects are recognized correctly. There are seven misclassifications obtained by the first method. Dinosaur images 2, 5, 10, 16, 17, 18 and 21 are wrongly classified as Mummy images 11, 16, 11, 21, 3, 16 and 21. The first method gave 70.83% correct classifications for the Dinosaurs group.

Table 2 contains the results obtained for the Mummy group of images used as query with both proposed algorithms.

Table 2. Mummy image retrieval result

	Method II		Method I	
Image	Similar	Similarity	Similar	Similarity
M_0	M_4	0.8098	M_22	-0.473274
M_1	M_5	0.8175	M_23	-0.473435
M_2	M_6	0.8638	M_6	-0.456423
M_3	M_7	0.8465	M_7	-0.449088
M_4	M_0	0.8098	M_18	-0.475947
M_5	M_1	0.8175	M_23	-0.474928
M_6	M_2	0.8638	M_2	-0.456423
M_7	M_3	0.8465	M_21	-0.448823
M_8	M_12	0.8213	M_22	-0.464342
M_9	M_13	0.7947	M_0	-0.485602
M_10	M_14	0.8489	M_14	-0.469516
M_11	M_15	0.8636	M_15	-0.448240
M_12	M_8	0.8213	M_18	-0.466055
M_13	M_9	0.7947	M_4	-0.486284
M_14	M_10	0.8489	M_10	-0.469516
M_15	M_11	0.8636	M_11	-0.448240
M_16	M_20	0.8793	M_20	-0.451026
M_17	M_21	0.8741	M_21	-0.444759
M_18	M_22	0.8662	M_22	-0.448271
M_19	M_23	0.8663	M_23	-0.454096
M_20	M_16	0.8793	M_16	-0.451026
M_21	M_17	0.8741	M_17	-0.444759
M_22	M_18	0.8662	M_18	-0.448271
M_23	M_19	0.8663	M_19	-0.454096

We obtained even better results for both methods for the Mummy group of images used as query images. The compared Mummy images are recognized correctly. Analyzing Table 2 we can conclude that we have 100% correct classification results obtained with both proposed methods for the Mummy query group.

Table 3 contains the results obtained for the Scull group of images used as query with both proposed algorithms.

Table 3. Scull image retrieval result

Image	Method II		Method I	
	Similar	Similarity	Similar	Similarity
S_0	S_7	0.7866	S_14	-0.492076
S_1	S_2	0.7869	S_16	-0.477059
S_2	S_1	0.7869	S_17	-0.477816
S_3	S_4	0.7853	S_18	-0.487566
S_4	S_5	0.7980	S_20	-0.480737
S_5	S_6	0.8294	S_20	-0.478156
S_6	S_5	0.8294	S_21	-0.475674
S_7	S_6	0.8002	S_22	-0.486436
S_8	S_15	0.8368	S_15	-0.476013
S_9	S_10	0.8483	S_10	-0.482384
S_10	S_11	0.8604	S_20	-0.474668
S_11	S_10	0.8604	M_17	-0.478968
S_12	S_11	0.8470	D_1	-0.478980
S_13	S_12	0.8367	S_21	-0.483276
S_14	S_15	0.8190	S_15	-0.478761
S_15	S_8	0.8368	S_8	-0.476013
S_16	S_17	0.7937	S_1	-0.477059
S_17	S_16	0.7937	S_2	-0.477816
S_18	S_19	0.7938	S_3	-0.487566
S_19	S_20	0.8010	S_4	-0.481828
S_20	S_21	0.8252	S_10	-0.474668
S_21	S_20	0.8252	S_6	-0.475674
S_22	S_21	0.8044	S_21	-0.484460
S_23	S_22	0.7932	S_16	-0.491731

The first column in Table 3 represents the query image, second column represents the most similar image retrieved by each method and the third column is the maximum similarity coefficient determined by the applied technique.

Analyzing Table 3 we can conclude that we have 100% correct classification results obtained with the second method as for the previous two query groups Dinosaur and Mummy. The compared objects are recognized correctly. There are only two misclassification obtained by the first method. Scull images 11 and 12 are wrongly classified as Mummy image 17 and Dinosaur image 1, respectively. The first method gave 91.67% correct classifications for the Scull group.

3.2   Results Obtained with Latent Semantic Indexing Method

In the first step, the input image was rescaled to 320×200 pixels. It means that the each image was represented by the 320×200=64 000 features. Moreover, in order to assume a color representation for the image retrieval, each image was represented by the RGB color model. In contrary to previous research [13-15], an alternative sparse coding of color images was implemented as a sequence of quantized FFT representations of RGB components. It means that the each image is represented by $3 \times 64\,000$ sparse features, which gives us 192 000 sparse features of an image in summary. Thanks to the LSI preprocessing quantization, these color image features remain sparse. In other words, both the quantization and the user defined coefficient $k=8$ significantly reduced memory requirements of LSI [14]. As a result, the document matrix of 72 images allocated only 0.82 MB of the computer memory. Moreover, LSI can be implemented very effectively by solving a partial symmetric eigen-problem (so called the SVD-free approach [13]). For this reason, LSI computations required less than 0.1 seconds.

Image retrieval results are presented by decreasing order of similarity. In all cases, the query image is situated in the upper left corner, see Figures 3-6. The similarity of the query image and the retrieved image is written in parentheses. In order to achieve well arranged results, only 9 most significant images are presented.

LSI image retrieval results seem to be very promising. Images were correctly classified in all cases except one case, see Figure 6. It gives us the probability of failure $1/81=1.23$ %. The remaining results are classified well: the most similar image is from the same group as the query.

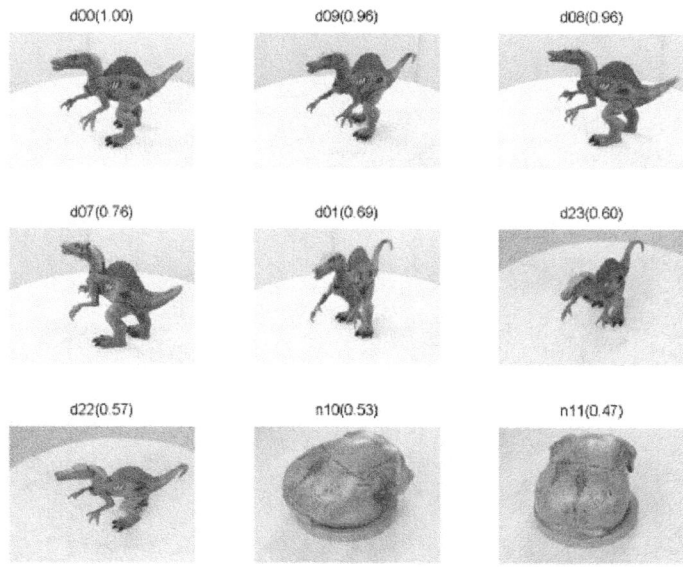

Fig. 3. An example of LSI image retrieval results: experiments with Dinosaur query image

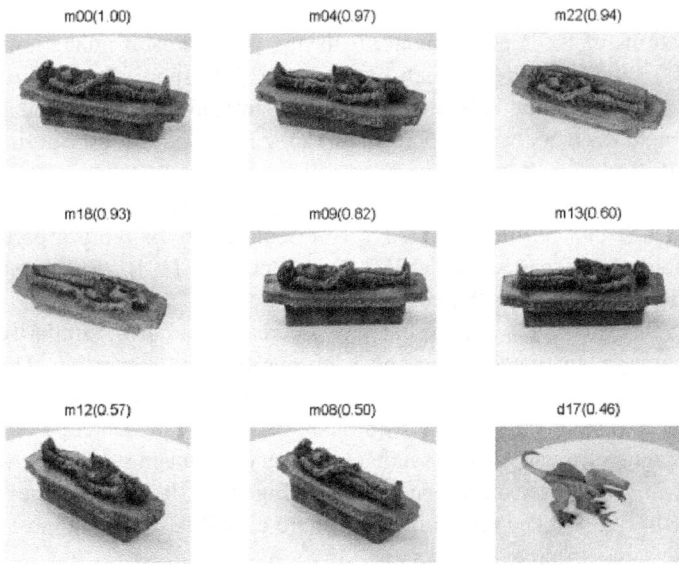

Fig. 4. An example of LSI image retrieval results: experiments with Mummy query image

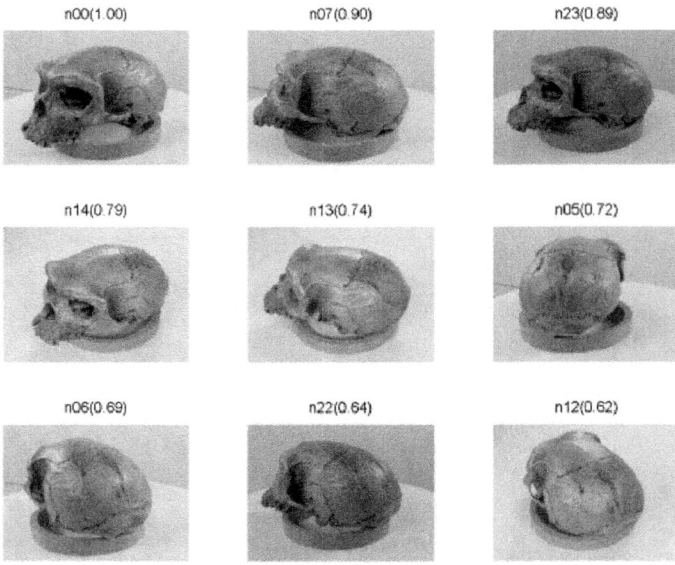

Fig. 5. An example of LSI image retrieval results: experiments with Scull query image

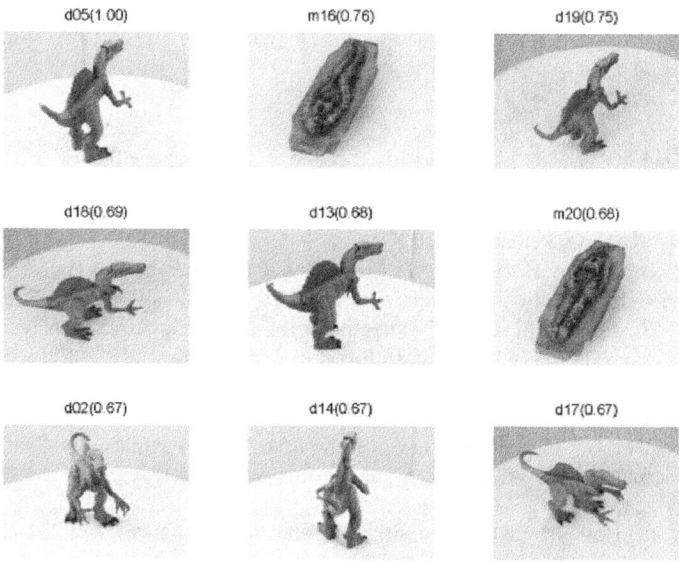

Fig. 6. An example of LSI image retrieval results: experiments with Dinosaur query image: failure recognition

4   Conclusion

Three approaches to the recognition and classification of different objects in various images are presented in this article. The results of the recognition test are promising and they show the ability of presented algorithms to successfully recognize various objects in real images. Of course, the quality of images and proper localization can influence the resulting errors by inaccurate localization. The proper localization will be subject to future work.

The proposed methods also show to be rotation, translation and scale invariant which open their potential application in wide range of areas.

We have applied two different methods to get proper object comparison and identification. The two following algorithms were tested and compared: the processing of the image with extracted edges and the processing of the whole image. As it can be observed from presented numerical results, both algorithms have shown compatible, accurate and comparable results. The advantage of the second proposed method is its simplicity, effectiveness, 100% correct classification results and practical implementation and realization.

An SVD-free sparse LSI algorithm with an alternative sparse representation of color images is presented as a tool for the object classification problem. In our experiments, the LSI algorithm was numerical stable. Results seem to be very robust: There is only one incorrectly detected case and it gives us 98.77% correct classification results.

Future research would concentrate on combining explicit mapping from presented low-level image features to semantic abstractions, which can be used for computer based interpretation of query images.

Acknowledgments. This work was partially supported by the Ministry of Education, Youth and Sport of the Czech Republic (Project No. 1M06047). Special thanks to Ruzica Vlajkovic and Milos Zeljkovic.

References

1. Engel, K., Hadwiger, M., Kniss, J., Rezk-Salama, C.: Real-time volume graphics, pp. 112–114. AK Peters Ltd. (2006)
2. Jähne, B., Scharr, H., Körkel, S.: Principles of filter design. In: Handbook of Computer Vision and Applications. Academic Press (1999)
3. Farid, H., Simoncelli, E.P.: Differentiation of discrete multi-dimensional signals. IEEE Transactions on Image Processing 13(4), 496–508 (2004)
4. Kroon, D.: Numerical Optimization of Kernel Based Image Derivatives, Short Paper University Twente (2009)
5. Scharr, H.: Optimal Second Order Derivative Filter Families For Transparent Motion Estimation. In: 15th European Signal Processing Conference, EUSIPCO 2007, Poznan, Poland, September 3-7 (2007)
6. Duda, R., Hart, P.: Pattern Classification and Scene Analysis, pp. 271–272. John Wiley and Sons
7. Gonzalez, R.C., Woods, R.E.: Digital Image Processing, 2nd edn. Prentice Hall (2001)
8. Singhal, A., Buckley, C., Mitra, M.: Pivoted Document Length Normalization, NSF grant supported study, Cornell University Ithaca NY
9. Garcia, E.: Patents on Duplicated Content and Re-Ranking Methods. In: SES 2005: Advanced Track Issues: The Patent Files, San Jose, August 8-11 (2005)
10. Tan, P.-N., Steinbach, M., Kumar, V.: Introduction to Data Mining, ch. 8, p. 500. Addison-Wesley (2005) ISBN: 0-321-32136-7
11. Zeljković, V., Praks, P.: A Comparative Study of Automated Iris Recognition Using the Biorthogonal Wavelets and the SVD-Free Latent Semantic Methods. In: ZNALOSTI 2007 Annual Conference on Knowledge Acquisition, Czecho-Slovak Knowledge Technology Conference, Ostrava, Czech Republic, February 21-23, pp. 143–154 (2007) ISBN: 978-80-248-1279-3
12. Berry, W.M., Dumais, S.T., O'Brien, G.W.: Using linear algebra for intelligent information retrieval. SIAM Review 37, 573–595 (1995)
13. Praks, P., Dvorský, J., Snášel, V., Černohorský, J.: On SVD-free latent semantic indexing for image retrieval for application in a hard industrial environment. In: IEEE ICIT 2003 (2003)
14. Praks, P., Kučera, R., Izquierdo, E.: The sparse image representation for automated image retrieval. In: IEEE ICIP 2008, pp. 25–28 (2008), doi:10.1109/ICIP.2008.4711682
15. Praus, P., Praks, P.: Hierarchical clustering of RGB surface water images based on MIA-LSI approach. Water SA 36(1), 143–150 (2010)
16. Krawiec, K., Bhanu, B.: Visual learning by evolutionary and coevolutionary feature synthesis. IEEE Trans. Evol. Comput. 11(5), 635–650 (2007)
17. Watchareeruetai, U., Takeuchi, Y., Matsumoto, T., Kudo, H., Ohnishi, N.: Redundancies in linear GP, canonical transformation, and its exploitation: a demonstration on image feature synthesis. Genet. Program Evolvable. 12, 49–77 (2011), doi:10.1007/s10710-010-9118-x

Optimal Interleaving for Robust Wireless JPEG 2000 Images and Video Transmission

Daniel Pascual Biosca and Max Agueh

LACSC - ECE Paris, 37 Quai de grenelle,
75015 Paris, France
{biosca,agueh}@ece.fr

Abstract. In this paper we study the impact of interleaving on JPEG2000 images and video transmission through wireless channels. Based on interleaving impact evaluation, we derive a lower bound limit for the successful images decoding rate in wireless environments. Since the successful decoding rate is of central importance to guarantee Quality of Service to wireless clients, we rely on the derived limit to evaluate the performance of near-optimal interleaved frames using a wireless JPEG 2000 based client/server application. This work is a step toward optimal interleaving for robust Wireless JPEG 2000 based images and video transmission.

Keywords: Interleaving, Wireless JPEG2000, Successful decoding rate, Forward Error Correction, Reed-Solomon codes.

1 Introduction

With the development of smart wireless fixed and mobile devices, efficient multimedia transmission over wireless error-prone channels becomes an important issue. Among existing images representation standards, JPEG 2000 1 is one of the most promising to address robust wireless images/video transmission challenges. Actually, JPEG 2000 defines an extension named JPWL [2] [3] (JPEG 2000 for wireless - 11th part of the standard) for reliable transmission of JPEG 2000 based codestreams over error-prone channels. Hence techniques such as Forward Error Correction (FEC) with Reed-Solomon (RS) codes, Unequal Error Protection (UEP) and data interleaving are proposed to increase the robustness of JPEG 2000 codestreams against transmission errors. Although, JPEG 2000 based FEC techniques has been intensively investigated in the literature [4][5][6][7], few works address JPEG 2000 codestreams interleaving issues. In [8], F. Frescura and G. Baruffa propose a backward-compatible JPEG 2000 virtual interleaving which improves the effectiveness of the RS codes. The proposed virtual interleaver guarantees the backward compatibility of JPEG 2000 frames by computing nonconsecutive parity bytes. However, as only parity bytes are interleaved, remaining part of the JPEG 2000 codestreams are still significantly sensitive to transmissions errors.

L. Atzori, J. Delgado, and D. Giusto (Eds.): MOBIMEDIA 2011, LNICST 79, pp. 217–226, 2012.
© Institute for Computer Sciences, Social Informatics and Telecommunications Engineering 2012

Since, JPEG 2000 codestreams headers and marker segments are the most important part of the codestreams, a specific emphasis should be taken to integrate them in a overall and more generic interleaving scheme.

In this work we study the impact of interleaving on JPEG 2000 images and video transmission over wireless networks. To the best of our knowledge the present work is the first to rely on interleaving to derive a lower bound limit for successful decoding rate for robust JPEG 2000 images/video streaming over wireless channel. Thus, a straightforward comparison to already implemented interleaving techniques is not possible.

2   Wireless JPEG 2000 Overview and Interleaving Framework

In this section, we present an overview of JPEG2000 Wireless standard and we provide an analysis of interleaved codeword error probability.

2.1   Wireless JPEG2000

Wireless JPEG2000 [2] [3] defines a set of 19 RS codes [2] to protect each part of JPEG 2000 codestreams against transmission errors. A $RS(n, k)$ code can correct up to $t = (n - k)/2$ or symbols. In JPEG 2000 codestreams, redundancy is allocated inside Error Protection Block (EPB) markers segments. A detailed description of JPWL codestream is available in [2]. In figure 1, we present the JPWL codestream structure considered in this work. This codestreams is constituted with K tile-parts. Main header is protected with N EPBs; The first tile-part is protected with L and M EPBs respectively for header and bitstream; last tile-part uses P and X EPBs respectively for its header and its bitstream protection. All EPBs are in packed mode.

2.2   Gilbert-Elliot Channel Model

The Gilbert-Elliot (GE) model is widely used to simulate the burst-error behavior of the wireless channels. The GE model considered in this work is a Markov chain of order 1 and is extensively presented in [9]. This GE model has two states: the state Good, where the channel symbol is correctly transmitted; and the Bad state, where the channel symbol is corrupted. The transition probability from Good state to Bad state is p_{gb}, which is generally low; and the transition probability from Bad state to Good state is p_{bg}, which is often high. The stationary values for the two states are given by:

$$\pi_B = SER = \frac{p_{bg}}{p_{bg} + p_{gb}}, \tag{1}$$

$$\pi_G = 1 - SER = \frac{p_{gb}}{p_{bg} + p_{gb}} \tag{2}$$

where SER is the Symbol Error Rate.

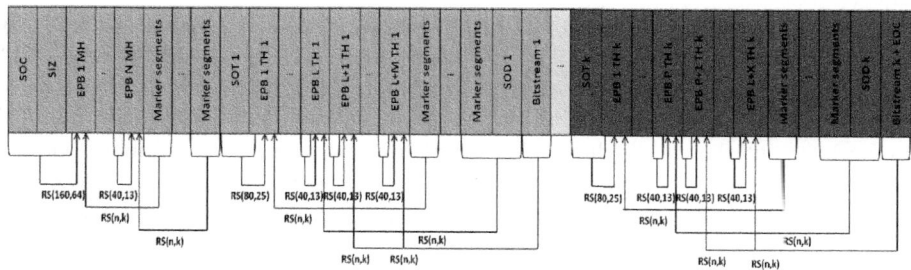

Fig. 1. JPWL codestream protected with EPBs

From [10] the transition probabilities can be expressed as:

$$p_{bg} = (1 - SER)(1 - \rho) \tag{3}$$

$$p_{gb} = SER(1 - \rho) \tag{4}$$

where $\rho = 1 - p_{gb} - p_{gb}$ is the correlation between two consecutive error symbols. Since error bursts may be very harmful for the error correction process, interleaving the protected data before transmitting it through the channel, helps to significantly decrease the decoding error rate. Hence, with interleaving, the correlation between two consecutive error symbols decreases by ρ^I, where I represents the interleaving depth. Then, channel parameters can be expressed as:

$$p_{bg}{}^I = (1 - SER) \cdot (1 - \rho^I) \tag{5}$$

$$p_{gb}{}^I = SER \cdot (1 - \rho^I) \tag{6}$$

As interleaving increases, the error distribution of the channel becomes more uniform, resulting in a memoryless Binary Symmetric Channel (BSC) with same SER. Indeed $\lim_{I \to \infty} p_{bg} = (1 - SER)$ and $\lim_{I \to \infty} p_{gb} = SER$.

2.3   Impact of Interleaving on Error Probability Reduction

In this section we investigate the impact of interleaving on error probability reduction at the decoder side. In the scenario considered, data is protected with RS codes and transmitted through a GE channel. From [10] the probability of having residual errors in a codeword after RS error correction in a GE channel is:

$$P_{cw}(n, k) = \sum_{m=t+1}^{n} P(m, n) \tag{7}$$

where $P(m, n)$ is the probability of having exactly m errors in n consecutive symbols. A detailed description of $P(m, n)$ is available in [10]. For infinite interleaving, the codeword error probability in a BSC channel [11] can be used:

$$P_{cw-bsc}(n, k) \leq \sum_{m=t+1}^{n} \binom{n}{m} SER^m (1 - SER)^{n-m} \tag{8}$$

Figure 2 presents the codeword error probability versus RS codes capability for different interleaving depths. We observe that increasing interleaving depth significantly reduces the codeword error probability. However, for RS codes with low error correction capability, interleaving is inefficient and may become harmful. This is because interleaving reduces the correlation between error symbols but also between error-free symbols. Since the SER remains constant, increasing the interleaving depth reduces the bursts length at the expense of increasing the number of bursts.

3   Successful Decoding Rate

We define the successful decoding rate S_{frame} as the percentage of JPEG 2000 images which are free of errors after error correction in the main header, in any of the tile-part headers, in the EPBs marker segment fields used to protect the bitstreams, and in the End Of Codestream (EOC) marker segment. Hence, we have:

$$S_{frame} \geq (1 - P_{main})(1 - P_{tile\,1})(1 - P_{bs\,1}) \cdots$$
$$(1 - P_{tile\,N})(1 - P_{bs\,N})(1 - P_{eoc}) \tag{9}$$

where P_{main}, P_{tile}, P_{bs} and P_{eoc} are respectively the probability of error in the main header, the tile-part headers, bitstreams and EOC marker segment.

3.1   Basis Assumption

Since Successful decoding rate is an important metric for our interleaving methodology, we first make the assumption that S_{frame} is only constituted of images with error free headers and markers segments. In other words we make the hypothesis that S_{frame} has a lower bound whose estimation is of central importance for practical implementation of JPEG 2000 frames interleaver. We then validate this assumption by simulation using JPEG 2000 codestreams.

   Actually, our hypothesis is justified by two reasons. First, residual errors in marker segment fields may look like valid values defined by the standard and thus could not be detected and corrected by the decoder. Hence, those errors may significantly reduce decoded images quality and this leads us to consider them as unsuccessfully decoded images. However, even if those undetected errors are not corrected by the decoder, the bad quality of resulting images will lead to straightforwardly discard these images using the method proposed in [12]. Secondly, the number of bytes to protect with an RS code, may not be multiple of the codeword length, thus byte padding is used up to fill the codeword. If by chance the residual errors fall only inside padding data, the decoding rate will not be affected.

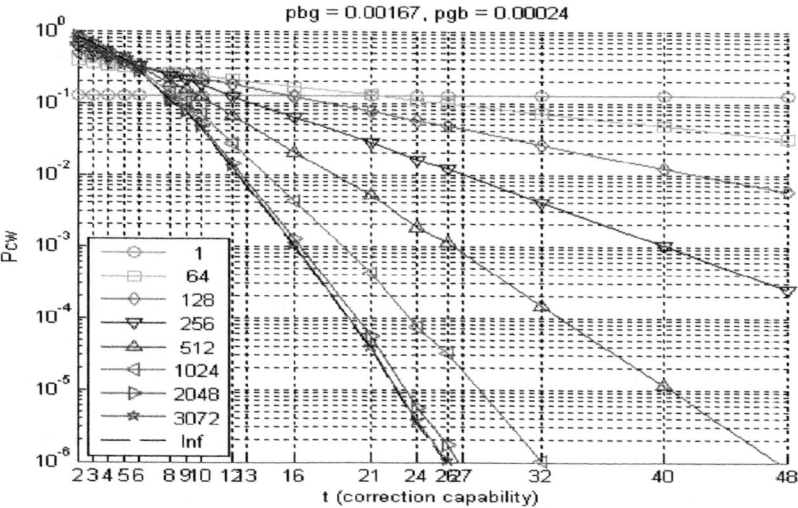

Fig. 2. Codeword error probability for RS codes in a GE channel with $p_{bg} = 0.00167$ and $p_{gb} = 0.00024$

3.2   Residual Error Probability Estimation

The probability of having residual errors in the main header is:

$$P_{main} = 1 - [(1 - P_{cw}(160, 64))(1 - P_{cw}(n_a, k_a))^a$$
$$(1 - P_{cw}(40, 13))(1 - P_{cw}(n_b, k_b))^b \cdots] \quad (10)$$

where a is the number of codewords in the first EPB protected with $RS(n_a, k_a)$, b is the number of codewords in the second EPB protected with $RS(n_b, k_b)$ and so on. In the same way, the probability of having residual errors in a tile-part header is:

$$P_{tile} = 1 - [(1 - P_{cw}(80, 25))(1 - P_{cw}(n_c, k_c))^c$$
$$(1 - P_{cw}(40, 13))(1 - P_{cw}(n_d, k_d))^d \cdots] \quad (11)$$

where c is the number of codewords in the first EPB protected with $RS(n_c, k_c)$, d is the number of codewords in the second EPB protected with $RS(n_d, k_d)$ and so on. The probability of having residual errors in the bitstream EPBs is:

$$P_{bs} = 1 - (1 - P_{cw}(40, 13))^{N_p} \quad (12)$$

where N_p is the number of EPBs used in the tile-part. Finally, the error probability for the EOC marker segment is given by:

$$P_{eoc} = P_{cw}(n_{last}, k_{last}) \quad (13)$$

3.3   Assumption Validation

In order to validate our basis assumption, we use Structural Similarity (SSIM) metric 13 to study the effect of residual errors in marker segments of a Lena 2k image. The characteristics of the lena.j2k images are: resolution 352x288; size off codeblocks 64x64; precinct 1; tile 1 (no offset used); component 1 ; resolution levels 6; quality layers 3 (compression rate 20, 10 and 5 respectively); JPEG 2000 data packets 18;

We observe from figure 3 and figure 4 that errors in headers are extremely harmful in terms of quality and successful decoding. Actually, JPEG 2000 images quality decreased significantly when transmission errors occur in the marker segments.

The current work is the first which investigates the JPEG 2000 marker segments sensitivy to wireless transmission errors. It's worth noting the proposed normalized residual error ratio allows comparison between different types of marker segments.

We notice from figure 3 and figure 4 that in the case of header or marker segment corruption, measured MSSIM is under 0.5 and successful decoding rate is under 50% (which is intolerable) whatever the marker. Our assumption which consists to consider only error free header and marker free decoded JPEG 2000 frames in S_{frame} estimation is valid.

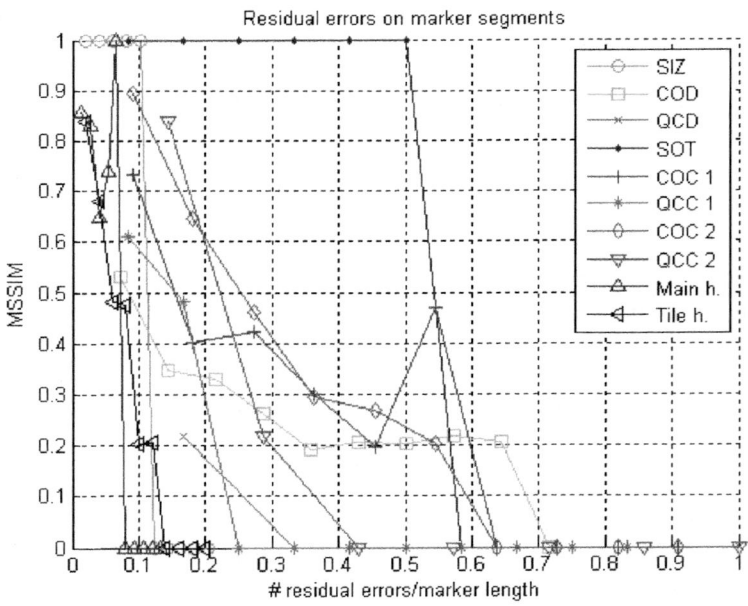

Fig. 3. Normalized residual errors ratio versus SSIM

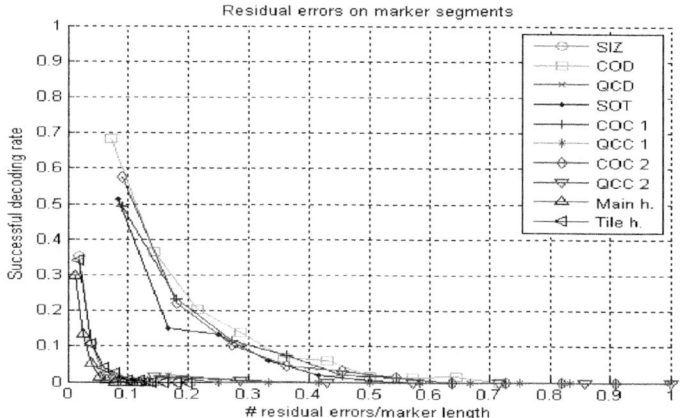

Fig. 4. Normalized residual errors ratio versus successful decoding rate

4   Wireless Performance of Interleaving on Our Wireless JPEG 2000 Transmission System

The video sequence used in this work is *speedway.mj2* video 14 which is constituted by 200 JPEG 2000 frames. The 352 x 288 video is transmitted through a GE channel using the JPWL based transmission system presented in [7]. RTP packet lengths of 512 and 768 are used . The packet traces are derived from real IEEE 802.11 wireless channel traces 15. JPEG 2000 frames marker segments are protected with the predefined RS codes. Equal Error Protection (EEP) is used to protect the whole codestream up to reaching the bandwidth constraint.

The generated GE channel characteristics are: $p_{bg} = 0.05227$ and $p_{gb} = 0.00024$ and the available bandwidth is 10 Mbps. In this scenario, S_{frame} is given by:

$$S_{frame} \geq \left(1 - P_{cw}(160, 64)\right)^3 \left(1 - P_{cw}(80, 25)\right)^2$$
$$\left(1 - P_{cw}(40, 13)\right)^{18} \left(1 - P_{cw}(n_{last}, k_{last})\right) \qquad (14)$$

In figure 5 and figure 6 the successful decoding rate is plotted for different interleaving depths (named as *Real*) along with the rate of frames without errors in the marker segments (named as *Minimum simulated*). We observe from figure 5 that the best results (more than 90% of successfully decoded images) are achieved for the interleaving depth overcome RTP packet length (here 512 bytes). However when RTP packet length increases the needed interleaving depth to achieve good performance seems to be a multiple of the RTP packet length. An interesting extension to this work could be to derive an optimal interleaving.

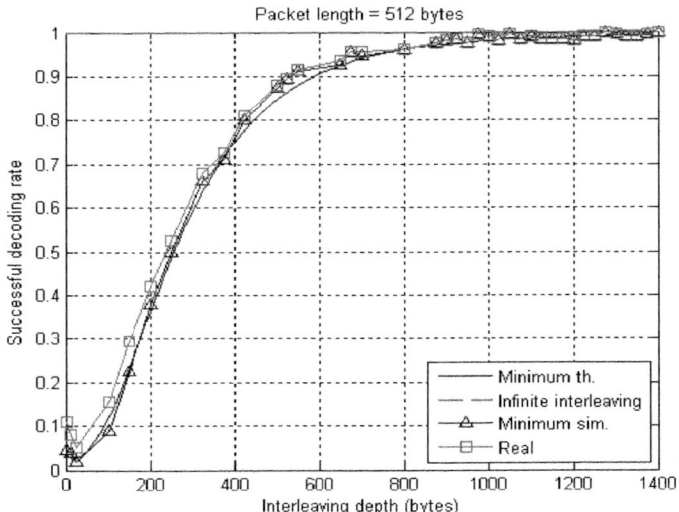

Fig. 5. Interleaving depth versus successful decoding rate – RTP packet length = 512 bytes

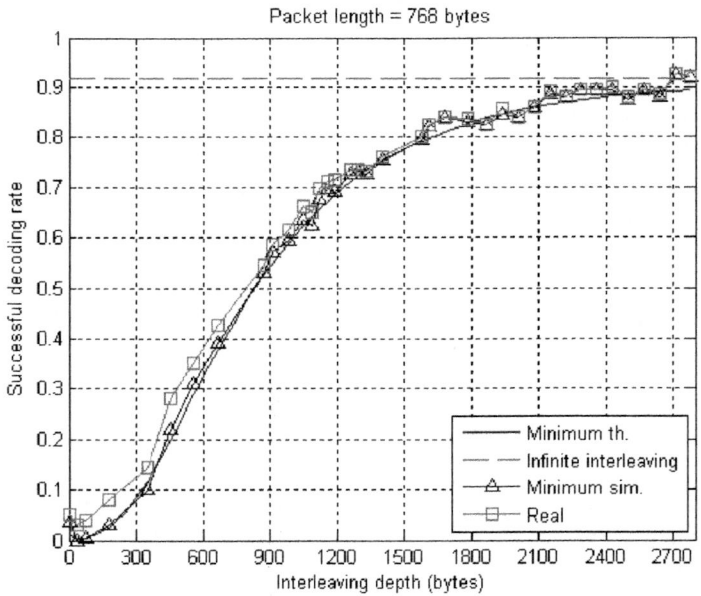

Fig. 6. Interleaving depth versus successful decoding rate – RTP packet length = 768 bytes

5   Conclusion

In this paper, we first investigate the impact of interleaving on robust wireless JPEG 2000 image and video transmission over wireless channels. Then, we derive a lower bound expression for successful decoded frames in wireless transmission of JPEG2000 images and video.

Our derived expression fits very well with JPEG 2000 based decoding images which are empirically estimated. We validate our expression using a wireless JPWL based client/server application. Since, successful decoding rate is significantly impacted by interleaving depth, our work could be considered as a valid step toward optimal interleaving for robust JPEG 2000 images and video transmission through wireless channels.

References

1. Information Technology-JPEG 2000-Image Coding System-Part 1: Core Coding System, ISO/IEC 15 444-1 (2000)
2. Information Technology-JPEG 2000-Image Coding System-Part 11: Wireless, ISO/IEC 15 444-1 (2005)
3. Nicholson, D., Lamy-Bergot, C., Naturel, X., Poulliat, C.: JPEG 2000 backward compatible error protection with Reed-Solomon codes. IEEE Transactions on Consumer Electronics 49(4), 855–860 (2003)
4. Agueh, M., Devaux, F.O., Diouris, J.F.: A Wireless Motion JPEG 2000 video streaming scheme with a priori channel coding. In: Proceeding of 13th European Wireless 2007 (EW 2007), Paris France (April 2007)
5. Guo, Z., Nishikawa, Y., Omaki, R.Y., Onoye, T., Shirakawa, I.: A Low-Complexity FEC Assignment Scheme for Motion JPEG 2000 over Wireless Network. IEEE Transactions on Consumer Electronics 52(1), 81–86 (2006)
6. Agueh, M., Diouris, J.F., Diop, M., Devaux, F.O.: Dynamic channel coding for efficient Motion JPEG 2000 streaming over MANET. In: Proceeding of Mobimedia 2007, Nafpaktos, Greece (August 2007)
7. Agueh, M., Diouris, J.F., Diop, M., Devaux, F.O., De Vleeschouwer, C., Macq, B.: Optimal JPWL Forward Error Correction rate allocation for robust JPEG 2000 images and video streaming over Mobile Ad-hoc Networks. EURASIP Journal on Advances in Signal 2008; Proc., Spec. Issue Wireless Video
8. Frescura, F., Baruffa, G.: Backward-Compatible Interleaving Technique for Robust JPEG 2000 Wireless Transmission. In: Atzori, L., Giusto, D.D., Leonardi, R., Pereira, F. (eds.) VLBV 2005. LNCS, vol. 3893, pp. 44–50. Springer, Heidelberg (2006), doi:10.1007/11738695_7
9. Elliot, O.: Estimates of error rates for codes on burst-noise channel. Bell SystemTechnical Journal 42, 1977–1997 (1963)
10. Yee, J.R., Weldon Jr., E.J.: Evaluation of the performance of error-correcting codes on a Gilbert channel. IEEE Trans. on Communications 43(8), 2316–2323 (1995)
11. Proakis, J.G.: Digital Communications, 3rd edn. McGraw-Hill, New York (2001)
12. Nishikawa, K., Munadi, K., Kiya, H.: No-Reference PSNR Estimation for Quality Monitoring of Motion JPEG 2000 Video Over Lossy Packet Networks. IEEE Transactions on Multimeda 10(4), 637–645 (2008)

13. Wang, Z., Bovik, A.C., Sheikh, H.R., Simoncelli, E.P.: Image Quality Assessment: From Error Visibility to Structural Similarity. IEEE Transactions on Image Processing 13(4), 600–612 (2004)
14. Speedway video sequences have been generated by UCL, http://euterpe.tele.ucl.ac.be/WCAM/public/Speedway%20Sequence
15. Loss patterns acquired during the WCAM Annecy 2004 measurement campaigns IST-2003-507204 WCAM. Wireless Cameras and Audio-Visual Seamless Networking (2004), http://www.ece.fr/~agueh/wcam_patterns

Interactive Image Viewing in Mobile Devices Based on JPEG XR

Bernardetta Saba, Cristian Perra, and Daniele D. Giusto

Department of Electrical and Electronic Engineering, University of Cagliari
Piazza D'Armi, 09123 Cagliari, Italy
{bernardetta.saba,cperra}@diee.unica.it, ddgiusto@unica.it

Abstract. Services for high definition image browsing on mobile devices require a careful design since the user experience is heavily depending on the network bandwidth, processing delay, display resolution, image quality. Modern applications require coding technologies providing tools for resolution and quality scalability, for accessing spatial regions of interest (ROI), for reducing the domain of the coding algorithm decomposing large images into tiles. Some state-of-the-art technologies satisfying these requirements are the JPEG2000 and the JPEG XR. This paper presents the design of an interactive high resolution image viewing architecture for mobile devices based on JPEG XR. Display resolution, resolution scalability, image tiling are investigated in order to optimize the coding parameters with the objective to improve the user experience. Experimental tests are performed on a set of large images and comparisons against accessing the images without parameter optimization are reported.

Keywords: JPEG XR, Interactive Access.

1 Introduction

The wide variety of devices for mobile imaging (PDA, Mobile Phone, and Smartphone) requires that the compressed image is capable to adapt to different viewing conditions. A big problem for the development of these applications is related to the high costs of storage and data transmission. The modern compression technologies offer a solution to the problems providing very efficient tools for high image compression rates.

The ISO/IEC JPEG [1] is universally accepted as the standard "de facto" in the field of lossy compression, despite it is very dated. JPEG is a compression algorithm based on DCT transform. Every image, after decomposition in 8x8 blocks, is sent to the encoder, which in turns consists of three simple steps: a FDCT, that converts the 8x8 block from the spatial to frequency domain, a quantizer that eliminates less important information and an entropy encoder to reduce the amount of images needed.

The JPEG 2000 standard [2] uses a different transform algorithm based on wavelet and an innovative system for entropy coding named EBCOT that provides better

L. Atzori, J. Delgado, and D. Giusto (Eds.): MOBIMEDIA 2011, LNICST 79, pp. 227–241, 2012.

compression performance than JPEG at high compression ratios. Image compression both binary and continuous-tone, random access to the bit-stream, processing domain encoded and robustness to transmission errors are some of JPGE 2000 tools. Nevertheless, the new algorithm has not been as successful as hoped, despite all the novel functionalities.

The search for efficient representation for digital photographs is always alive and the ISO/IEC committee has recently completed the standardization process of the new JPEG XR image compression algorithm [3].

JPEG XR has been designed to manage the dynamics of modern acquisition sensors, offering a range of new features and benefits to the needs of digital photography. JPEG XR has been designed to reduce the limitations of the available formats that do not maximize the quality of stored data and fail to reach optimal performance according to the used devices. It allows representing multiple images into a single file, to decode only a part of an image, and crop, reduce in quality, tilt or rotate without having to decode the file. This compression algorithm has been made to minimize the encoder and decoder algorithmic complexity so as to obtain a minimal memory footprint. This is mainly due to the fact that the operations that are performed in encoding and decoding are very simple (basically it is a set of sum operations). This simplicity makes it very attractive for mobile imaging applications.

JPEG XR can process very large images by dividing them into tiles, each tile can be independently decoded. An analysis of JPEG XR coding efficiency can be found in [5]. It was evinced that for low bitrate, small tile sizes damage compression efficiency. Experimental results show that the goodness of tile size is closely linked to the bitrate. In particular, it is strongly recommended to use tile sizes 512x512 and above, while it is not advisable to use tile sizes below 64x64 when compression efficiency is very relevant.

A new architecture for JPEG XR encoding is described in [6]. The paper shows that there is no dependency information in the intra-macroblock; so every process of JPEG XR encoding (PCT/POT, Quantization, Prediction, Adaptive Scanning and Entropy Coding) may be pipelined. In the previous architectures the entropy coding was implemented in order to manage the dependency intra-blocks sequentially.

The significant progress in many aspects of digital technology, especially within the area of image acquisition, data storage, printing and display, have led to the creation of a large number of multimedia applications relating to digital images, such as image browsing.

It is in applications like these that the scalability plays a fundamental role. There are two types of scalability [7]: spatial and quality scalability. In the first scenario, it possible to zoom in and out the original image; in the second scenario it is possible to see the image at different quality levels.

A solution to navigate through the entire collection of large image databases is presented in [8]: a multi-modes and integrated image retrieval method consisting in combining direct search and browsing retrieval paradigms to obtain a third model called seamless interaction. In this way, the client can specify their needs with a simple query, resulting in a gain both in efficiency and in system speed.

A JPEG 2000 interactive browsing based on a client-server Vmedia architecture is presented in [9]. In this architecture, the user is allowed to choose and eventually change the region of interest (ROI) to decode. This means that only a portion of the data stream needs to be read, reducing the amount of data transferred. The Vmedia protocol is able to give a preview of the ROI, that becomes sharper as the remaining information arrive. There is the possibility to reuse the data delivered when the user decide to change its ROI, resulting in a reduction of load time. This approach is very useful for large images such as painting databases.

An image browsing application based on JPEG-XR is presented in [10]. Region of interest specified by the user are progressively downloaded and displayed with an approach aiming to minimize the transferred information between the request and the image presentation. Server side images are stored in frequency mode order and exploiting partitioning of images into tiles. This architecture provides user experience comparable to image browsing based on JPEG 2000. Nevertheless, results have shown that PSNR image quality using JPEG 2000 is higher than JPEG XR if multiple embedded bitrates are used.

A visual attention system for image browsing applications with large image database on mobile devices has been described in [11]. This algorithm defines a selective model of attention objects (face and text) within the image, based on eye movements, to show a specific region of interest. This method is also implemented in [12] to dynamically adjust the content of the image to the display size of different users.

A path for image browsing and searching for optimal path, based on the visual attention model [11,12] is defined in [13]. A set of new approaches to automate tasks of image browsing on mobile devices is also proposed. In [14] the visual attention model is examined spatially. There is a proposal of a new method for image browsing. In this case the detect ROI, first is decreasing ranking and then readapted on small displays.

An algorithm to identify the sufficient display resolution for image browsing is detailed in [15]. The authors use Kullback-Leibler (K-L) distance to determine data loss caused by down-scaling an image. The experiments, based on visual attention and spatial contrast masking, show that find the correct display resolution does not depend on users. Two algorithms are compared, those based on vision approach and those based on non-vision approach and although both achieve good performance, the first method gives better results than the second.

A technique to extract objects from large images and view them on small displays is described in [16]. The goal is to derive human faces in an image, thus making them the ROI image. The image is then segmented and will be measured saliency. So, there will be an adaption of the ROI and will start the image browsing. It is required a compatible aspect ratio to show ROI on small displays. Before, the search path was done manually; now is being made with an automated method of browse image.

This paper presents the design of an interactive high resolution image viewing architecture for mobile devices based on JPEG XR. Display resolution, resolution scalability, image tiling are investigated in order to optimize the coding parameters with the objective to improve the user experience. Experimental tests are performed on a set of large images and comparisons against accessing the images without parameter optimization are reported.

2   JPEG XR Overview

JPEG XR image transform [3] is based on a two-level hierarchical lapped transform [4], that is a concatenation of two flexible and independent transformation operators: FCT (Forward Core Transform) and OT (Overlap Transform).

In order to execute the transformation, an image JPEGXR is divided into 4x4 macroblocks, for each color plane. Every of them consist of 16 4x4 non-overlapping blocks, on each of which is applied the FCT transform, that produces 1 DC and 15AC coefficients of first stage for any block.

The DC coefficients of all blocks are grouped together in a 4x4 block and then a FCT transform is applied again.

So, this produces other 16 coefficients: 1 DC and 15 AC coefficients of second stage. Finally, this coefficients are mapped in the first pixel of each block that is a part of the macroblock. Fig. 1 shows the coefficients mapping of the two stages of FCT, that produce 240HP coefficients of first stage and 15LP and 1 DC coefficients of second stage.

1	2	3	4	5	6	7	8	9	10	11	12	13	14	15	16
17	18	19	20	21	22	23	24	25	26	27	28	29	30	31	32
33	34	35	36	37	38	39	40	41	42	43	44	45	46	47	48
49	50	51	52	53	54	55	56	57	58	59	60	61	62	63	64
65	66	67	68	69	70	71	72	73	74	75	76	77	78	79	80
81	82	83	84	85	86	87	88	89	90	91	92	93	94	95	96
97	98	99	100	101	102	103	104	105	106	107	108	109	110	111	112
113	114	115	116	117	118	119	120	121	122	123	124	125	126	127	128
129	130	131	132	133	134	135	136	137	138	139	140	141	142	143	144
145	146	147	148	149	150	151	152	153	154	155	156	157	158	159	160
161	162	163	164	165	166	167	168	169	170	171	172	173	174	175	176
177	178	179	180	181	182	183	184	185	186	187	188	189	190	191	192
193	194	195	196	197	198	199	200	201	202	203	204	205	206	207	208
209	210	211	212	213	214	215	216	217	218	219	220	221	222	223	224
225	226	227	228	229	230	231	232	233	234	235	236	237	238	239	240
241	242	243	244	245	246	247	248	249	250	251	252	253	254	255	256

DC coefficient   LP coefficient   HP coefficient

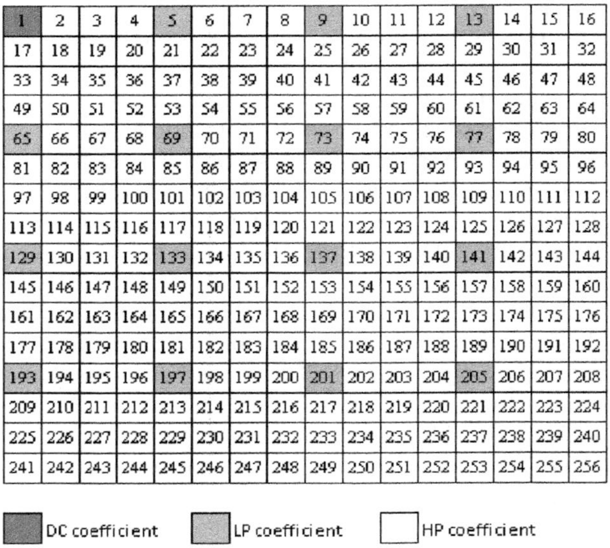

Fig. 1. JPEG-XR macroblock coefficient mapping

The FCT Transform, the core of the transform, consists of three elementary 2x2 filter operations:

- 2x2 Hadamard Transform: T2x2h;
- 1D rotate: TOdd;
- 2D rotate: TOddOdd.

FCT is applied to each 4x4 block in two different stage, as shown in Fig. 2. Each stage provides four 2x2 transform which may be done simultaneously or in a random sequence inside the stage. Anyhow, the second stage transform can initiate only if the first stage transform is completed.

The first stage of FCT includes four 2x2 Hadamard Transform (T2x2h): earlier is applied to corners (a, b, c, d) and then to the centers coefficients (e, f, g, h) of a 4x4 block; afterwards the T2x2h is applied to upper and lower edges (i, l, m, n) and finally to the right and left edges (o, p,q, r).

The second stage continues with a T2x2h for even-even basis (A, B, C, D), with a 1D rotation for even-odd basis (E, F, G, H) and odd-even basis (I, L, M, N) respectively and lastly with a 2D rotation for odd-odd basis (O, P, Q, R).

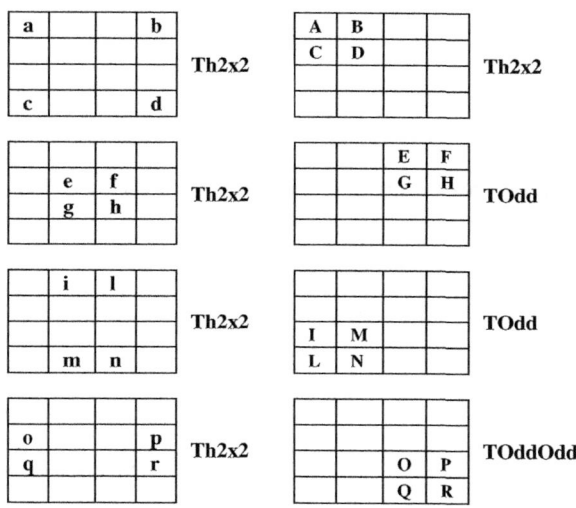

Fig. 2. Forward Core Transform steps

After the two stages, coefficients are re-ordered as shown in Fig.3.

i	Array[i]	i	Array[i]
0	0	8	1
1	8	9	11
2	4	10	15
3	6	11	13
4	2	12	9
5	10	13	3
6	14	14	7
7	12	15	5

Fig. 3. Forward Permutation

Every FCT operation is preceded by an optional filter OT (Fig.4), which is applied to a 4x4 areas, between the stacked two-dimensional blocks. (If OT_mode=0, no overlap operator is applied; if OT_mode=1 only the first level overlap is applied, otherwise if OT_mode=2 both level overlaps are performed). The overlap filter is designed to limit the blocking artifacts.

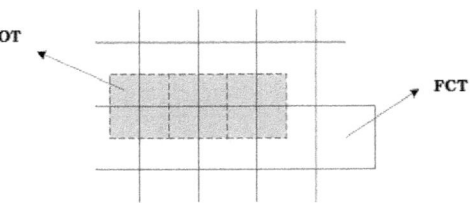

Fig. 4. Regions of support for the 4x4 FCT and OT operators [3]

The JPEG-XR decoder uses a block transform, called ICT, in which the stages are inverted. Every filter operations within the stages use its own inverse transform, and it is preceded by the inverse permutation function. The decoding process is summarize in Fig. 5.

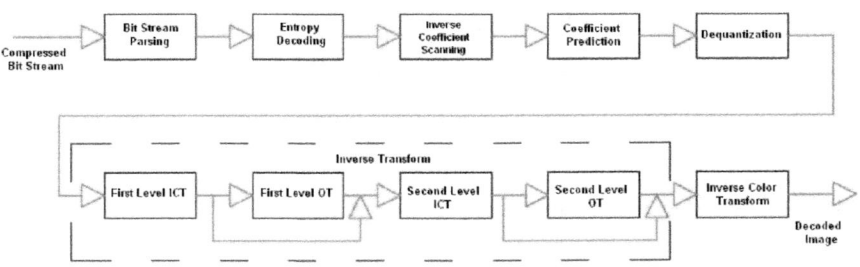

Fig. 5. Decoding process diagram [19]

The JPEG XR defines two approaches for access to the bit-stream: spatial and frequency. In both case, the bit-stream is composed by an image header followed by progressive tiles. In the first case the bit-stream of each tile is arranged in a macroblock order; in the second case the bit-stream of each tile is transmitted in multiple tile packets. It was considered the layout in the frequency mode, where the bit-stream of each tile is set up as a hierarchy of bands, as shown in Fig.6.

The tile coefficients are positioned in the following order: DC, LP, HP and FLEX band. FLEX provides additional information to HP band. The FLEX band may be not present.

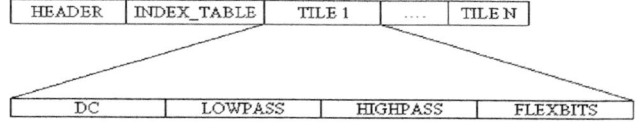

Fig. 6. Layout of JPEG XR bit-stream in frequency mode [3]

The group formed by DC, LP and HP sub-bands, produces different resolutions of the image information; while FLEX, if exists, can be used for progressive decoding.

The first scale of decoding can be obtained decoding the part of bit-stream that corresponds to DC, LP and HP coefficients.

It is then possible decode FLEX to produce the complete image.

A resolution equal to 1:16 is obtained decoding only the DC band. To obtain a 1:4 resolution of the image, only DC and LP bands need to be decoded.

3   Tiling

The process of partitioning a source image into rectangular non-overlapping blocks is called tiling. The JPEG 2000 standard compress each tile as an independent image from the other tiles. It implies that decoding any region of interest (ROI) inside a tile requires only the coded codestream of that tile. The tile are processed through the wavelet transform, (that can be reversible (RCT) or irreversible (ICT), to obtain different subbands, that are respectively LL,HL, LH and HH.

The JPEG XR standard is slightly different. A source image is divided into a grid of macroblock-aligned tiles, that enable fast local access. JPEG XR encoder can choose among three overlap filter operation, as described above: 1) non-overlapping that implies that no overlapping operation is performed; 2) one-level overlapping filter; 3) two-level overlapping filter. It is also possible decide whether to handle the tile boundaries in soft mode or in hard mode. In the first case, the overlapping filter is enabled within the tile and over the edges between the tiles. In the second case, the overlapping operation is applied only within the tile. In order to reconstruct a specific ROI, if the adjacent macroblock is placed in a distinct tile, it is necessary decode only a part of that tile.

Moreover, a structure for an optimized tile construction exists[17] and uses 256x256 ROI to reduce the overhead required to decode an image.

In the optimized case we can find one of these possibilities:

you want to decode a ROI that lies between two vertical tile;

or you want to decode a ROI that lies between two horizontal tile.

In the first condition, the ROI should be on a small case separated from the edge of the horizontal tile. In this way it is not necessary decode the tile that are above.

In the second occurrence the ROI should be crushed into two neighbors vertical tile boundaries for the same reason of the previous case.

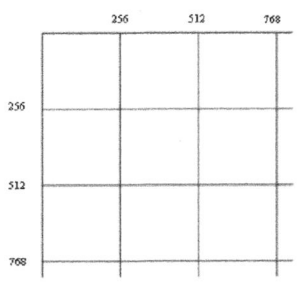
(a) JPEG 2000 e JPEG XR regular tiling

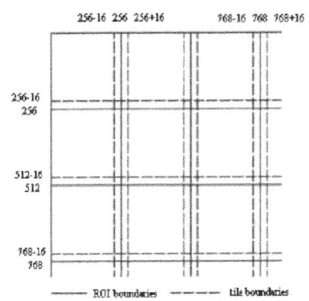
(b) JPEG XR Optimized Tile structure [17]

Fig. 7. Regular tiling (a) and optimized tiling (b)

4   Proposed Architecture

The proposed architecture is shown in Fig. 8. It is a client-server architecture for remote browsing applications for the packets exchange over the Web. The server has been implemented in C++ language and it is responsible for access management, allocation and release of resources. It stores a database of JPEG XR images and for each connection to a client user data are transferred by the means of the HTTP protocol.

At the server side JPEG XR images are stored with an optimized tile decomposition [17], in order to allow fast local access by the client device. An indexing tools analyses the JPEG XR codestream in order to extract the offset and the size of each subband. Indexes are stored in the index table for fast retrieval of the required information for the reconstruction of a ROI requested by the client. Index table is managed by the functional Index block in Fig. 8. This block has also the purpose of keeping track of the information already available at the client side and avoiding retransmission of the same data. The HTTP server receive incoming request and provide the chunks of data required by the HTTP client.

The client constructs the request on the basis of the current user view. In particular the display resolution and the level of zooms are used for determining what are the needed chunks of information and the corresponding indexes.

The proposed concept is slightly different from the concept used in the interactive protocol JPIP. In fact in the proposed architecture the computation of required chunk indexes is performed at client side, minimizing the information transferred for the request to a vector of index corresponding to the chunk number needed for display the current view to the user.

Moreover it is not needed to have an arbitrary ROI access but simplifying as much as possible the user interaction with the process of viewing high quality high resolution large images with mobile devices.

HTTP client receives incoming packet from the HTTP server containing JPEGXR image subband that are stored into the local cache. The block composer prepare the JPEG XR file merging the required chunks into a correct JPEG XR format in order to be decoded by the JPEG XR block.

The viewport block keep tracks of the visible portion of the image which is larger than the visualization device.

The user interacts with the display with classical image viewing operation such as zooming and panning.

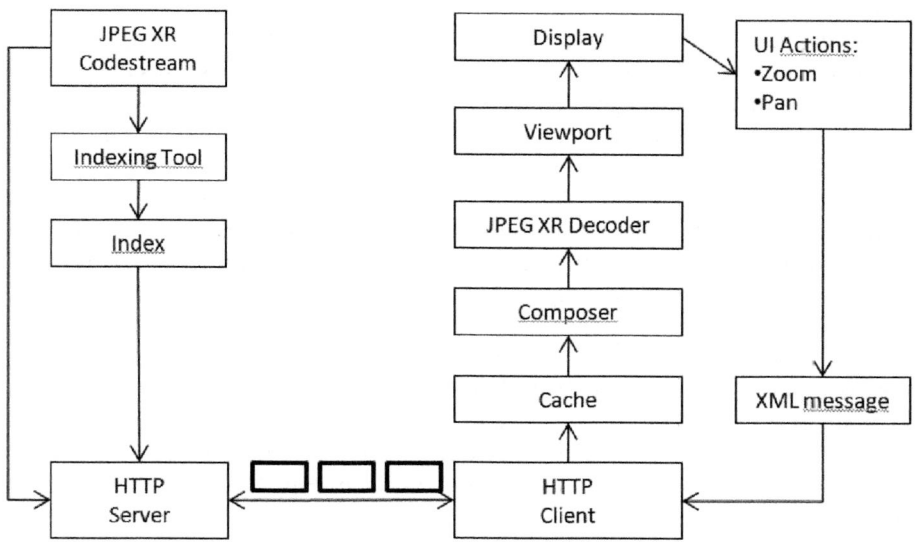

Fig. 8. Reference Architecture

5   Experimental Tests

In this section, we describe the tests performed and their results. Tests were conducted using the JPEG XR reference software and the KAKADU software [20]. The use case is as follows: once the desired ROI at a given resolution has been selected, the client must be able to zoom in and out the image that appears in his own device, or make horizontal or vertical scrolls of the current view of the image.

It is possible to request only some subset of the available sub-band coefficients. In fact, in the frequency mode, each sub-band can independently be decoded. All tiles of a particular sub-band are merged together in a unique data packet.

This allows creating smaller image previews using a resolution that fits the device used to load the image.

The client can request only the DC level transmission, if a low quality of image suffice. It is possible to increase the image quality, transmitting progressively all available sub-bands.

DC+LP+HP+FLEX sub-bands transmission ensures maximum image quality.

Removing irrelevant information obviously causes a loss of data image that must be quantified in some way. There are subjective and objective techniques. The subjective metric depends on the experience of the observers, while the objective metric consists to calculate the difference in statistical distribution of pixel values in digital images; in this way it is possible to quantify the distortion of the compressed image with respect to the original. The PSNR is the most widely used image objective metric.

The bitrate vs PSNR shows the efficiency of a compression algorithm: high bitrates provide a higher quality of the compressed data.

Several tests were performed using the three different overlapping filters and distinguishing between hard, soft and optimized mode. These results are then compared with those obtained with JPEG 2000 tests.

Image request at difference resolution level have been analyzed in order to report the bit-rate of the data transferred at each requests.

The following figures show the experimental results. Each figure contains only the average of the results of the most significant images used as a test set.

Experimental tests were carried out by defining three mode of encoding JPEG XR (soft, hard and optimized) and using the three OT mode (0,1,2). Subbands considered are: 1) only DC, 2) DC+LP, 3) DC+LP+HP, 4) DC+LP+HP+FLEX.

JPEG2000 codestream were produced in relation to JPEG XR bitrate, to make a comparison between the two coding algorithms.

In the first JPEG XR case (Fig. 9 (a-b-c)) the solution without any level of overlap was considered; in the second JPEG XR case (Fig. 10 (a-b-c)) only one overlapping filter was applied, while in the latter JPEG XR case (Fig. 11 (a-b-c)) both stages of overlapping have been taken into account. In all three cases there was a distinction between soft (a), hard (b) and optimized (c) mode, in order to compare this three methods.

As expected, experimental tests, in terms of bpp, show that in the case with L = 0 hard, soft and optimized mode show similar results. In the case with L= 1 soft and optimized mode show similar results, while the bpp slightly decreases in hard mode.

Finally, in the case with L= 2 optimized mode performance in terms of bpp presents results that are intermediate between those obtained with hard and soft mode.

In all experimental JPEG XR tests, therefore, hard tiles outperform soft tiles and optimized tile; but the results obtained with the tile optimization, in terms of overhead, are very acceptable if compared with those of JPEG 2000.

Fig. 12 shows the JPEG 2000 results obtained in the case of image subdivision in tiles of 256x256 size, without any type of optimization.

It is possible to notice that the JPEG 2000 bpp are very similar to those obtained with JPEG XR experiments, but in JPEG2000 we can see a gradual increase of bpp as bitrate increased, even if only DC.

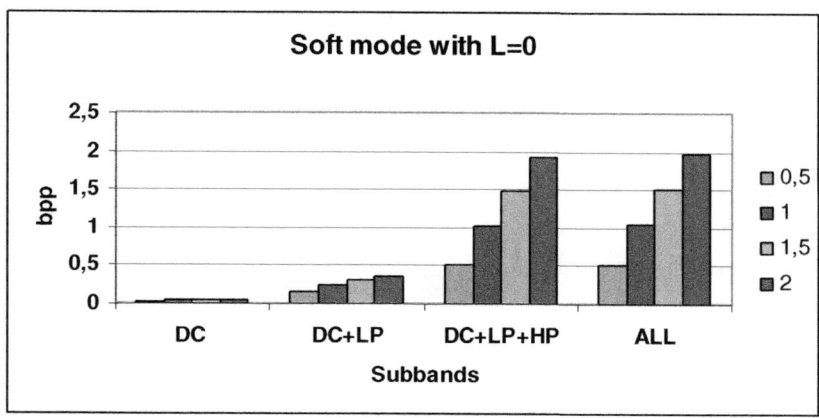

Fig. 9 (a). Bitrate for each JPEG XR subband for four target bitrates (soft mode, L=0)

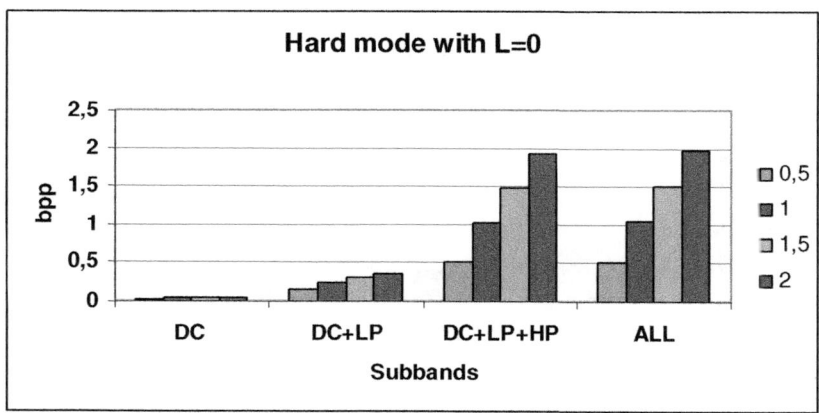

Fig. 9 (b). Bitrate for each JPEG XR subband for four target bitrates (hard mode, L=0)

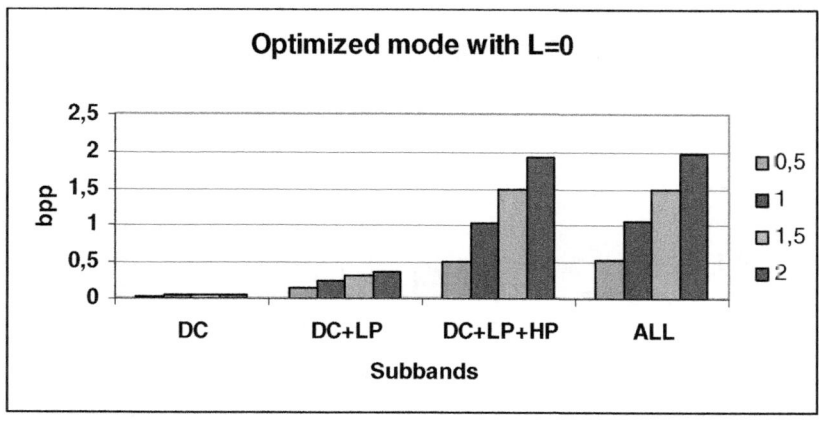

Fig. 9 (c). Bitrate for each JPEG XR subband for four target bitrates (optimized mode, L=0)

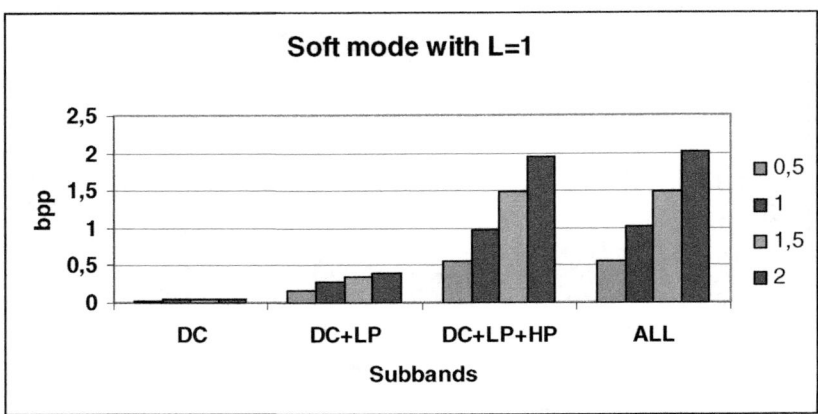

Fig. 10 (a). Bitrate for each JPEG XR subband for four target bitrates (soft mode, L=1)

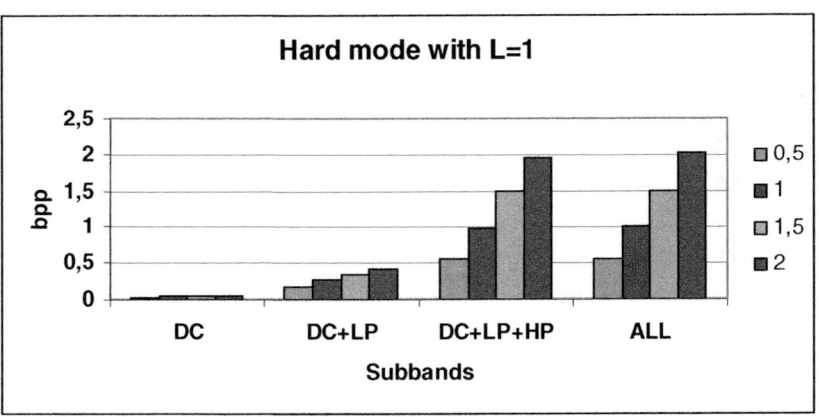

Fig. 10 (b). Bitrate for each JPEG XR subband for four target bitrates (hard mode, L=1)

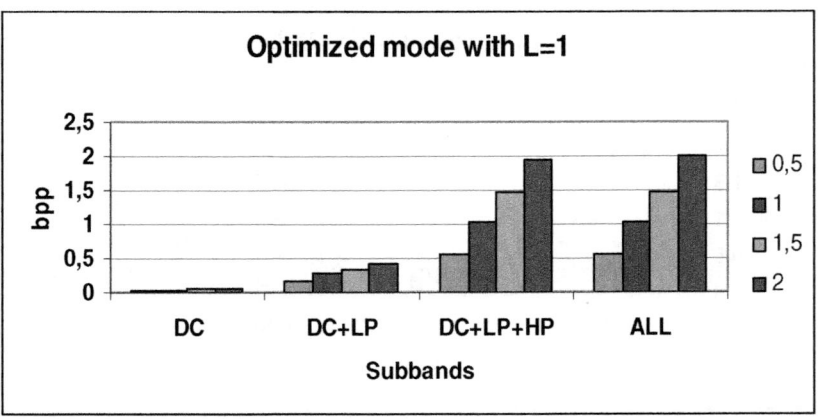

Fig. 10 (c). Bitrate for each JPEG XR subband for four target bitrates (optimized mode, L=1)

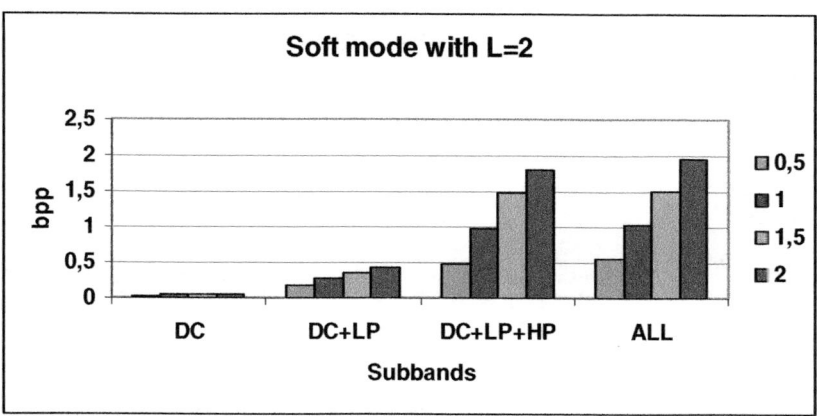

Fig. 11 (a). Bitrate for each JPEG XR subband for four target bitrates (soft mode, L=2)

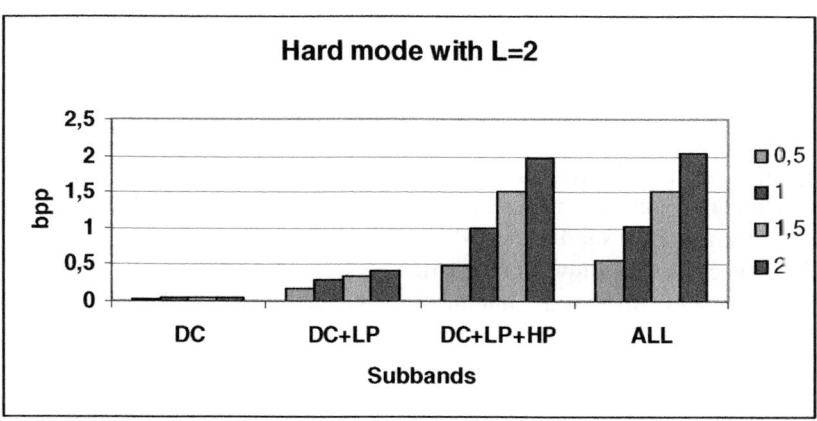

Fig. 11 (b). Bitrate for each JPEG XR subband for four target bitrates (hard mode, L=2)

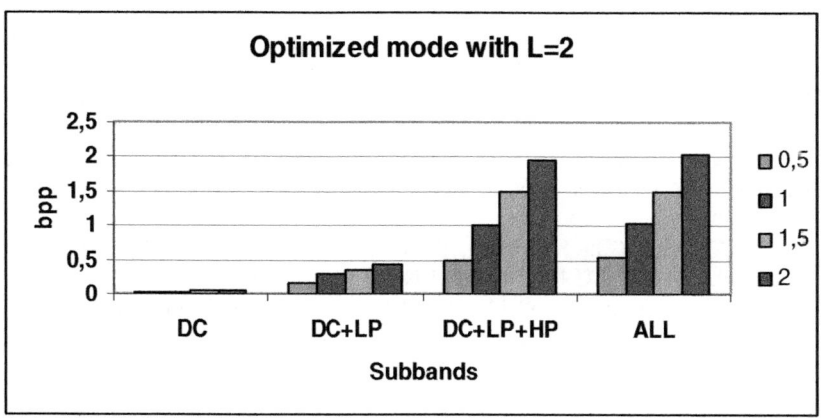

Fig. 11 (c). Bitrate for each JPEG XR subband for four target bitrates (optimized mode, L=2)

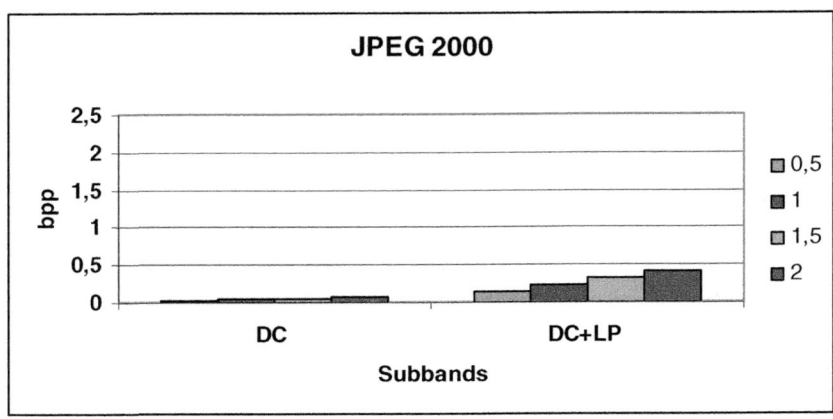

Fig. 12. Bitrate for each JPEG 2000 subband for four target bitrates

6   Conclusions

An architecture for remote browsing of large images coded using fast local access means provided by JPEG XR has been presented. In order to minimize the transferred information, the JPEG XR coded file format should make use of the frequency mode order and partitioning of images into tiles. The main goal is transmitting only some subset of the available sub-band coefficients.

This is necessary to allow an interactive access to portion of images (ROI), that are downloaded and displayed, minimizing the amount of data transferred and maintaining an acceptable image quality for the user experience.

References

1. Wallace, G.K.: The JPEG still picture compression standard. Communications of the ACM 34(4), 31–44 (1991)
2. Skodras, A., Christopoulos, C., Ebrahimi, T.: The JPEG 2000 still image compression standard. IEEE Signal Processing Magazine 18(5), 36–58 (2001)
3. ITU-T Rec. T.832 | ISO/IEC 29199-2, Information technology – JPEG-XR image coding system – Image coding specification, http://www.itu.int/rec/T-REC-T.832
4. Malvar, H.S.: Efficient signal coding with hierarchical lapped transforms. In: Proc. IEEE Int. Conf. on Acoustics, Speech, and Signal Processing, Albuquerque, pp. 1519–1522 (April 1990)
5. Tu, C., Sullivan, G.J., Srinivasan, S.: Effect of tile size on JPEG XR compression efficiency, ISO/IEC JTC 1/SC 29/WG 1 N4980, Japan (April 2009)
6. Hattori, K., Tsutsui, H., Ochi, H., Nakamura, Y.: A High-Throughput Pipelined Architecture for JPEG-XR Encoding. In: IEEE/ACM/IFIP 7th Workshop on Embedded Systems for Real-Time Multimedia, ESTIMedia 2009 (2009)
7. Jain, A., Panchanathan, S.: Scalable Compression for Image Browsing. IEEE Consumer Electronics 40(3) (August 1994)

8. Pecenovic, Z., Do, M.N., Vetterli, M., Pu, P.: Integrated Browsing and Searching of Large Image Collections. In: Laurini, R. (ed.) VISUAL 2000. LNCS, vol. 1929, pp. 279–289. Springer, Heidelberg (2000)

9. Li, J., Sun, H.-H.: On interactive browsing of large images. IEEE Transactions on Multimedia 5(4), 581–590 (2003)

10. Perra, C., Giusto, D.D.: An image browsing application based on JPEG-XR. In: International Workshop on Content-Based Multimedia Indexing, June 18-20, pp. 396–401 (2008)

11. Fan, X., Xie, X., Ma, W.Y., Zhang, H.J., Zhou, H.Q.: Visual attention based image browsing on mobile devices. In: Proceedings of ICME, Baltimore, MD, vol. I, pp. 53–56 (2003)

12. Chen, L.Q., Xie, X., et al.: A visual attention model for adapting images on small displays. ACM Multimedia Systems Journal 9(4), 353–364 (2003)

13. Xie, X., Liu, H., Ma, W.-Y., Zhang, H.-J.: Browsing large pictures under limited display sizes. IEEE Transactions on Multimedia 8(4), 707–715 (2006)

14. Hasan, M.A., Kim, C.: An Automatic Image Browsing Technique for Small Display Users. In: ICACT 2009: Proceedings of the 11th International Conference on Advanced Communication Technology, vol. 3 (2009)

15. Fan, X., Xie, X., Ma, W.-Y.: Detecting the sufficient display resolution for image browsing. In: 7th International Conference on Mobile Data Management (MDM 2006), p. 22 (2006)

16. Liu, H., Jiang, S., Huang, Q., Xu, C., Gao, W.: Region-based visual attention analysis with its application in image browsing on small displays. In: Proceedings of the 15th International Conference on Multimedia, pp. 305–308 (2007)

17. Tu, C., Sullivan, G.J., Srinivasan, S.: Optimizing JPEG XR tile structure for fast local access, Microsoft Corporation, Redmond, USA

18. Christopoulos, C., Skodras, A., Ebrahimi, T.: The JPEG 2000 Still Image Coding: An Overview. IEEE Transactions on Consumer Electronics 46(4), 1103–1127 (2000)

19. Dufaux, F., Sullivan, G.J., Ebrahimi, T.: The JPEG-XR image coding standard. IEEE Signal Processing Magazine 26(6), 195–199, 204–204 (2009)

20. Kakadu Software Home, http://www.kakadusoftware.com

The Use of Selected Transforms
to Improve the Accuracy of Face Recognition
for Images with Uneven Illumination

Tomasz Marcin Orzechowski[1], Andrzej Dziech[1],
Tomasz Lukanko[1], and Tomasz Rusc[2]

[1] Department of Telecommunications, AGH University of Science and Technology,
Al. Mickiewicza 30, 30-059 Krakow, Poland
[2] Institute of Physics, The Jan Kochanowski University of Humanities and Sciences
ul. Zeromskiego 5, 25-369 Kielce, Poland
tomeko@agh.edu.pl, dziech@kt.agh.edu.pl,
lukanko@student.agh.edu.pl, tomasz.rusc@ujk.edu.pl

Abstract. This paper presents new methods of the illumination normalization in images preprocessed for face recognition system. The main problem in statistical methods of face recognition is illumination. Different lighting conditions between photos taken indoor and outdoor may drastically decrease the level of correct classification. Variations of the illumination lie mostly in low-frequency band, so it is possible to use several transforms operating on frequency domain of an image. This approach is to truncate appropriate number of coefficients in frequency domain to minimize variations under different lighting conditions. This paper presents methods using transforms such as: Two Dimensional Discrete Cosine Transform type II (2D-DCT-II) and two Periodic Piecewise-Linear Transforms, such as: Periodic Haar piecewise Linear Transform (PHL) and Periodic Walsh piecewise-Linear Transform PWL. The main advantage of this approach is that, it does not require any modeling steps and it can be implemented in real-time face recognition systems.

Keywords: Discrete Transforms, Discrete Cosine Transform, DCT, Periodic Walsh piecewise Linear Transform, PWL, Periodic Haar piecewise Linear Transform, PHL, Image preprocessing, Face recognition, Illumination reduction, Illumination normalization.

1 Introduction

Variable illumination is one of the most important problems in face recognition. Lighting condition change between indoor and outdoor environments, so one of main aims for face recognition researchers in wise understood computer vision is to create face recognition system insensitive for illumination changing.

L. Atzori, J. Delgado, and D. Giusto (Eds.): MOBIMEDIA 2011, LNICST 79, pp. 242–251, 2012.
© Institute for Computer Sciences, Social Informatics and Telecommunications Engineering 2012

The article deals with spectral analysis of an image, so all modifications are done on the transformed image into the frequency domain. The approach presented in this paper can be presented in the following three steps:

- Image values normalization;
- Transformation of image into logarithm and frequency domain and truncation appropriate number of coefficients;
- Computing the inverse transform and truncate a histogram of an image.

2   Image Values Normalization

Four following steps in normalization process could be distinguished:

- Subtraction the minimal value of an image from all pixels;
- Computing the difference between minimum and maximum value of an image;
- Computing the division of results of step I and II;
- Multiplication results of step III by 255.

To normalize an image, it is necessary to compute the minimal value of the whole image. Then subtract this value from all pixels. This operation leads to obtain '0' as a minimal value in whole image.

Next, it is necessary to compute value compound of subtraction maximum and minimum value of the whole image. Then, divide the results of both steps: I and II.

The last step is just to multiply the results of the second step by 255. It is necessary to get images normalized. The new image contains values between 0 and 255. All these operations can be represented by the following formula:

$$[IMG_{norm}] = \frac{[IMG] - [IMG_{min}]}{[IMG_{max}] - [IMG_{min}]} * 255 \tag{1}$$

Where,

$[IMG]$ represents an image matrix

$[IMG_{min}]$ represents a matrix, which has the same size as an image matrix and all values of this matrix are equal to the minimum value of an image matrix.

$[IMG_{max}]$ represents a matrix, which has the same size as an image matrix and all values of this matrix are equal to the maximum value of an image matrix.

$[IMG_{norm}]$ represents a normalized image matrix.

3   Discrete Transforms

In this section, all used transforms will be presented in detail: DCT, PHL and PWL.

3.1   Discrete Cosine Transform (DCT)

DCT is a Fourier related transform, similar to Discrete Fourier Transform.

There are eight types of this transform. The most common is type II, which is described by equation 2.

$$X_{DCT}(k,l) = \sum_{m=0}^{M-1}\left[\sum_{n=0}^{N-1} x(m,n)\beta(l)\cos\left(\frac{\pi l}{N}\left(n+\frac{1}{2}\right)\right)\right]\alpha(k)\cos\left(\frac{\pi k}{M}\left(m+\frac{1}{2}\right)\right)$$

$$(2)$$

Where,

m, n represent rows and columns of image.

$$\alpha(k) = \begin{cases} \sqrt{\dfrac{1}{M}}, k = 0 \\[4mm] \sqrt{\dfrac{2}{M}}, \quad k = 1 \dots M-1 \end{cases}$$

$$(3)$$

$$\beta(l) = \begin{cases} \sqrt{\dfrac{1}{N}}, \quad l = 0 \\[4mm] \sqrt{\dfrac{2}{N}}, \quad l = 1 \dots N-1 \end{cases}$$

$$(4)$$

The Two Dimensional Cosine Transform – II is a compound of two series one-dimensional cosine transform [2][3][4].

3.2   Periodic Walsh piecewise-Linear Transform (PWL)

The Periodic Walsh piecewise-Linear PWL functions, which are the basis functions of the PWL transform, are obtained by integrating periodic Walsh functions as follows [4][5]:

$$PWL(0,t) = 1, \qquad t \in (-\infty, +\infty)$$

$$(5)$$

$$PWL(i,t) = \frac{2^{k+1}}{T}\int_{mT}^{t+mT} Wal(i,\tau)d\tau$$

$$(6)$$

Where $i = 1,2,\dots,N-1$, $k = 1,2,\dots,\log_2 N$, $m = 0,1,2,\dots$,k is the group index of PWL functions and m is the number of period.

The matrix form of the forward and inverse PWL transforms may be formulated as follows:

Forward transform: $[C(N) = [-2^{-(k+1)}][PWL(N)][X(N)]$

$$(7)$$

Inverse transform: $[X(N)] = [IPWL(N)][C(N)]$

$$(8)$$

Where:

    [C(N)]   is a vector of PWL coefficients;
    [X(N)]   is a vector of sampled signal;
    [PWL(N)]   is a matrix of forward transform;
    [IPWL(N)]   is a matrix of inverse transform;
    [-2-(k+1)]   is a diagonal matrix of normalization.

3.3   Periodic Haar piecewise Linear Transform (PHL)

The set of Periodic Haar piecewise-Linear (PHL) functions [5] is obtained by integrating the well-known set of Haar functions. The set of PHL functions is defined by:

$$PHL(0, t) = 1, \qquad t \in (-\infty, +\infty) \tag{9}$$

$$PHL(1, t) = \left[\frac{2}{T} \int_{mT}^{t+mT} Har(i, \tau) d\tau \right] + \frac{1}{2} \tag{10}$$

$$PHL(i + 1, t) = \frac{2^{k+1}}{T} \int_{mT}^{t+mT} Har(i + 1, \tau) d\tau \tag{11}$$

Where $i = 1, 2, \dots, N - 2$, $k = 1, 2, \dots, (\log_2 N) - 1$, $m = 0, 1, 2, \dots$, k is the group index of PWL functions and m is the number of period. The matrix form of the forward and inverse PWL transforms may be formulated as follows:

Forward transform: $[C(N) = \left[\frac{-1}{2^{-(k+1)}} \right] [PHL(N)][X(N)]$ (12)

Inverse transform: $[X(N)] = [IPHL(N)][C(N)]$ (13)

Where,

    [C(N)]   is a vector of PWL coefficients;
    [X(N)]   is a vector of sampled signal;
    [PHL(N)]   is a matrix of forward transform;
    [IPHL(N)]   is a matrix of inverse transform;
    $[\frac{-1}{2^{-(k+1)}}]$ is a diagonal matrix of normalization.

4   Algorithm of Normalization

The normalization process described in this section bases on the approach presented in [9][10] allows reducing the shadow effect.

Figure 1 presents an example image, where half of the face is covered by the shadow, whilst Figures 2,3,4 present results of transforms: PHL, DCT, PWL made on the example image.

Fig. 1. The example image, where half of the face is covered by the shadow

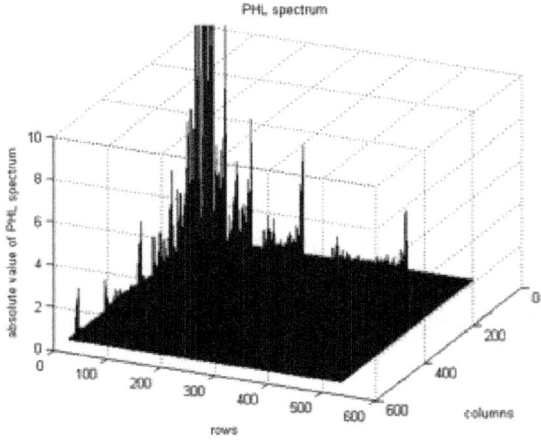

Fig. 2. Spectrum of Periodic Haar Piecewise-Linear Transform (PHL)

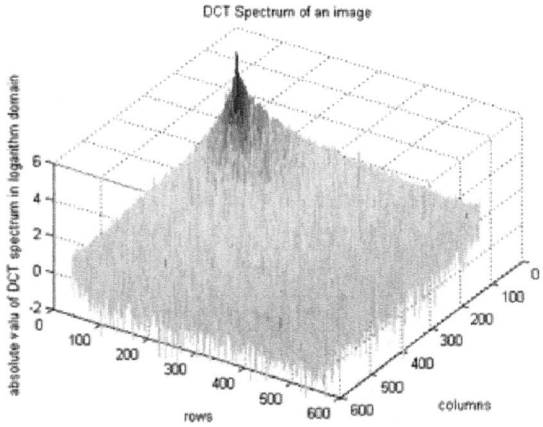

Fig. 3. Spectrum of Two Dimensional Discrete Cosine Transform (DCT)

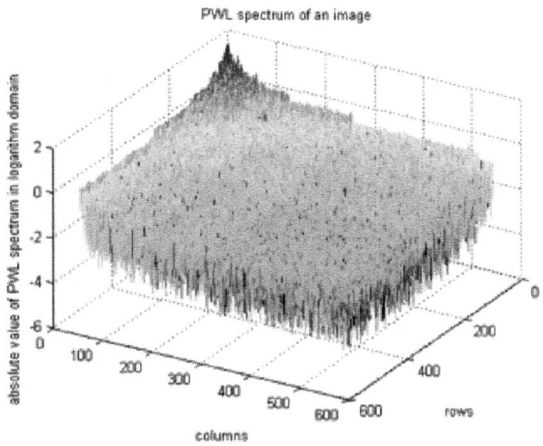

Fig. 4. Spectrum of Periodic Walsh Piecewise-Linear Transform (PWL)

It is easy to see that the most significant coefficients are placed in the upper corner. All values quickly diminish with the distance from the corner. This fact concerns all presented types of transforms.

Three discrete transforms can be used in presented algorithm. The Discrete Cosine Transform is popular and commonly used in many cases, also in illumination reduction algorithms.. However Periodic Walsh Piecewise-Linear Transform and Periodic Haar Piecewise-Linear Transform have been never us in such case. Illumination reduction is a new issue where these two transforms can be implemented.

It is possible to specify some steps in algorithm of illumination reduction:

Step 1. The first step of the normalization process is to read consecutive coefficients in specific way [7][8].

Approach presented in Figure 5 shows the way to read particular coefficients of an image. It is necessary to start reading coefficients from the biggest to the smallest. Starting reading in the left upper corner allows obtaining firstly the most significant coefficients. Smaller (low-frequency) values will be read after bigger (high-frequency) values of the transform. Discarding the low- frequency transform's coefficients in the logarithm domain is identical to compensate the illumination variations.

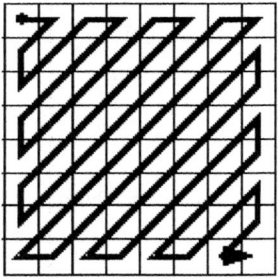

Fig. 5. Scheme how to read transform's coefficients (zigzag scan)

Step 2. The first coefficient describes overall illumination of an image. It is worth to change it in specific way. Firstly the average value of the image in gray scale must be computed. Then logarithm of the average value must be multiplied by a square root of amount of all pixels in the image.

All this computation could be presented by a given equation:

$$IMG(1,1) = \log(\alpha) \cdot \sqrt{M \cdot N} \tag{14}$$

Where α is an average level of gray scale of an image, M and N describe size of an image.

Step 3. The last proposed step is to change values to zero of some low-frequency coefficients, which are also strongly connected with illumination [1][2].

A particular image determines how many coefficients should be changed to zero value. It depends on quality of an image, resolution, etc.

5   Results

Figure 6 presents example of results of all described methods in this paper. In the figure 6a) there is presented original image where half of the face is covered in the shadow. Figures 6b), 6c), 6d) present results of illumination reduction.

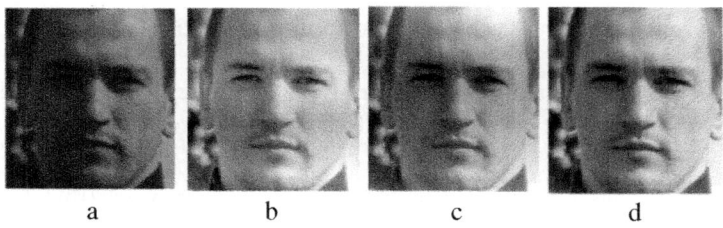

a   b   c   d

Fig. 6. a) original image, b) normalization using DCT, c) normalization using PHL, d) normalization using PWL

Presented methods are appropriate to decrease differences between indoor and outdoor light conditions.

Figure 7 presents another example of results. A problem presented in the Figure 7a) is dark image. Figures 7b), 7c), 7d) present results of using described transforms.

As it is presented in the Figure 7d) *Periodic Walsh Piecewise-Linear Transform* gives better results than well-known *Two Dimensional Discrete Cosine Transform*.

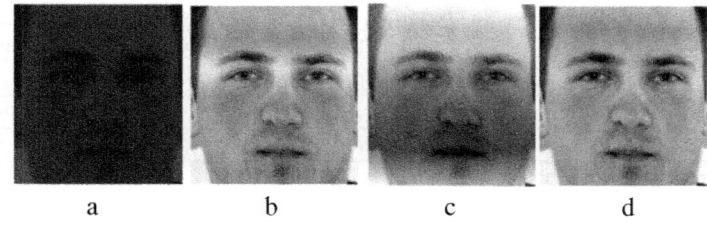

a   b   c   d

Fig. 7. a) original image, b) normalization using DCT, c) normalization using PHL, d) normalization using PWL

Results of Illumination Reduction Algorithm in Face Recognition System
Article [6] presents works about statistical face recognition algorithms.

Tests described in [6] were repeated with illumination reduction algorithm. Tests consider two statistical algorithms Principle Component Analysis and Linear Discriminant Analysis and also a special fusion of these two algorithms presented in [6].

Following tests were carried out and described in this article:

- Single classifier (without normalization)
- Fusion of LDA & PCA (without normalization)
- Single classifier (normalization)
- Fusion of LDA & PCA (without normalization)

Tables 1-3 present results of correct classification in face recognition system. Tests were done on AT&T database. Data set was not divided in any subsets.

The column "First Place" describes the percentage of correct classification in particular algorithms in the way that correct person was in the nearest image, whilst the column "First Six Places" describes the percentage of correct classification in a particular algorithm in way that correct person was in one of six nearest images.

Tests prove that in this case PWL transform method gives better results than DCT method.

It must be noticed that results depends on database and it is possible that in different data base other transform e.g. PHL transform gives better results.

Table 1. Results of correct classification in face recognition system (PWL transform)

	Single classifier without normalization		Fusion of LDA & PCA without normalization	Single classifier with normalization		Fusion of LDA & PCA without normalization
	PCA	LDA	PCA & LDA	PCA	LDA	PCA & LDA
First Place	80,25%	82%	86%	82.75%	84.75%	88,25%
First six places	89%	90,25%	91,25%	91%	92,25%	93%

Table 2. Results of correct classification in face recognition system (DCT transform)

	Single classifier without normalization		Fusion of LDA & PCA without normalization	Single classifier with normalization		Fusion of LDA & PCA without normalization
	PCA	LDA	PCA & LDA	PCA	LDA	PCA & LDA
First Place	80,25%	82%	86%	82.5%	84%	87,25%
First six places	89%	90,25%	91,25%	90,75%	91,75%	92%

Table 3. Results of correct classification in face recognition system (PHL transform)

	Single classifier without normalization		Fusion of LDA & PCA without normalization	Single classifier with normalization		Fusion of LDA & PCA without normalization
	PCA	LDA	PCA & LDA	PCA	LDA	PCA & LDA
First Place	80,25%	82%	86%	79%	80%	81%
First six places	89%	90,25%	91,25%	83%	83,5%	85%

6   Conclusions

The aim of works was to examine commonly used algorithms of reduction of the illumination influence and try to implements new methods of illumination reduction for face recognition systems.

Three methods, presented in the article, which use transforms such as: Two Dimensional Discrete Cosine Transform type II (2D-DCT-II) and Periodic Piecewise-Linear Transforms: Periodic Haar piecewise Linear Transform (PHL) and Periodic Walsh piecewise-Linear Transform (PWL) were implemented and tested.

To compare this works to other i.e. article [1] it is worth to notice this article describes spectral modifications of an image, but using only Discrete Cosine Transform. Article [2] is a comparative study among several preprocessing algorithms, but the idea of using Periodic Haar piecewise Linear Transform (PHL) and Periodic Walsh piecewise-Linear Transform (PWL), presented in this article is new.

Tests prove that the use of PHL and PWL transforms gives better results than 2D-DCT in many cases. The tests are fully reproducible.

Methods presented in the article can be used as a powerful addition to face recognition systems. It is worth to add that presented algorithms are fast, so presented methods are implemented in real-time face recognition system.

The plans for future are to integrate these normalization methods with race recognition system to increase the level of correct classification.

Acknowledgments. This work has been financed by the European Regional Development Fund under the Innovative Economy Operational Programme, INSIGMA project no. POIG.01.01.02-00-062/09.

References

1. Chen, W., Er, M.J., Wu, S.: Illumination Compensation and Normalization for Robust Face Recognition Using Discrete Cosine Transform in Logarithm Domain. IEEE Trans. on Systems, Man, and Cybernetics, part B: Cybernetics 36(2), 458–466 (2006)
2. del Solar, J.R., Quinteros, J.: Illumination Compensation and Normalization in Eigenspace-based Face Recognition: A comparative study of different pre-processing approaches
3. Zieliński, T.: Cyfrowe przetwarzanie sygnałów – in Polish (ang. Digital Signal Processing) WKŁ Warszawa, pp. 663–664 (2009)
4. Dziech, A., Ślusarczyk, P., Tibken, B.: Methods of Image Compression by PHL Transform. Journal of Intelligent and Robotic Systems 39, 447–458 (2004)
5. Dziech, A., Belgassem, F., Nern, H.J.: Image Data Compression using Zonal Sampling and Piecewise-Linear Transforms. Journal of Intelligent and Robotic Systems 28, 61–68 (2000)
6. Dziech, A., Orzechowski, T., Łukańko, T.: Testing fusion of LDA and PCA algorithms for face recognition with images preprocessed with Two-Dimensional Discrete Cosine Transform. In: Proc. of The 15th WSEAS Conference on CIRCUITS, Corfu, Greece (2011)
7. Mendonça, A., Magalhães, M.: Illumination Normalization Methods for Face Recognitino. In: Proc. of the 20th Brazilian Symposium on Computer Graphics and Image Processing, SIBGRAPI 2007 (2007)
8. Savvides, M., Kumar, V.: Illumination Normalization using Logarithm Transforms for Face Authentication. In: Kittler, J., Nixon, M.S. (eds.) AVBPA 2003. LNCS, vol. 2688, pp. 549–556. Springer, Heidelberg (2003)
9. Adini, Y., Moses, Y., Ullman, S.: Neurobiol. Dept., Weizmann Inst. of Sci., Rehovot, Face recognition: the problem of compensating for changes in illumination direction. IEEE Transactions on Pattern Analysis and Machine Intelligence, 721–732 (July 1997)
10. Sanderson, C., Paliwal, K.K.: Fast features for face authentication under illumination direction change. Pattern Recognition

Objective Evaluation of WebP Image Compression Efficiency

Maurizio Pintus, Giaime Ginesu, Luigi Atzori, and Daniele D. Giusto

Department of Electronic Engineering, University of Cagliari, Italy
m.pintus@gmail.com, {g.ginesu,l.atzori}@diee.unica.it,
ddgiusto@unica.it

Abstract. Performances of multimedia coding techniques are still improving in terms of compression ratio, coding features, and robustness against errors even if at a slower pace with respect to what we were used to up a decade ago. One of the latest codec which is expected to improve on the state of the art is the WebP algorithm released by Google. With the intent to evaluate the extent of this improvement, in this paper we provide an objective evaluation of the compression efficiency of WebP, by comparing it with alternative algorithms. From the results it appears that the performance of the proposed codec is in line with that of the alternative methods, without achieving any major improvement and lacking several features.

Keywords: Image compression, lossy coding, codec assessment.

1   Introduction

The compression of still images has undergone a significant improvement in the past decades. In the nineties, compression ratios experienced an increase from a 2:1 – 3:1 factor, with lossless entropy coders, to 20:1 and more thanks to the lossy JPEG standard. A decade later, a further 20% increase has been achieved though JPEG 2000. Improvements also involved the development and support for advanced features, such as progressive and lossless to lossy coding, multi-channel and HDR support or region of interest coding. Nowadays, the research and development community is mainly focused on moving pictures, as an extension of still image coding, whereas compression efficiency is improving at a slower pace. The success of a new compression technology then depends on both its performances and features, and is deeply influenced by other commercial factors, such as the presence of patents or licensing royalties and the support in major software packets. Nonetheless, each time such new technology is submitted to the attention of the community, there is the need to evaluate it. The evaluation is performed through comparative studies with existing technologies to test the compression efficiency achieved by the proposed coding algorithm, its computational complexity, and any additional functionalities. An example of such activity can be found in [1].

L. Atzori, J. Delgado, and D. Giusto (Eds.): MOBIMEDIA 2011, LNICST 79, pp. 252–265, 2012.
© Institute for Computer Sciences, Social Informatics and Telecommunications Engineering 2012

Compression efficiency expresses the ability of the coding algorithm to maximize the visual quality of a compressed image versus the number of bits used to represent it. Subjective evaluation consists in collecting quality statistics from a sample of users feedbacks and are expensive and time consuming. On the other hand, objective quality assessment makes use of computer algorithms in order to automatically estimate the perceived visual quality.

In this paper, we focus on the compression efficiency evaluation of the new image format from Google, WebP [2]. Released in late September 2010, WebP is a lossy compression algorithm to be used on photographic images, which features predictive coding and exploits variable block sizes. It is reported to offer 39.8% more byte-size efficiency than JPEG for the same quality. Performance evaluation is accomplished by comparing the results of WebP with three state of the art image compression formats (JPEG, JPEG 2000, JPEG XR) in terms of two objective quality metrics (PSNR and SSIM). The paper is organized as follows: in Section 2 an overview of the competing coding algorithms and the objective quality metrics is provided, Section 3 provides the results, while in Section 4 conclusions are drawn.

2   Background

This section briefly illustrates the coding algorithms under analysis and the quality metrics used in the assessment.

2.1   Coding Algorithms

JPEG dates back to 1990, when the International Standard Organization created the Joint Photographic Experts Group with the task of developing an international compression standard for still pictures. The resulting standard was published in 1993 under the reference ISO/IEC 10918. JPEG compression can be described in six main steps: 8×8 pixels block decomposition, discrete cosine transform, thresholding and quantization, zig-zag scan, run length coding and variable length coding. JPEG compression can be either lossy or lossless.

JPEG 2000 has gained the status of international standard in 2000 as ISO/IEC 15444. The discrete cosine transform was replaced by a newly designed wavelet-based method. New features include: multiple resolution representation, random code-stream access and processing, also called Region of Interest and side channel spatial information. A more accurate description of the JPEG 2000 characteristics can be found in [3].

JPEG XR is based on a technology originally developed and patented by Microsoft. The codec was first announced as Windows Media Photo in 2006 and then renamed to HD Photo in the same year. Thanks to the collaboration with the Joint Photographic Experts Group, HD Photo gained the status of international standard ISO/IEC 29199 under the name JPEG XR. Differences between JPEG XR and JPEG include: 2-level hierarchical transformation within 16x16 macroblock regions, lossless integer transform employing a lifting scheme, optional overlap prefiltering step before each of its 4x4 transform stages, prediction of coefficient values across transform blocks applied to the DC and AC, adaptive reordering and Huffman coding for the coefficients, variable coefficients quantization step sizes inside the same color plane of the image.

WebP is a new format for lossy image compression developed by Google in 2010. It is based on the VP8 video codec with a Resource Interchange File Format (RIFF) container. VP8 innovations include an alternate or constructed reference frame, consisting of image data that is encoded into the bitstream but never displayed. It serves to improve the encoding of subsequent frames by providing an additional predictor than previously transmitted frames [4]. Intra prediction, actually used in the case of image compression, is mostly taken from H.264. Loop filtering is used to remove blocking artifacts introduced by quantization of DCT coefficients from block transforms. WebP uses VP8 intra predictive coding. Images are divided into blocks of pixels of variable sizes, whose values are predicted using the values in neighboring block, so that only the difference (residual) is encoded. Residuals are DCT and Hadamard transformed, quantized and entropy-coded through a non-adaptive arithmetic coder [5]. WebP does not currently provide for alpha channel or HDR support, nor lossless or lossy-to-lossless compression.

2.2 Quality Metrics

PSNR is considered as the most recognized and least complex quality metric. However, its output does not correlate well with the image quality degradation since as perceived by the Human Visual System (HVS).

Structure SIMilarity (SSIM) [7] defines the quality degradation as the product of luminance, contrast, and structural errors affecting the image structure. The structural error is defined as the residual error in the image after its normalization with respect to luminance and contrast. The general form of the SSIM between signal x and y is defined as:

$$SSIM\ (x, y) = [l(x, y)]^\alpha \cdot [c(x, y)]^\beta \cdot [s(x, y)]^\gamma, \tag{1}$$

where α, β and γ are parameters that define the relative importance of the three components. If $\alpha = \beta = \gamma = 1$, the resulting SSIM index is given by:

$$SSIM = \frac{\left(2\mu_x\mu_y + C_1\right)\left(2\sigma_{xy} + C_2\right)}{\left(\mu_x^2 + \mu_y^2 + C_1\right)\left(\sigma_x + \sigma_y + C_2\right)}. \tag{2}$$

Although its sensitivity to relative translations, scaling and rotations of images, the SSIM index is quite simple and it performs quite well across a wide variety of image and distortion types. It is able to improve on the traditional PSNR by providing results with are more correlated with the image quality as perceived by the Human Visual System.

3   Assessment

3.1   System Description

All tests were done according to the process described in Fig. 1. All images from each dataset were compressed using the chosen codecs (JPEG, JPEG 2000, JPEG XR and WebP) and the original images were compared with the co-decoded images, on the basis of two quality metrics (PSNR and SSIM). Since not all the codecs provide a tool to directly set the compression ratio, but rather allow for controlling the resulting quality, a first set of coding trials has been performed. Then, the trials from the competing algorithms with the closest compression ratio have been matched. The presented results are the quality index averaged over all the images for each considered dataset. When reporting average results, the lowest bitrate displayed for each codec is the highest bitrate produced through the compression of all images at its lowest quality level. Similarly, the highest bitrate displayed for each codec is the lowest bitrate produced through the compression of all images at its highest quality level. For example, given the chosen output bitrate range from 0 to 8 bpp, with a step of 0.1 bpp, if there were 3 images in the dataset and the bitrates of the images compressed with the lowest quality were 0.15, 0.18 and 0.25, the lowest bitrate displayed in the plot would be 0.3. If the bitrates of the images compressed with the highest quality were 3.05, 3.5 and 4, the highest bitrate displayed in the plot would be 3.

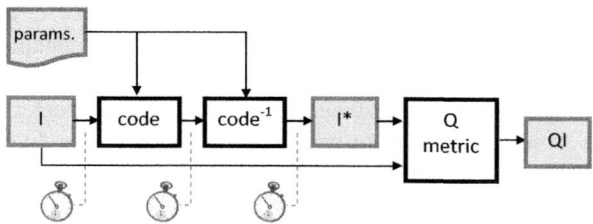

Fig. 1. Performance assessment process

For the experiments, both 24 bpp RGB and 8 bpp luminance images were used from 3 standard datasets. The Canon dataset [8] consists in 18 images with two different resolutions: 512×480 pixels and 512×512 pixels. The Kodak dataset [9] is made of 23 images with 768×512 pixels resolution.

Finally, the "The new test images" dataset contains 15 high-resolution images, ranging from 3008×2000 to 7216×5412 pixels [10]. Because of the limitations of the WebP implementation, images from this dataset have been subsampled by a 2 factor. Table 1 summarizes the software that was used for the compression and conversion between different image formats and the quality measurement.

Table 1. Software tools used for the assessment

Task		Tool
Conversion between lossless formats		GIMP [11]
JPEG codec		Convert Image [12]
JPEG 2000 codec		ImageMagick [13]
JPEG XR	coding	jpg2wdp [14]
	decoding	XnView [15]
WebP	coding	solution that can be found in [16]
	decoding	WebPConvert [17]
SSIM		The SSIM Index for Image Quality Assessment [18]

3.2 Results

Results are first shown as PSNR and SSIM for each greylevel dataset; average values are reported for each dataset, while bpp values are shown in logarithmic scale.

As expected, JPEG achieves the worst performance; the old standard is unable to achieve very high compression ratios and results in lower quality at low bitrates when compared to the other algorithms.

Considering the Canon (Fig. 2) and the Kodak (Fig. 3) datasets, JPEG 2000, JPEG XR and WebP show a similar behavior with slight differences. According to both PSNR and SSIM, WebP performs slightly better than the competing techniques for bitrates from 0.1 bpp to about 0.6 bpp for the Canon dataset and around 1 bpp for the Kodak dataset. For bitrates higher than 1 bpp, JPEG XR outperforms WebP, which provides even lower quality with respect to JPEG 2000. It has to be noted that the WebP codec is unable to run in the entire bitrate range where the other codecs operate. This is a drawback attributable to the used codec and not to the coding algorithm. However, the average values reported in the graph are only computed for bitrates available for all images in the dataset.

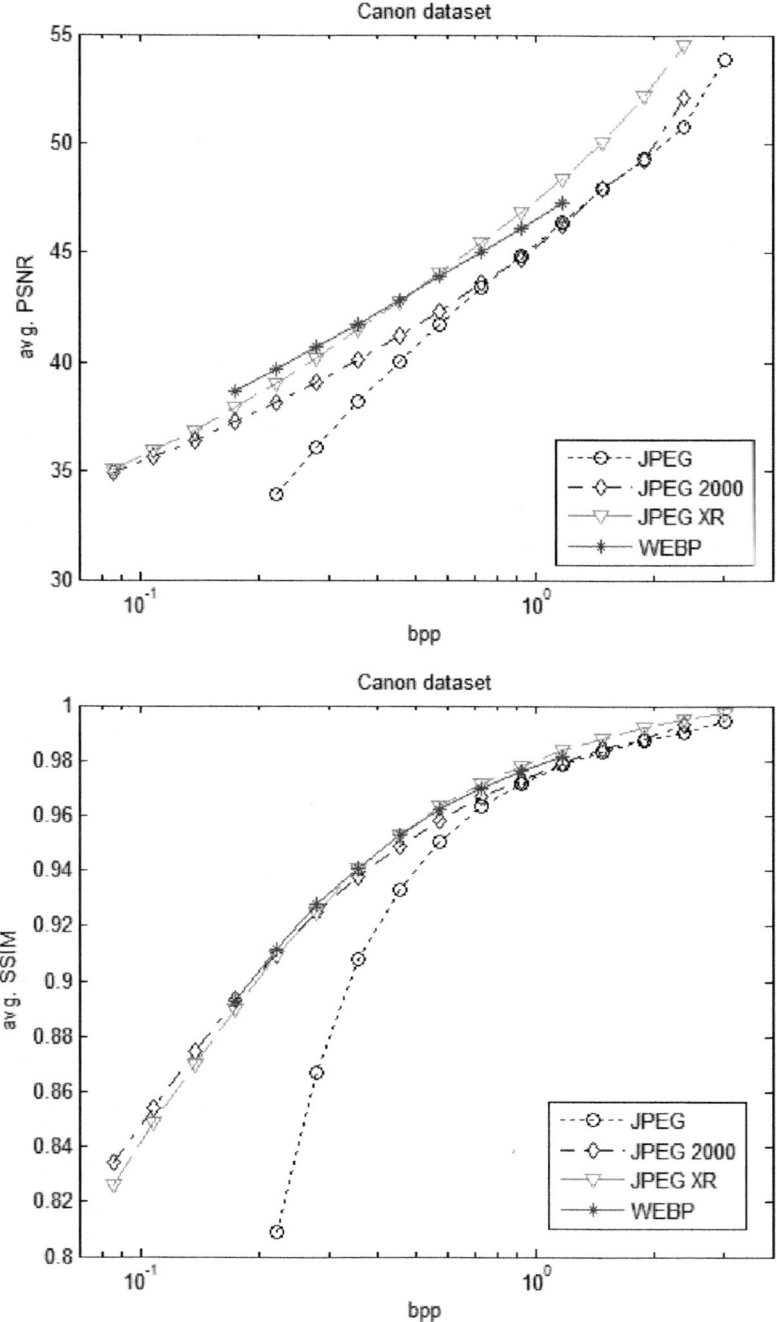

Fig. 2. PSNR and SSIM average results for the Canon dataset

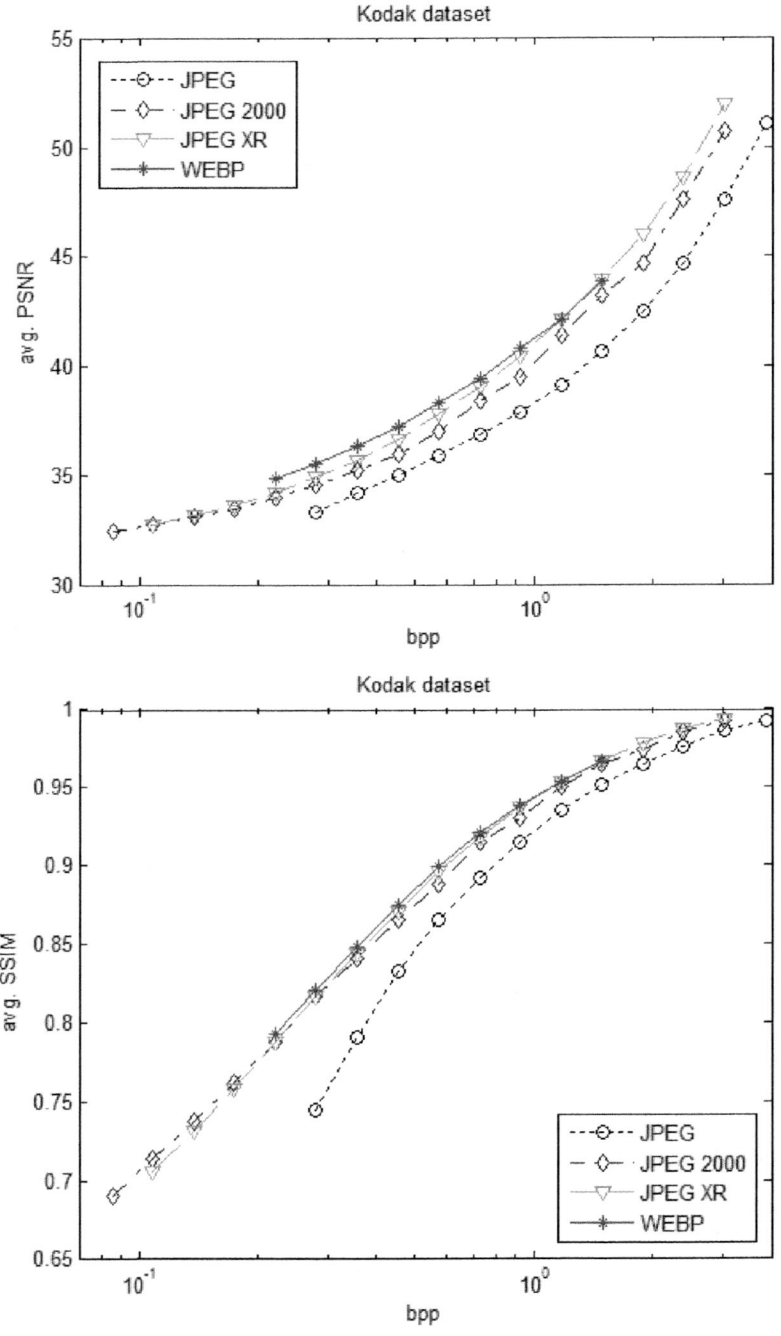

Fig. 3. PSNR and SSIM average results for the Kodak dataset

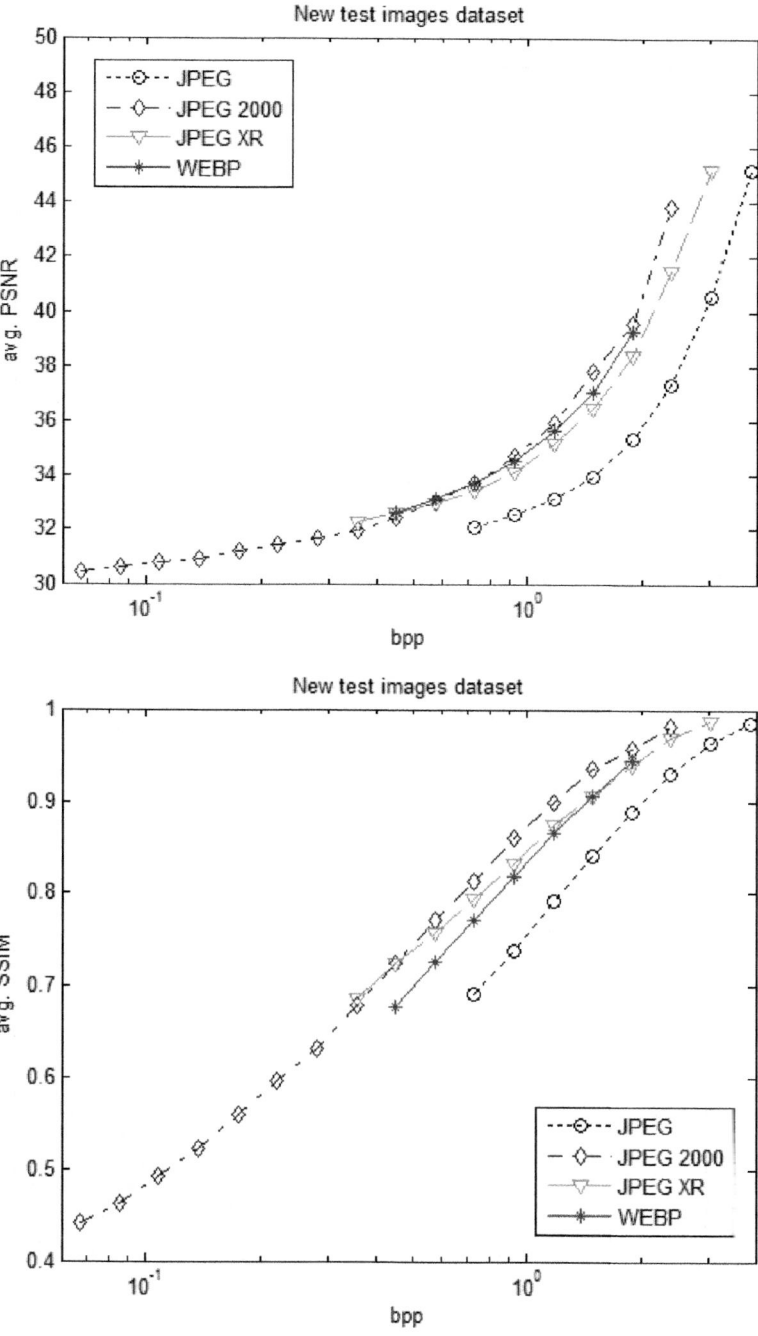

Fig. 4. PSNR and SSIM average results for the "new test images" dataset

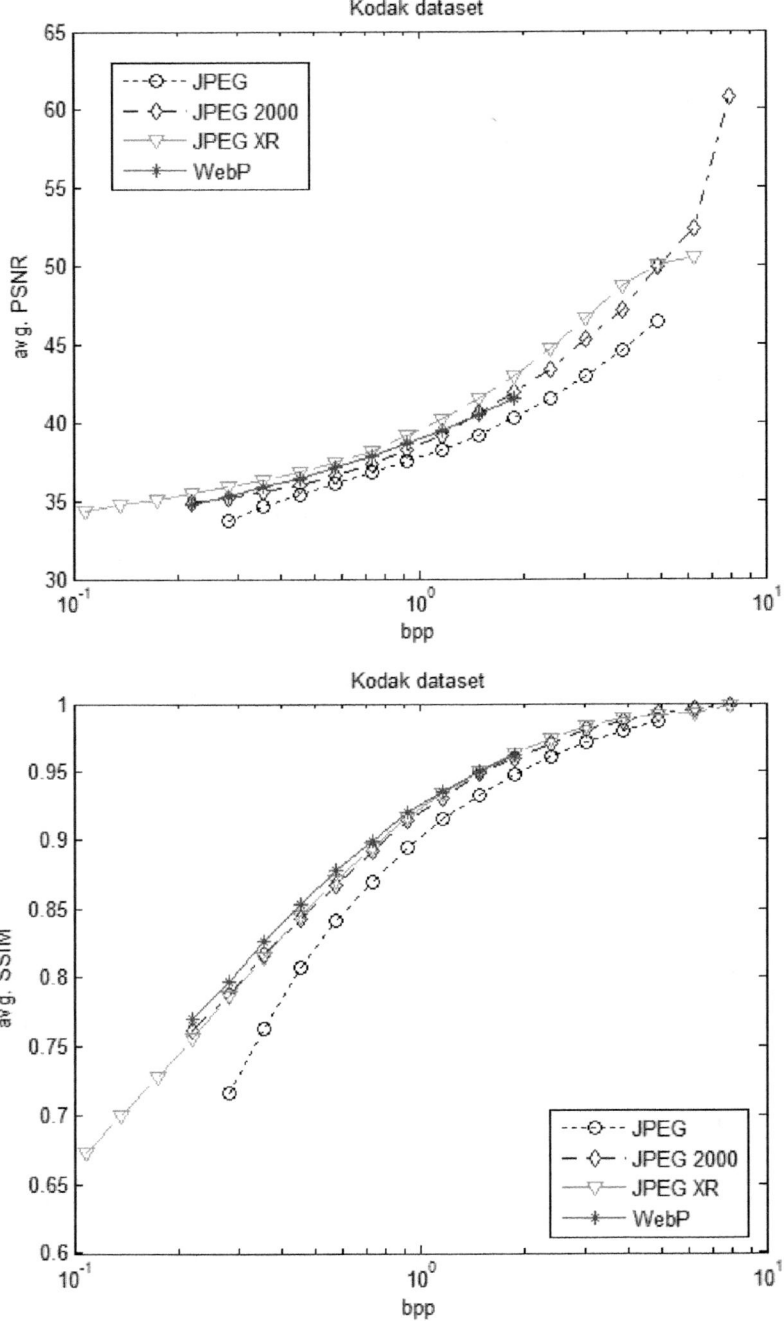

Fig. 5. Average RGB PSNR and SSIM for the Kodak dataset

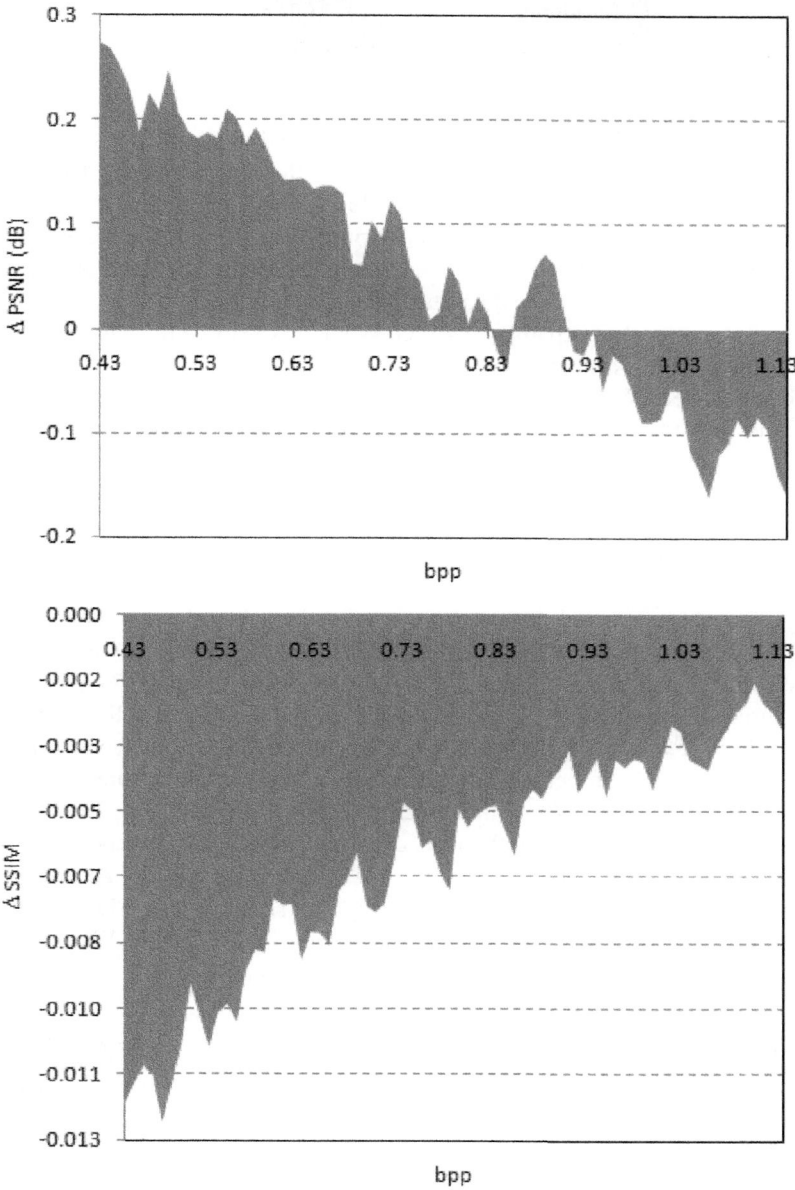

Fig. 6. Comparison between WebP and JPEG XR in terms of the average PSNR and SSIM difference

WebP performance seems to get worse with high-resolution and synthetic images. The average results from the "The new test images" dataset (Fig. 4) show that JPEG 2000 is the best among the competing algorithms in terms of both PSNR and SSIM, whereas WebP prevails over JPEG XR in terms of PSNR but not in terms of SSIM.

The dataset under consideration is made of high-resolution and highly detailed images, two of whom are non-natural scenes (computer graphic and 2D math plot respectively). As in the previous cases, the JPEG 2000 codec implementation is able to reach much higher compression ratios than WebP.

RGB results are not dissimilar from those presented in the previous figures. Fig. 5 presents the average PSNR and SSIM values for the Kodak dataset. Except for the JPEG XR superiority in the PSNR results, which is not in accordance with the SSIM quality, the greylevel results are mostly confirmed. The JPEG XR achievements could be explained with the superior color management offered by such standard. Such figures are common to the three datasets used for this study.

In Fig. 6 an overall comparison between WebP and JPEG XR is shown. PSNR and SSIM results have been averaged over all images from all datasets and only the difference between WebP and JPEG XR is shown. The limited bitrate range has been chosen in accordance with the available measurements. Regarding PSNR, WebP seems to perform better than JPEG XR at low bitrates. Such a superiority is as much as 0.25 dB at 0.43 bpp and linearly decreases to fade at 0.9 bpp, with the inversion of such trend at higher bitrates. On the other hand, SSIM results are in favor of JPEG XR, starting from a 0.011 gain at 0.44 bpp and decreasing as the compression ratio decreases.

Figs. 7-9 show a visual comparison between the selected algorithms for bitrates lower than 1 bpp. As expected, JPEG images present evident blocking artifacts and are perceptibly worse than those from competing algorithms. JPEG 2000, XR and WebP produce comparable outputs, although JPEG 2000 and XR seem to better preserve high-frequency components, while WebP sacrifices some details in favor of visual appearance. Such behavior can be observed by looking at either the petals in Fig. 7 (JPEG XR vs WebP) or the concrete wall in Fig. 8 (JPEG 2000 vs WebP). Some details are lost in the WebP image, even though the average appearance is clear, resembling the effect of bilateral filtering. This effect is not equally acceptable in the case of complex or synthetic images. Fig. 9 shows that JPEG 2000 better approximates the original "zone_plate" signal, followed by JPEG XR. In this case, WebP anisotropic filtering effect results into unnatural edges.

Fig. 7. Portion of image 7 (from the Kodak dataset) compressed at 0.26 bpp

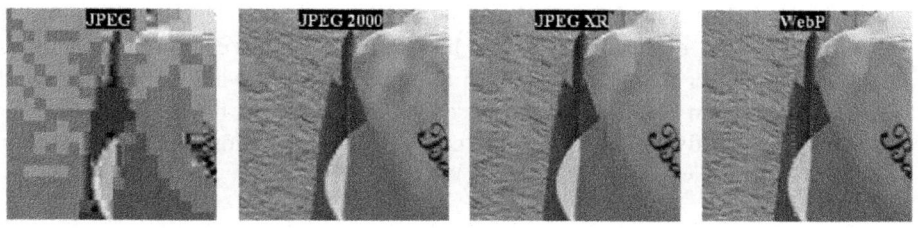

Fig. 8. Portion of image 3 (from the Kodak dataset) compressed at 0.15 bpp

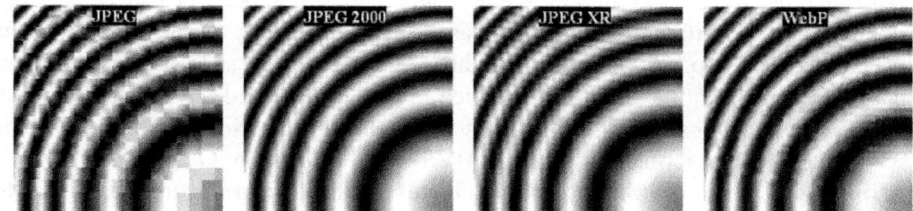

Fig. 9. Portion of the image "zone_plate" (from "The new test images" dataset) compressed at 0.95 bpp

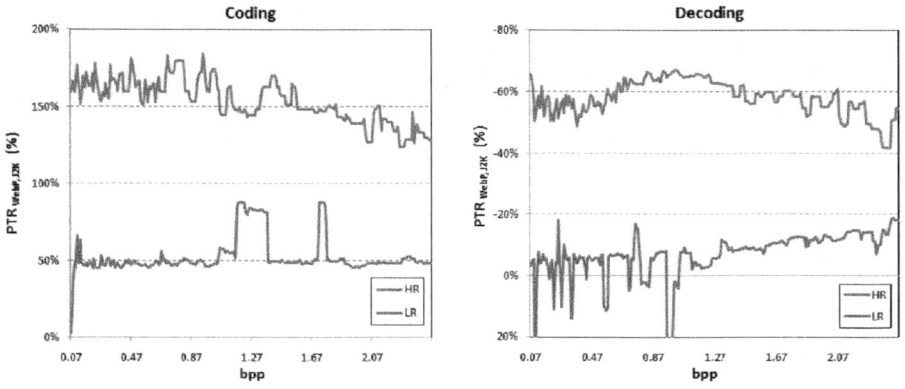

Fig. 10. Average processing time rate between WebP and JPEG 2000 as described in eq. 3 for the coding and decoding phase

Finally, Fig. 10 reports on the average processing time rate between WebP and JPEG 2000 for the coding and decoding of two groups of 10 images (LR: 512×480 and HR: 1500×1500) as:

$$PTR_{WebP,J2K} = (t_{WebP} - t_{J2K})/t_{J2K} \%. \qquad (3)$$

A consumer PC with Intel Core2 Duo T5800 CPU @ 2.00 GHz and 4 GB RAM has been used for the experiments. It can be noted that the WebP implementation is significantly slower than JPEG 2000 in the coding phase, while it outperforms the recent compression standard in the decoding. It must also be observed that such difference is relatively small in the case of low-resolution images, which are frequently used in web applications.

4   Conclusions

An objective assessment of the WebP image codec has been provided. The new compression algorithm from Google has been evaluated in terms of compression efficiency by comparing its experimental results with other state of the art codecs. The WebP algorithm shows good efficiency with natural images for bitrates up to about 1 bpp. In this scenario, WebP is often slightly superior than the competing techniques, both in terms of PSNR and HVS-based evaluation. On the other hand, its performance drops with high resolution/highly detailed or synthetic images and for bitrates higher than 1 bpp. Visual comparison reveals a blurry effect probably due to loop filtering, which can be pleasing or unnatural depending on the scene. Processing time evaluation shows that the current implementation of WebP is fairly inefficient in the coding phase and reasonably fast in the decoding of low-resolution images.

From such considerations, and with some caution due to experimenting through a single implementation, WebP classifies among the current state of the art algorithms for image compression, without providing any important innovations or performance boost. Moreover, WebP does not support several features that are common nowadays, such as transparency or lossy to lossless compression. Given such limitations, WebP should really provide further technical improvements, given the aim of its authors to replace the old JPEG, a task already failed by other standard competitors.

References

[1] De Simone, F., et al.: Subjective evaluation of JPEG XR image compression. In: SPIE Optics and Photonics, Applications of Digital Image Processing XXXII, San Diego, CA, vol. 7443 (2009)
[2] WebP homepage, http://code.google.com/speed/webp/
[3] JPEG 2000 Part I Final Committee Draft v. 1.0 (March 2000)
[4] Inside WebM Technology: The VP8 Alternate Reference Frame, http://blog.webmproject.org/2010/05/inside-webm-technology-vp8-alternate.html
[5] WebP code, http://code.google.com/speed/webp/index.html
[6] Chalmers, A., et al.: Image Quality Metrics. In: ACM SIGGRAPH, New Orleans, USA, July 23-28 (2000)
[7] Wang, Z., et al.: Image quality assessment: from error visibility to structural similarity. IEEE Trans. on Image Processing 13(4), 600–612 (2004)
[8] Canon dataset, http://www.cipr.rpi.edu/resource/stills/canon.html

[9]   Kodak dataset,
        http://www.cipr.rpi.edu/resource/stills/kodak.html
[10] The new test images dataset,
        http://www.imagecompression.info/test_images/
[11] GIMP, http://www.gimp.org/
[12] Convert Image,
        http://www.softinterface.com/Convert-Image%5CConvert-JPG.htm
[13] ImageMagick, http://www.imagemagick.org/script/index.php
[14] Herila, B.: jpg2wdp, http://www.bherila.net/2010/01/easily-batch-
        convert-jpeg-to-hd-photo-with-jpg2wdp/
[15] XnView, http://www.xnview.com/
[16] WebP Discussion,
        https://groups.google.com/a/webmproject.org/group/webp-
        discuss/browse_thread/thread/b8b20e010565dc09/174256299ece6ec2
[17] WebP for.NET, http://webp.codeplex.com/
[18] Wang, Z., et al.: The SSIM Index implementation,
        http://www.ece.uwaterloo.ca/~z70wang/research/ssim/

Implementing Mobile Applications with the MIPAMS Content Management Platform

Xavier Maroñas, Silvia Llorente, Eva Rodríguez, and Jaime Delgado

Distributed Multimedia Applications Group (DMAG), Departament d'Arquitectura de
Computadors (DAC), Universitat Politècnica de Catalunya (UPC),
C/Jordi Girona, 1-3, 08034 Barcelona
{xmaronas,silviall,evar,jaime.delgado}@ac.upc.edu

Abstract. New mobile devices (pda's, tablets) permit the implementation of
new business models as they are always connected and provide multimedia
capabilities for capturing images, videos, music or even conversations.
Together with an architecture for the secure management and distribution of
multimedia content called MIPAMS, we propose a mobile business model with
the implementation of a mobile application based on iOS (Apple operating
system for mobile devices) for publishing added value content captured with a
mobile device.

Keywords: Mobile Applications, MIPAMS, iOS, Android.

1 Introduction

This paper describes how to use MIPAMS (Multimedia Information Protection And
Management System) [1], a service-oriented DRM platform developed at the authors'
research group, Distributed Multimedia Applications Group (DMAG) [2] to
implement fully functional mobile applications for the management and distribution
of multimedia content in a secure way, respecting intellectual property rights
governing them.

The solution presented in this paper implements part of the ideas described in [3],
where several business scenarios based on MIPAMS were presented. Specifically, we
focus on the scenario designed for mobile devices, which applies to an electronic
publishing scenario, but other applications could be implemented as we will describe
in the future work section.

The paper is organized as follows. First, MIPAMS architecture and its modules are
briefly described. Then, we show a use case where a mobile application connects to
MIPAMS architecture for implementing an electronic publishing scenario for mobile
devices. Afterwards, some mobile operating systems are presented, together with a
description of how new applications can be developed over those platforms. Then, we
describe how we can connect MIPAMS with applications developed for iOS [4].
Finally, some conclusions and future work are presented.

L. Atzori, J. Delgado, and D. Giusto (Eds.): MOBIMEDIA 2011, LNICST 79, pp. 266–280, 2012.
© Institute for Computer Sciences, Social Informatics and Telecommunications Engineering 2012

2   Implementing Mobile Applications for an Electronic Publishing Scenario

In [3] we presented a scenario where mobile devices and applications were especially relevant for electronic publishing. It represented an electronic publishing scenario where mobile devices were used to generate "publishable" content that could have some added value. In that paper, some other scenarios were defined, showing the different ways in which an architecture for the secure management and distribution of multimedia content, known as MIPAMS, could be used to implement those functionalities in a secure way. A brief overview of this architecture is given in subsection 2.1.

The scenario described was considering the principle that mobile devices are not the most suitable devices for capturing and distributing high quality images or videos. However, they are always available and, possibly the most important, always connected. So, the most important feature for our scenario is the "opportunity" of the multimedia content being captured; that is, it was taken at the right place at the right moment. This is particularly important for unforeseen events such as natural disasters, accidents or celebrities found in an unexpected or funny situation, as this kind of events cannot be later reproduced and official mass media are not present at the instant when the event happens. In such cases, the multimedia content could even have monetary value for the author, who may try to sell or at least ask for attribution of the images or videos taken.

Starting from that scenario, this paper details a specific use case using a mobile scenario with Apple devices integrated with the MIPAMS platform. Thus, we provide the tools for a business model to facilitate exploitation of content generated in those devices.

2.1   Use of a Back-End Existing Architecture to Support Electronic Publishing with Mobile Devices

This section describes MIPAMS (Multimedia Information Protection And Management System), a service-oriented content management platform, developed by the DMAG (Distributed Multimedia Applications Group) [2]. It is mainly intended for applications where management of rights over digital multimedia content is required. This architecture will act as a back-end for the mobile scenario described before.

The MIPAMS architecture is based on the flexible web services approach, as it consists of several modules and services that provide an important part of the functionality needed for governing and protecting multimedia content. The main advantage of having service-oriented DRM functionality relies on the possibility of decoupling it into different subsystems depending on the needs of the application that is going to be implemented, while being able to share the same common services between different applications with different requirements, thus reducing costs. This also permits its use in innovative scenarios very different from the ones originally intended, like Social Networks [5][6] or Electronic Health Record management and exchange [7]. Nowadays, users of both applications are really concerned about privacy. MIPAMS and more specifically the License, Protection and Authorization services, can be the key elements to improve the protection of users' privacy in both, online social networks and personal health records systems.

Figure 1 depicts the MIPAMS architecture, for which we provide next a general overview of its components and the different services being offered.

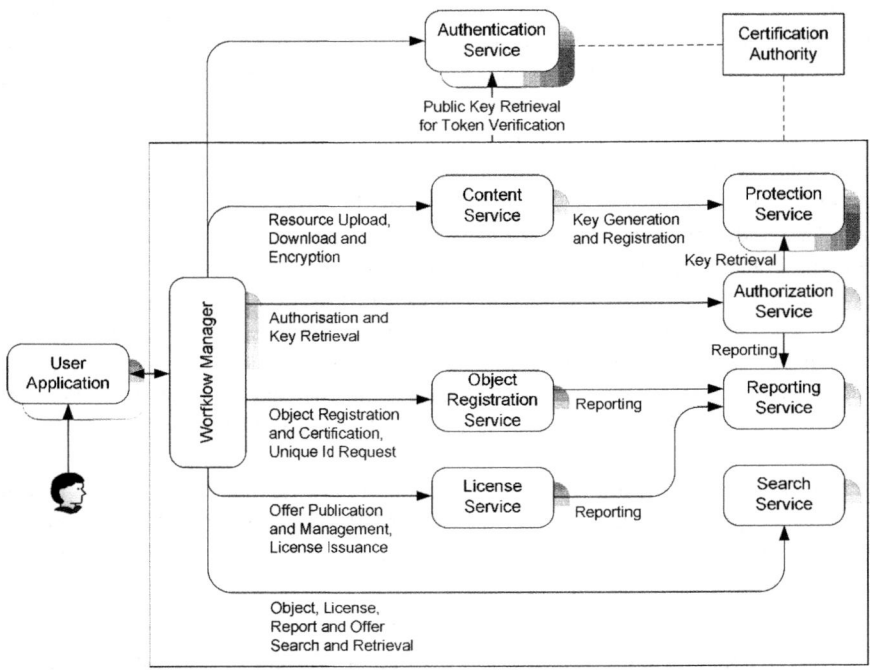

Fig. 1. MIPAMS Architecture overview

Content Service (CS) enables applications to upload and download digital resources such as audio or video files, text documents, etc. Those resources can be optionally encrypted under request. If encryption is selected, the protection keys will be first requested to the Protection Service (PS) and then registered back into PS, once encryption is performed. Moreover, one can request unique identifiers to the Object Registration Service (ORS) in order to uniquely identify uploaded content in the system. Unique identifiers can be also obtained from external sources; it just depends on the business scenario implemented using MIPAMS.

Object Registration Service (ORS) permits the registration of digital representations (i.e. digital objects) of multimedia content (comprising content and metadata). This information is packaged using the MPEG-21 Digital Item XML-based format [8]. ORS registers and digitally signs objects so that they can be later checked for authenticity and integrity. It also provides unique identifier for uploading contents into CS, as already explained.

License Service (LS) deals with rights offers and license issuance over digital objects. Rights offers include the rights and conditions that can be acquired by users over some digital content. They are defined by contents' rights holders, which include content creators. License issuance refers to the creation of a license according to a rights purchase or any other situation where a rights holder grants a set of rights to other user. Licenses are expressed using MPEG-21 Rights Expression Language [9].

Authorization Service (AS) checks whether a user owns any appropriate license that grants him the right to perform a requested action (e.g., play) over a digital object. The authorization is based on the license based authorization mechanism defined in [9]. This module shares license repository with LS. After positive authorization and if content is encrypted, AS requests corresponding encryption keys to PS and returns them to the requesting application. This is the only means for decrypting protected content.

Protection Service (PS), as introduced before, generates encryption keys upon request, registers encryption keys associated to uniquely identified content and provides the encryption keys for protected content to the AS.

The User Application (UA) is the player, edition tool, browser or any other means managed by the user to deal with the digital content, for instance registering or accessing it. It may have an internal trusted module to locally enforce DRM features.

Workflow Manager (WM) may be an integral part of the UA or otherwise be located in the server part (e.g. web portal, brokerage service) to reduce the UA complexity. It controls access to the rest of services inside MIPAMS architecture, like license issuance, authorization, content upload, etc.

Search Service (SS) enables applications to perform accurate searches amongst metadata in the MIPAMS system. It can be used for searching content, licenses, offers or reports or a combination of them.

Reporting Service (RS) collects usage reports regarding object registration, license issuance and positive authorizations. Those reports may be used for computing statistics as well as for billing or tracking purposes. From the information stored it is possible to generate standards-based representations like MPEG-21 Event Reports [10].

Authentication Service (ATS) is needed to authenticate the identity of users. It generates SAML (Security Assertion Markup Language)-based tokens [11] that identify MIPAMS users. Any service in the MIPAMS architecture will require a token argument to be provided in order to authenticate users. Tokens are digitally signed by the ATS, so that they can be checked for authenticity and integrity by the receiving service. Moreover, the ATS deals with user registration and management.

Finally, there is a need for having a recognized Certification Authority (CA), which issues credentials for the different Components and Actors in the system, as X.509 [12] certificates and private keys for the different architectural components.

As far as we are concerned, there is no other modular and standards based architecture that provides all functionality offered by MIPAMS. The most similar initiatives are Chillout [13], developed by the Digital Media Project (DMP) [14], and Convergence European project [15]. There is also a new MPEG standard under development, Multimedia service platform technologies (MSPT), also known as MPEG-M. The aim of MPEG-M Part 4 [16] is to define elementary services that can be combined to provide complex services. MPEG-M follows the same principles as MIPAMS where each module could be seen as an elementary service, although MIPAMS is already implemented and in use by several research projects. To provide a complex service, MIPAMS implements a workflow manager which performs the appropriate calls to each elementary service, that is, a MIPAMS module.

3   Use Case

The use case presented in this section defines how MIPAMS modules interact with the electronic publishing scenario described in section 2. In such cases, the multimedia content captured by the mobile device may even have economical relevance, as they could be published in online newspapers, the gossip news or even broadcasted on television. Therefore, the author may register the content for different purposes: to try to get some revenues or just for later attribution.

Figure 2 depicts MIPAMS modules involved in the electronic publishing scenario and their interaction with the Mobile User (the person using the mobile device) and the Mobile Application (a specific application for registering content). Figure 2 shows the registration of the content, image or video into the registration portal, including how the author can create some offers over the content to get some revenues. It is worth noting that one could create an offer that provides the content for free, but the authorship and some limiting conditions may still apply.

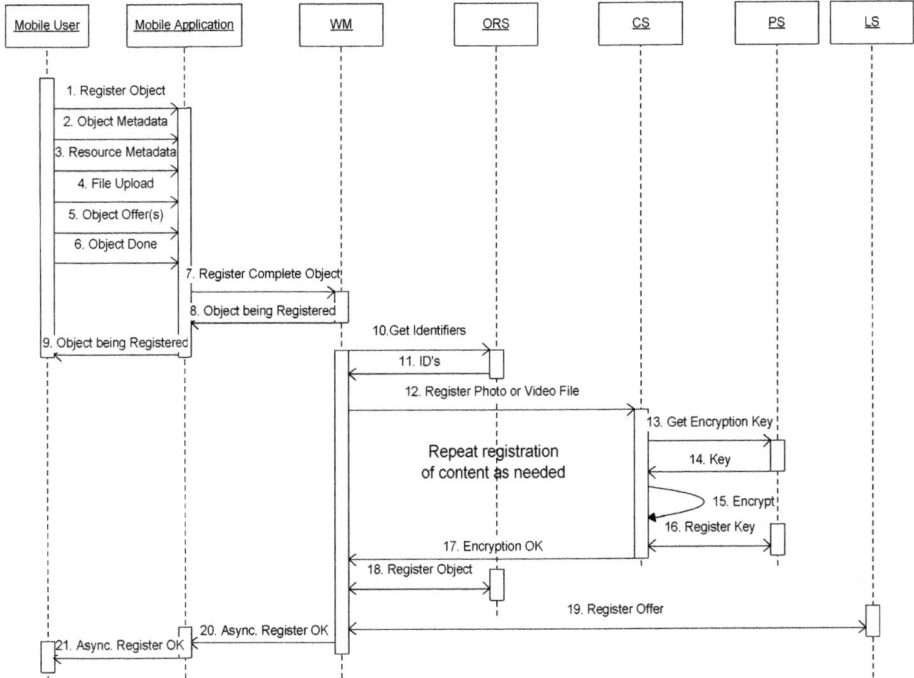

Fig. 2. Registration of image or video captured with the mobile device, including offers

Figure 2 illustrates the content registration process, including also offer creation, to facilitate the registration in a one-step-process from a Mobile Application. The specific steps involved in this process are the following:

1. User starts registration of content locally in an application for mobile devices (Mobile Application, MA). Some information is predefined to facilitate the registration process.
2. User fills a form with all metadata associated to the complete digital object. Some fields, like author, can be automatically filled by the MA.
3. User fills several forms (one for each image or video, that is, each resource) with metadata associated to each resource.
4. User selects the file containing each resource. This file is accessible by the mobile device (inside any local storage).
5. User defines the offers applying to the object. User has to insert the different sale conditions they offer for the registered object. These conditions include what can be done with the content (play, print, etc.) together with some conditions (territory, number of times one can perform the action or payment conditions). Again, license templates or predefined forms can be used to facilitate the task.
6. User indicates that all object information has been inserted and the registration information has to be sent to the server.
7. MA sends all information to the Workflow Manager (WM) module.
8. WM sends an immediate response to the MA. Later on, if the registration is successful, the Mobile User will be informed through the MA.
9. MA shows an indication that registration is in progress to the Mobile User.
10. WM requests identifiers to the Object Registration Service (ORS).
11. ORS sends the identifiers requested to the WM, one for the object and one for each resource file (even if they have not been already uploaded).
12. WM sends resource to the Content Service (CS).
13. Since the user has requested encryption of the resource, CS asks for encryption keys to the Protection Service (PS).
14. PS returns the keys for encryption algorithm and key length specified by CS.
15. CS encrypts and stores the file with the given key.
16. CS registers the encryption key in the PS for permitting later decryption.
17. CS sends WM notification of correct content storage and encryption. Steps 10 to 17 are repeated for each resource uploaded by the user. If no resources are uploaded, these steps can be done later.
18. When all resources are properly uploaded and encrypted, WM requests ORS the registration of the complete object, which is digitally signed to guarantee digital object integrity. The format used for storing the object is the MPEG-21 Digital Item.
19. WM registers the offer provided by the Mobile User. If there is more than one offer, this step will be repeated as many times as needed.
20. WM sends asynchronous notification of object registration to the MA.
21. MA informs the user that the object has been properly registered. In any case, Mobile User always can search her registered objects to check if everything is correct. This is especially useful when the mobile device connection is lost for some reason.

Once there are offers over a registered object, other users are able to purchase it. At that moment, a license is created based on the selected offer and the purchasing user can access the purchased content. For the moment, content download is only possible from an application installed in a laptop or PC, but with the major introduction of tablets between users, other possible applications and business scenarios could apply. Tablets are devices between 7 and 10 inches of screen that are not as powerful as PCs but they are more than a simple mobile phone. So, for tablets one could use the mobile application for registering content due to its quickness but also content download and consumption could be done combining both functionalities, the ones for mobile phones and the ones for PC.

Nevertheless, there are some other tasks that could be done from the mobile device. After registering and creating offers over some content, the author may look for her registered objects or check if any of her objects has been purchased, requesting the event reports associated to them. Figure 3 illustrates the interaction between the Mobile User and the different MIPAMS modules. In fact, there is one service dedicated to searches inside MIPAMS, which is the Search Service. This service accesses the necessary databases (Objects, Reports, etc.) to get the requested information.

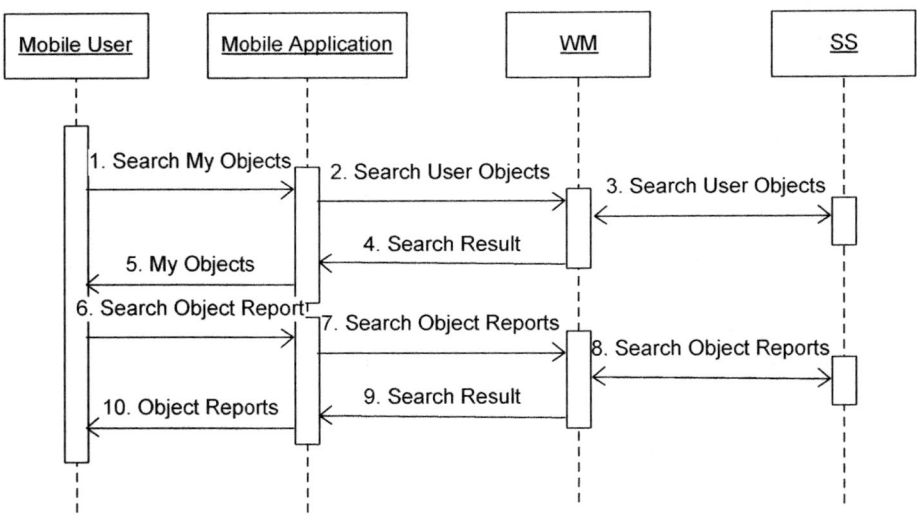

Fig. 3. Search of object reports

Figure 3 illustrates how a user can look for her registered objects and request the activity reports (mainly purchases) over them. The following steps are involved:

1. Mobile User searches her objects through the Mobile Application.
2. MA sends the request to the WM.
3. WM contacts the SS, which searches the objects registered by the specified author. SS sends the search result to the WM.

4. WM returns results to the MA.
5. MA presents objects information to the Mobile User.
6. The Mobile User selects an object from the ones returned by the previous search in the MA and requests its reports.
7. MA sends a report request to WM indicating which user's object wants to query.
8. WM sends a report request to SS indicating which user's object wants to query. SS returns Event Reports accomplishing the criteria to the WM.
9. WM returns Event Reports to the MA.
10. MA shows results to the Mobile User.

There are other possibilities in the use cases presented. For instance, the registration process may be implemented as a simple one-step process using predefined offers, where the right and some basic conditions are already defined and the user only has to select them. Once content is registered, users may always create other offers from a web portal. For instance, a user gives play permission to a low quality version of the content and later on offers a high quality version for broadcasting at a higher price.

Another improvement for mobile devices may be the minimization of metadata associated to the different resources. This will facilitate registration as the messages sent will be smaller and thus the connection time will be less, too.

Apart from the mobile application, users will always be able to access content from a PC, and update offers and metadata as required.

Finally, a major concern for the mobile user may be the connection bandwidth and speed. So, if she does not have a quick connection, uploading a photo or a video may not be possible. In that case, she may just register basic metadata and offers and upload the digital resource later, by means of WiFi connection, providing better connection capabilities and usually supported by current mobile devices.

4   Implementing Electronic Publishing Scenario with Different Mobile Platforms

The implementation of the electronic publishing scenario presented in sections 2 and 3 could be done on different Mobile Platforms, including iOS [4], Android [17], BlackBerry OS [18] or Windows Phone [19] as most of them provide direct connectivity over HTTP/HTTPS. With this basic connection support, it is possible to connect with web services implementing the SOAP protocol [20].

For each platform, there are several possibilities for establishing the connection between the mobile device and a remote web service, as presented in the used case defined in section 3.

The first and possibly the most expensive one in terms of development time is the implementation of a dedicated SOAP client working over the HTTP/HTTPS support provided by the mobile platform.

The second one is the use of an external library that makes the implementation of a SOAP client easier, as provides the basic functionality for creating a SOAP request message and receiving a SOAP response message. Almost every mentioned mobile operating system provides a library with this kind of protocol support.

Finally, there is the possibility of creating another Front End using REST [21] services, as there is a major support of REST services on mobile devices as they have less communication overhead than SOAP.

In the following subsections an overview of Android and iOS systems is provided before the description of a mobile application running on iOS.

4.1   Android Overview

Android [22] is an open-source software stack for mobile devices (phones and other devices like tablets) that includes an operating system, middleware and key applications. The Android Open Source Project (AOSP) [17], led by Google, is tasked with the maintenance and further development of Android. The Android SDK [23] provides the tools and APIs necessary to begin developing applications on the Android platform using the Java programming language. Many device manufacturers have brought to market devices running Android, and they are readily available around the world.

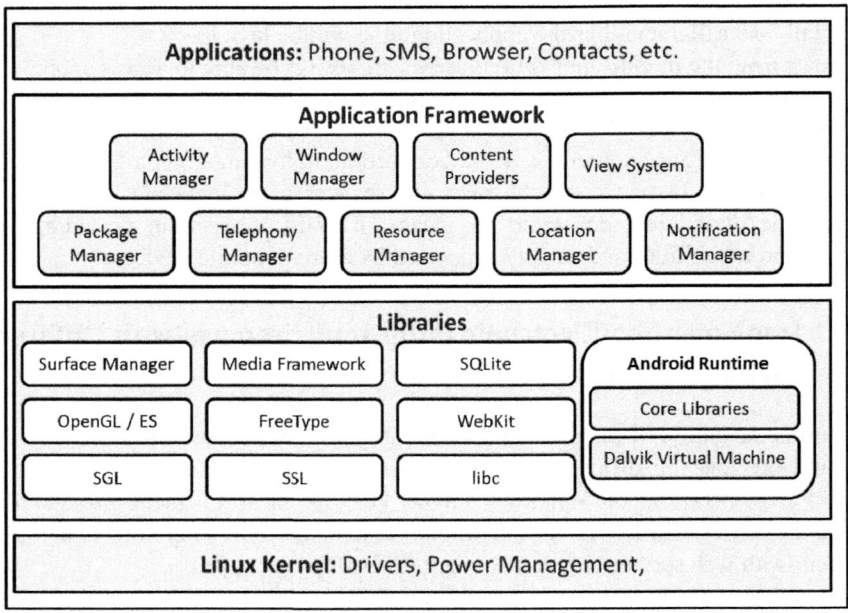

Fig. 4. Android Architecture overview

Android relies on Linux version 2.6 for core system services such as security, memory management, process management, network stack, and driver model. The kernel also acts as an abstraction layer between the hardware and the rest of the software stack. Figure 4 shows an overview of the Android architecture, which is organized in different layers. See [24] for a detailed description on how to implement a mobile application with Android.

Android latest version is 3.1, made public in May 2011 [25]. It introduces many changes for both, users and developers, like new user interface, new 2D and 3D graphics, etc. It is specifically optimized for devices with larger screen sizes, particularly tablets that currently are the new trend in smart always connected devices.

4.2   iOS Overview

iOS [4] is the official mobile platform for Apple [26] mobile devices, including the iPhone, iPad and iPod Touch. This platform includes both the operating system and the frameworks and libraries needed for developing iOS compatible applications. Also, the XCode developer tools package [27] provides the IDE, instruments, compilers and simulators with which build applications. All these stuff is only available inside the iOS developer program [28] in which you have to be registered to access the information, forums, tools, SDK, etc.

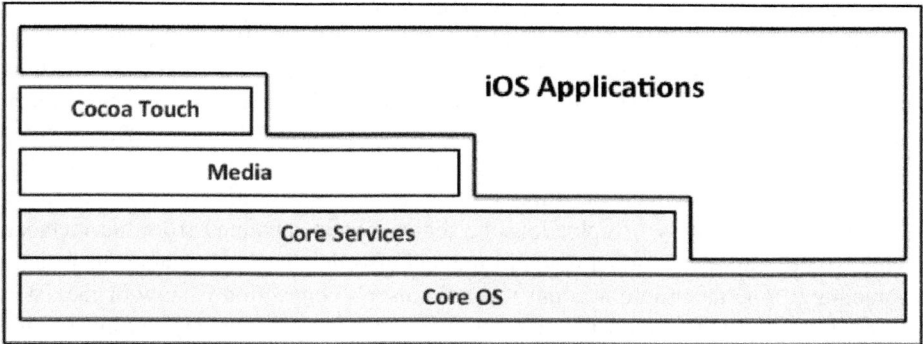

Fig. 5. iOS Architecture overview

iOS platform is based on the Objective-C programming language [29], used by Apple both in Mac OS X and iOS development. iOS can be divided in 4 basic technology layers as described in Figure 5. Applications are built not only on top of these layers but using any of them depending on the requirements. However, most of things can be done using the "Cocoa Touch" level and other existing high-level frameworks.

iOS 4 is the latest major version, and introduces more than 100 changes from the previous one including the possibility of grouping applications using folders, change the wallpaper image, compatibility with external keyboards using Bluetooth connections, etc. But, one of the most significant improvements on the system was the introduction of the "multitasking" feature, which makes possible for the developers to keep the application alive even when users close it.

4.3   iPhone Implementation

This section describes the details of the development and the distribution of a MIPAMS client in a mobile environment. In the development subsection there are also some screenshots of a preliminary interface design of the application. In [30][31] there are described previous mobile applications implemented by the authors, that give an idea of the evolution of the mobile devices in short time.

Development. Something to be considered when implementing a mobile application are the target device capabilities. Although there are many different devices that use the same OS, all of them have some common features: the reduced size of the screen and the limited computational capacity. Even the way in which user interacts with the device is very different from a PC environment. For all these reasons the application has to be as simple as possible in terms of screen interfaces and steps. We propose a one-step content registration and the use of predefined offers to make the registration process easier and faster. We can achieve those tasks by sending just one compacted message, reducing both the time and the amount of data sent.

Although both the content metadata and license introduced by the user are restricted, the current implementation of MIPAMS includes a web portal, which can be accessed from any web browser. Using this portal, a user may change object's information afterwards. That makes sense in a scenario where the user wants to upload the minimum object information to make it accessible as quickly as possible.

Another way to reduce the time waiting for the registration response from the server is the possibility of uploading the digital object separated from the metadata. This would reduce the global upload time and also the amount of data sent which is something to be taken into account when the user's connection is slow or inexistent. In addition, the registration process could be done asynchronously, letting the user interact with the other options of the application or even close it just a moment after the registration message is sent. In this case, the user would receive a message from the server with the registration result when the process is finished.

We are developing a first version of the described mobile application using iOS technologies. The first conceptual interface screenshots are the ones shown in Figure 6. In this version, only the content registration (metadata and licenses) and content upload is available. The search of other's objects and offers and the option of searching reports will be implemented later. As it is a mobile application, it would not include all the features that the MIPAMS portal has, but there are some other operations that could be considered like the possibility of including a simple license editor in the mobile device.

All these points make sense when talking about devices like phones or PDAs, but there are some other mobile devices that have to be taken in account. Both iOS and Android have a version of the operating system designed for tablets. It is not clear if it has to be a native application for this kind of devices or it is better to use an adapted web application.

Image 1 in Figure 6 shows the different options provided to the user, content registration and search and reporting. Image 2 refers to the location of the content to be uploaded, followed by Image 3 where some basic metadata can be filled. Images 4 and 5 refer to the offers part of the application. Image 4 shows the offers menu, where some predefined offers can be selected. Image 5 shows the information the user has to fill in case she decides to create a new offer. Finally, Image 6 shows the object registration summary, showing the basic information that will be sent to the server for registering the object, which includes object information together with offers created.

At the time of writing, we have already developed some of the steps mentioned above. We have implemented some of the Workflow Manager (WM) functions in a servlet. Doing so, we reduce to the minimum the information sent and received by the mobile device, and the time waiting for the response.

The first service we have connected is the Authentication one. As this application is supposed to work in a trusted and secure environment, the user will not be able to use it without being authenticated in the MIPAMS system.

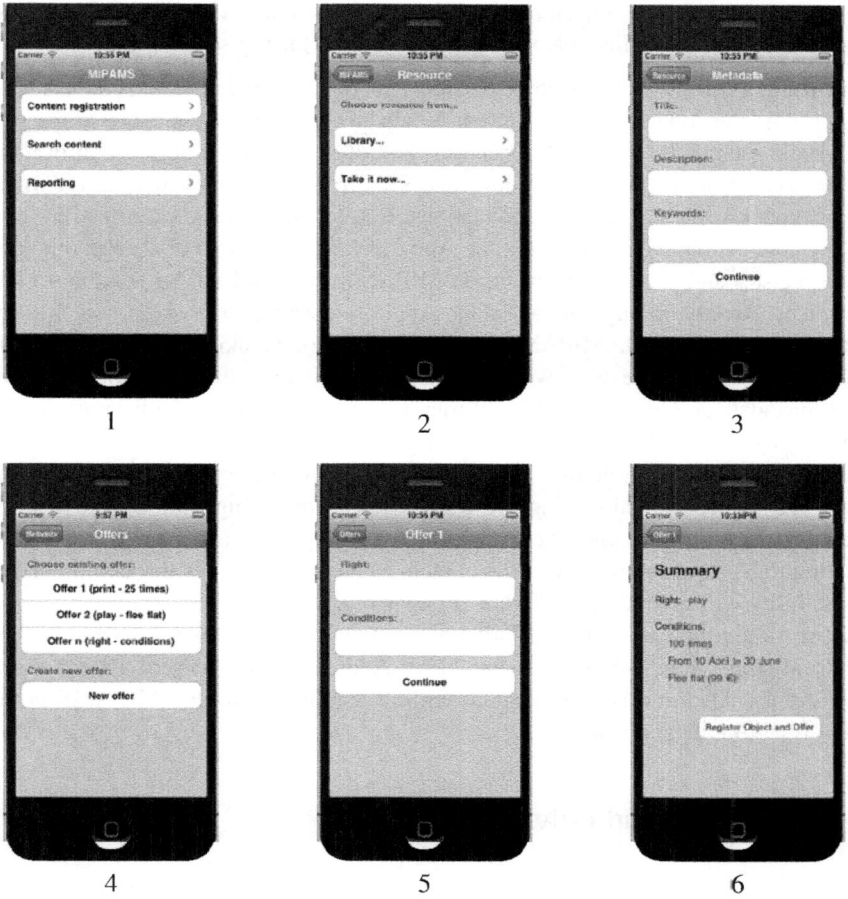

Fig. 6. MIPAMS mobile client screenshots

This first service has been connected using an OpenSource library called, ASIHTTP [32]. This framework lets us establish an HTTP connection in a very simple way. The application just has to know the destination URL and the parameters to send; then the library will manage the connection, the request and the response, including the possible errors. Also, it can work both synchronous and asynchronously, returning control to the mobile application while it is sending a request and waiting for the response. The library implements some protocol methods to alert the application whether it receives the response or identifies connection errors.

After this connection, the application is ready to continue with the registration process. To do so, we maintain the connection architecture used for login. A new method has been developed in the WM servlet, responsible of the Object Creation process (both the metadata and the digital resource upload). With that method, the mobile application is able to implement a one-step registration, sending the object metadata and digital resource at the same time, using an asynchronous connection. The servlet receives the request and makes the needed steps to register the object metadata, obtaining a unique id and uploading the resource. After that, WM sends the response message to the mobile application, which shows the user the registration result.

Distribution. One of the most restrictive things in Apple mobile devices is the business model used on the applications distribution. When an application is uploaded to the App Store, Apple retains the 30% of the benefits, and does not let the application use any other payment system. That constraint reduces the possibilities of creating a parallel business model inside this kind of applications. But, for the specific case of MIPAMS, the client application distributed using the App Store would be free of charge, allowing users to register and upload content inside MIPAMS. It would not be possible to buy any content directly through it, avoiding Apple's distribution charges. So, for the business model presented, MIPAMS application is not breaking Apple's rules but complementing them. The user will just use the application to upload or query the content, metadata and offers, but the real transactions for acquiring the licenses will be done through the MIPAMS web, possibly using a PC or a tablet.

Furthermore, MIPAMS mobile client would not allow the user to buy or download content. The content download and use would imply a much higher complexity in the application development and the possibility that Apple refuses it. It would be interesting to have some preview of the content as the user searches through them, but this possibility will be studied for future versions of software.

It is worth noting that Android Market also retains a 30% of benefits when selling an Android App. Thus, the business model for Android would be very similar to the one for Apple. Also in this case, content will be purchased through MIPAMS portal instead of the mobile application.

5   Conclusions and Future Work

This paper has presented a business model for the development of mobile applications for the secure management and distribution of content captured with the mobile device itself. This model lies in the MIPAMS Architecture, a standards-based architecture where different modules provide the protection, rights definition and object registration functionality needed for secure content distribution. We have also presented a use case in section 3 to clarify how a mobile application connecting with MIPAMS should be implemented, describing the different steps involved in object registration, metadata introduction and how to associate offers to the content registered using the mobile platform. Section 4 describes the implementation of a mobile application to connect with MIPAMS over iOS operating system. Some screenshots and some possibilities for implementation and distribution of the mobile application have been described.

For complementing the use case a first version of the application has been presented. Nevertheless, we are thinking of other kind of mobile devices, tablets, which have a bigger screen and more rendering capabilities that smaller mobile devices, like iPod or iPhone. A more complex application may be developed and distributed for tablets.

Moreover, we are also evaluating the use of Android operating system for the secure management and distribution of multimedia content with MIPAMS. The use of web services and the lessons learnt from the iOS implementation will facilitate this task.

Acknowledgements. This work has been partially supported by the Spanish government through the project MCM-LC (TEC 2008-06692-C02-01).

References

1. Delgado, J., Torres, V., Llorente, S., Rodríguez, E.: Rights management in architectures for distributed multimedia content applications. Trustworthy Internet. Springer, Heidelberg (2011)
2. Distributed Multimedia Applications Group (DMAG), http://dmag.ac.upc.edu
3. Delgado, J., Llorente, S., Rodríguez, E., Torres-Padrosa, V.: A Mobile Scenario for Electronic Publishing based on the MIPAMS Architecture. In: 15th International Conference on Electronic Publishing Digital Publication and Mobile Technologies (June 2011)
4. iOS technologies, http://developer.apple.com/technologies/ios/
5. Delgado, J., Rodríguez, E., Llorente, S.: User's Privacy in Applications provided through Social Networks. In: 2nd ACM Workshop on Social Media, WSM 2010 (2010); ISBN: 978-1-60558-933-6
6. Llorente, S., Rodríguez, E., Delgado, J.: Secure Management of Social Networks Applications Data. In: Proceedings of the 8th International Workshop for Technical, Economic and Legal Aspects of Business Models for Virtual Goods (2010)
7. Rodriguez, E., Delgado, J., Alcalde, G.: Protection of patients' privacy by means of standard technologies. In: Proceedings of the 9th International Workshop for Technical, Economic and Legal Aspects of Business Models for Virtual Goods (2011)
8. ISO/IEC. ISO/IEC IS 21000:2 – Part 2: Digital Item Declaration (2005)
9. ISO/IEC. ISO/IEC IS 21000:5 – Part 5: Rights Expression Language (2004)
10. ISO/IEC. ISO/IEC IS 21000:15 – Part 15: Event Reporting (2006)
11. OASIS. Security Assertion Markup Language (SAML) (2005), http://saml.xml.org/
12. Internet Engineering Task Force (IETF): RFC 3280 Internet X.509 Public Key Infrastructure Certificate and Certificate Revocation List (CRL) Profile (April 2002)

13. Digital Media Project. Chillout, http://chillout.dmpf.org/
14. Digital Media Project, http://www.dmpf.org/
15. European FP7 Project CONVERGENCE, http://www.ict-convergence.eu/partners/
16. ISO/IEC. ISO/IEC DIS 23006-4 Multimedia service platform technologies – Part 4: Elementary Services (2011)
17. Android Open Source Project (AOSP), http://source.android.com/about/philosophy.html
18. Blackberry overview, http://us.blackberry.com/ataglance/
19. Windows phone, http://www.microsoft.com/windowsphone/es-es/default.aspx
20. Simple Object Access Protocol (SOAP), http://www.w3.org/TR/soap/
21. Fielding, R. T.: Representational State Transfer (REST), 5th chapter of Thesis dissertation Architectural Styles and the Design of Network-based Software Architectures, http://www.ics.uci.edu/~fielding/pubs/dissertation/top.htm
22. Android Developers, http://developer.android.com
23. Android SDK, http://developer.android.com/sdk/index.html
24. Chang, G., Tan, C., Li, G., Zhu, C.: Developing Mobile Applications on the Android Platform. In: Jiang, X., Ma, M.Y., Chen, C.W. (eds.) WMMP 2008. LNCS, vol. 5960, pp. 264–286. Springer, Heidelberg (2010)
25. Android 3.1 Highlights, http://developer.android.com/sdk/android-3.1-highlights.html
26. Apple, http://www.apple.com/
27. XCode tools, http://developer.apple.com/xcode/
28. iOS developer center, http://developer.apple.com/devcenter/ios/index.action
29. Objective C programming language overview, http://developer.apple.com/library/mac/#documentation/Cocoa/Conceptual/ObjectiveC/Introduction/introObjectiveC.html
30. Llorente, S., Delgado, J., Maroñas, X.: Implementing mobile DRM with MPEG-21 and OMA. In: WOSIS 2007 Proceedings. INSTICC Press (2007)
31. Llorente, S., Delgado, J., Maroñas, X., Barrio, R.: Experiencing Digital Rights Management in Mobile Environments. In: AXMEDIS 2008 Proceedings. IEEE Computer Society (2008)
32. ASIHTTP library, http://allseeing-i.com/ASIHTTPRequest/

Virtual Device: Media Service Fitness, Selection and Composition Considering Composition Interactivity and Synchronization

Niall Murray, Brian Lee, A.K. Karunakar, Yuansong Qiao, and Enda Fallon

Software Research Institute, Athlone Institute of Technology, Ireland
{nmurray,ysqiao,akkarunakar}@research.ait.ie,
{blee,efallon}@ait.ie

Abstract. The virtual device enables seamless use of application services residing on different devices in the vicinity of the user. In a pervasive environment, numerous service combinations can be selected to undertake a task. Current works aim to determine the best possible media services for composition by considering user preferences, environment capabilities and similarity between requested and available services. Previously, the authors considered all of above as well as potential local and remote content sources and destination devices. Here this is extended by considering end-to-end service latency to determine service fitness. The end-to-end delay of a service instance is important to consider as it directly affects the interactivity of the system. Services are selected for composition based on our fitness model. We model and simulate this issue and explain the results of our experimentation.

Keywords: Virtual device, Atomic service fitness, service composition. media service selection.

1   Introduction

Today users are surrounded by technology that is heterogeneous, pervasive, and variable [1]. The virtual device combines media services from different devices to support multi-modal communication. Logically it can be viewed as a single device, but each of the services that constitute the virtual device are from different devices. Examples of devices that could provide such services include: small handheld multifunctional devices with limited processing and display capabilities, enhanced single function dedicated devices or powerful multifunctional multimedia systems (e.g. PDA, PCs, HDTV's, network speakers and surround sound systems).

Considering devices in a pervasive environment, multiple instances of the same service types are likely to exist. The suitability (or fitness) of individual services is calculated to distinguish between instances. By selecting and composing the blue-chip services of different devices, the virtual device supports user task satisfaction beyond what a user companion device can offer. A service is defined as an indivisible, self-contained application unit that performs a processing function on multimedia content [2].

L. Atzori, J. Delgado, and D. Giusto (Eds.): MOBIMEDIA 2011, LNICST 79, pp. 281–294, 2012.
© Institute for Computer Sciences, Social Informatics and Telecommunications Engineering 2012

The virtual device supports the user task by combining the best service(s) of different devices within the context of the same session. Devices independent of their network connection or platform provide media services (see Fig. 1) to their peers for composition. In general, devices that provide one-to-many services support one service, that is the premium service associated with that device. For example, consider a HDTV that supports video and audio (many TV's currently support many other media services). It's logical to assume that the premium service of this device is its video display considering that people have manually configured surround sound systems in their homes to complement their TV box. Consequently by selecting the premium services from different devices, it is possible to optimally support the user task considering available services. Fig. 1 also includes what we define as content manipulation services, which is outside of scope of this work.

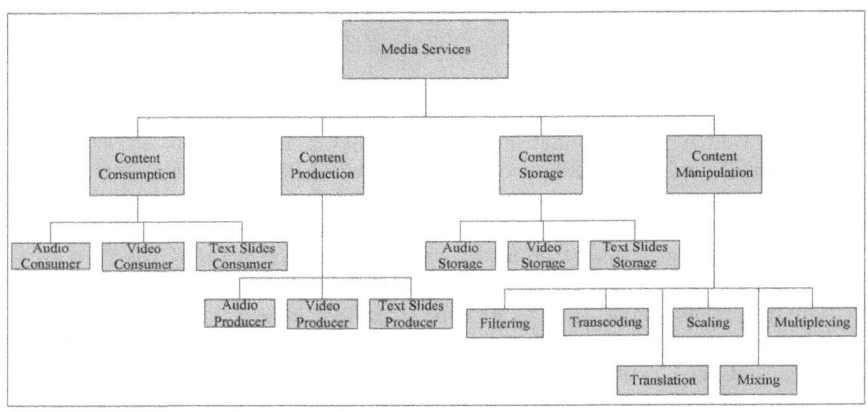

Fig. 1. Multimedia Content Service types [2]

If the goal virtual device is to provide a satisfactory Quality of Service for the user, low delay tolerant scenarios like real-time communication using this type of cluster application must be interactive. It should give the appearance that the selected services of different devices are indeed from a single device. If varying degrees of end-to-end delay exist between the selected services, the virtual device will appear disjointed, with audio/video/other multimedia skews degrading the user experience.

Concurrent services require strict functional and/or timely synchronization [3]. Synchronization ensures that the delivery of media is played out in a timely manner but it can lead to increased latency in a system. The nature of the virtual device, like any other cluster-to-cluster application (3DTI, collaborative workspaces etc.), means that different services available on different devices will have different end-to-end delays. This is because "different sensory streams employ their own protocols and adaptation algorithms in response to the bandwidth dynamics according to their diverse quality-of-service (QoS) requirements" [4]. We consider system end-to-end delay as being effected by network delays [5], network jitter [5] and end system jitters [5].

We model this multi-source, multi-destination service composition challenge using graph theory as per [6][7][8][9].

The goal is to achieve the maximum quality media service composition by considering context. If this selection process is not performed properly, the search will generate non optimized results, causing a low QoS perception from the consumer point of view [10]. Simply choosing the most powerful device or service does not always result in the most efficient and most favourable user experience [2]. It may not be the user's preferred mode of communication. We propose the consideration of estimated end-to-end service delay along with local and remote content consumption and production services, user preferences and device resources. Considering service end-to-end delay results in services being selected that provide greater interactivity in the system. Synchronization schemes like [11] can then be applied to streams with the lowest end-to-end delay. Also as previously proposed [2], considering remote services enables early elimination of unusable services and facilitates reduced media processing costs and delays. Finally, this work provides a method to overcome the dependency [12] on service users' feedback.

The overall contributions of this paper are:

1. Defining the service selection and composition problem to include service end-to-end delay, along with our previous contribution of local and remote services, user preferences and device resources.
2. We model and define the fitness of a service as a function of service end-to-end delay, availability, encodings and potential remote services.
3. We present the results of our experiments displaying the benefits of using the proposed algorithm.

This paper is organized as follows: In section 2 we outline related work. Section 3 defines the problem with the aid of a sample scenario. Section 4 describes how atomic services are modelled in terms of end-to-end delay, availability, compression formats supported and service type. Section 5 presents the algorithms for service composition. Section 6 discusses and explains experimentation and simulation results. Finally in Section 7 this paper is concluded, our contributions are reviewed and our future work explained.

2   Related Work

A broad range of related work involving task based service composition in pervasive environments and MANETS, content delivery and adaptation, the connection of devices in pervasive environments exists.

Similar to this work, Sousa et al. in [1] describe an approach to finding the best match between the user's requirements and the environment's capabilities. Hossain et al. in [13] determine the best possible composition as a function of gain (the extent of which a media service satisfies a user in a particular context) and cost of the service. In [14], Karmouch et al. define service composition in MANETS as a distributed constraint satisfaction problem. Similar to us they use a QoS model based on the work of Pertuttan et al. [15]. Pertuttan's QoS model refers to a degree of match between the user task requirements and the quality of the service composition. This work is based on a per service quality assessment considering the user task. We borrow facets of, and extend Pertuttan's model with end-to-end service delay, device capability and consideration of the remote services. Mukhtar et al. define an approach for task composition considering user preferences and device capabilities [16]. In [8], they use graph theory to define the user task. In [12] Atrey et al. use how regularly a service has been composed with another service, to determine a reputation for that service. In [17] Jiang et al. address service composition based on the prospect of minimum disruption. None of these works consider the end-to-end service delay to select local atomic services.

In relation to selection of the coordinator device, Karmouch et al. in [18] implement a broker based distributed service composition protocol which extends the work of Chakraborty et al. [19]. Basu et al. [7] define graph based approaches to distributed application composition approaches in MANETS.

[20][21][22] propose different approaches for connecting devices in pervasive environments. In [22], Schuster et al. provide a service orientated architecture for virtual device composition utilizing middleware on all devices. In [20], Senthivel et al. construct ad hoc service overlay networks (SON) based on service requests. In [21], Park et al. propose an interoperability framework based on the JXTA protocol. In [23], Ibrahim et al. survey middleware approaches to service composition and define service composition as a four step process: translation, generation, evaluation and execution. In [9], Kalasapur et al. propose a SOA based middleware platform which also incorporates graph theory. In [24], Lee et al. propose an approach based on a virtual device software manager, a middleware manager and hardware adaptation. In [25], Grigoras et al. address MANET formation based on device constraints like bandwidth and battery power.

Much service composition research has focused on media delivery. In [26], Gu et al. propose SpiderNet which provides Statistical QoS assurances for service composition. In [27], Jafarpour et al. strategically place content adaptation nodes in an overlay network to reduce costs in terms of communication and computation. In [28], Qian et al. contribute a multimedia delivery algorithm to calculate the lowest delay path. Xu et al. [29] propose a distributed Storage-assisted Data-driven overlay network to support P2P Video-on-Demand services.

In [30] Nahrstedt et al. introduce and discuss challenges with web services based approaches to multimedia delivery. SPovNet [31] is an overlay based solution that facilitates the spontaneous deployment of distributed network applications and services. In [32], Kim et al. discuss an emerging trend of media orientated service composition with SON's and outline challenges. They also dicuss virtualized resource components as a futuristic solution. In [33], Buford et al. suggest an Internet-scale P2P Overlay to facilitate expanding the capability of a device. Huang et al. in [4][34] propose synchronization solutions for 3DTI where the challenges are similar to those of the virtual device. Boronat et al. in [5] provide a detailed survey on inter stream and group synchronization schemes, whilst in [11] propose a RTP/RTCP based synchronization scheme for cluster-to-cluster applications. In [35] Murray et al. compare SIP and the Advanced Multimedia System as approaches to support future multimedia systems. In [36] Eid et al. incorporate multiple modalities (audio, video, graphics, haptics and scent data) in their Admux framework. In [37], Huang et al. uses prediction to improve interactivity of group synchronization with haptic media. In [38], Ghinea et al. aim to determine perceived synchronization boundaries of olfactory data. In [39] Nunome et al. assess group synchronization schemes for audio and video in wireless ad hoc networks and conclude that a new group synchronization scheme is required. The webinos project [40] aim is to provide a service platform where applications can be used and shared across devices.

To the best of our knowledge, the use of end-to-end delay of potential services has not been a driver to calculate fitness, select and compose local atomic services.

3   Problem Definition

We envisage public and private environments consisting of multiple mobile and stationary nodes that provide content related services. Fig. 2 shows a number of devices providing one-to-many services. All nodes are Internet connected devices where resident services can be invoked by peer nodes. Consider the following scenario defined as real time communication between distributed compositions. Brian is talking with John on his virtual device enabled smart phone and is walking from his office to his car. John is working from home and is using the microphone from his personal computer and surround sound speaker system to converse with Brian. Also in John's study are a PC connected camera and a large display screen. Once Brian sits in his car and places the phone in the holder, his call is automatically transferred from his handheld device to the services available in his car. He is now using the microphone embedded in the sun visor, car radio speakers and can see John on the LCD panel integrated into the car. John can also now see Brian because of the camera built into the steering wheel.

Local services from Brian's perspective are the services resident on devices within his vicinity that he uses to communicate with John (e.g. the microphone embedded in the visor). Remote services from Brian's perspective are the services available to John for communication (e.g. microphone in his personal computer).

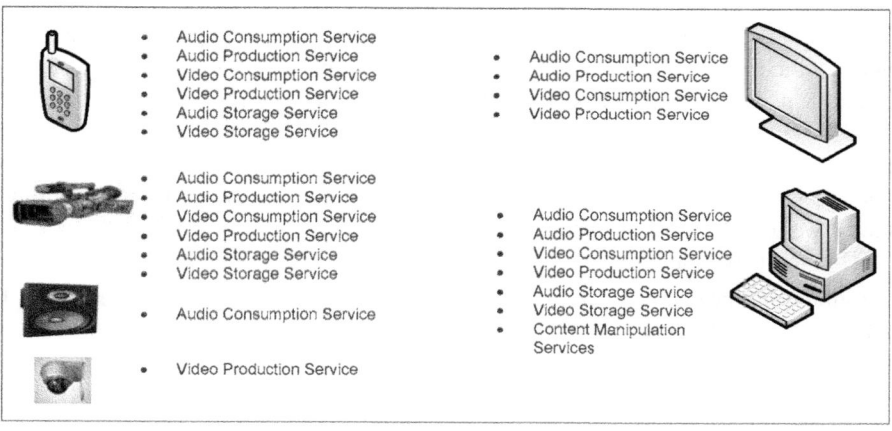

Fig. 2. Shows a number of different devices providing one or many services that could be used to in service composition to create the virtual device

Considering this scenario and Fig. 2, many of the devices in a user's vicinity may provide similar functional services. However differences may exist between these services in terms of capabilities, availabilities, user preferences, usage costs and end-to-end delays. In the scenario outlined, the user may want to use many types of services (e.g. video and audio consumption and production services) in a communication session. Consequently there are multiple constraints in terms of what the user requires, and multiple choices in terms of devices providing different services to solve these problems. Determining the quality of compositions is a variant of a 0-1 Knapsack problem, called multiple dimensional, multiple choice 0-1 Knapsack. Multiple dimensions refer to the multiple constraints and multiple choices refer to choosing one among a set of similar items. In optimizations research this problem has been proven to be NP-complete [41].

Our work aims to achieve the best possible service composition by selecting the fittest instances of the individual service types by considering end-to-end service delay, device resources, user preferences and remote services.

4   Service Composition System

Composing services from distributed devices requires a number of steps. Many of these steps are outside the scope of this document and are outlined in [2]. [2] also provides descriptions of all possible entities required for such a system. It is assumed an Internet scale network exists. Devices provide their content production, storage, consumption and manipulation services for execution by peers. Atomic services provide content processing functionality as per Fig. 1. The end-to-end service delay is the time taken to deliver content from a local content production service to a remote content consumption service or from a remote content production service to a local content consumption service.

4.1   Modelling

This network can be represented by a bipartite digraph, G. This graph is composed of nodes which represent devices that provide one-to-many services. Links in the graph reflect connections between these devices. The disjoint sets in the graph represent the local services and remote services described in section 3. The resources of a device are the quantifiable non-functional characteristics of a device. A scoring of device resources is executed with respect to its peers. Further information on modelling of the network and resources can be obtained in [2].

The services available on a particular node $S = \{S_1, S_2,...S_k\}$ are the content consumption, production, storage and manipulation services as per Fig. 1. The services supported by a device n are denoted by $S_i(n)$ where $ST(s)$ represents the

$$S_i(n) = \{ST, E, A, D\}, \text{ where } 0 \leq i \leq k \tag{1}$$

service type, E represents the content compression formats supported and $A(s_i)$ represents the availability status of the service [2]. It is assumed that all nodes in the network can exchange information. $D(s_i)$ represents the end-to-end delay value from a remote site to a local consumption service instance or from a local production service to the remote site. D can be calculated using approaches discussed in [4][11]. End-to-end service delay values are used in algorithm 1 to calculate the fitness of local service instances with respect to their peers. In [2] a specification of a user task is defined. Local and remote content consumption and production services (S_L and S_r respectively), availability, device resources and user preference are considered. Similar to [6][7][8] this specification is detailed using graph theory.

In [2], the contribution was using the existence of remote services and their encoding formats to determine the fitness of local services. The novel aspect of this work is use of the end-to-end delay between potential local services S_L and potential remote services S_r to determine the fitness of local services. The goal is selection of the best possible composition set providing the user with the most interactive user experience.

4.2   Local Atomic Service Fitness Calculation Considering End-to-End Delay

A three step suitability function is used to calculate the fitness of local atomic services of the same type. Per service type ranking tables are generated which reflect the fitness of a service instance with respect to other services instances of the same type. In [2], the work of Pertutten et al [15] was modified and extended to produce local atomic service type ranking tables based on Atomic Service Value ($ASV_t(s_i)$) and Final Service Value ($FSV_t(s_i)$). $ASV_t(s_i)$ considers the devices resources, user inputted weightings, atomic service similarity with requested service, user preferences and availability. As shown in Fig 3, discovery of local services is a prerequisite for this step.

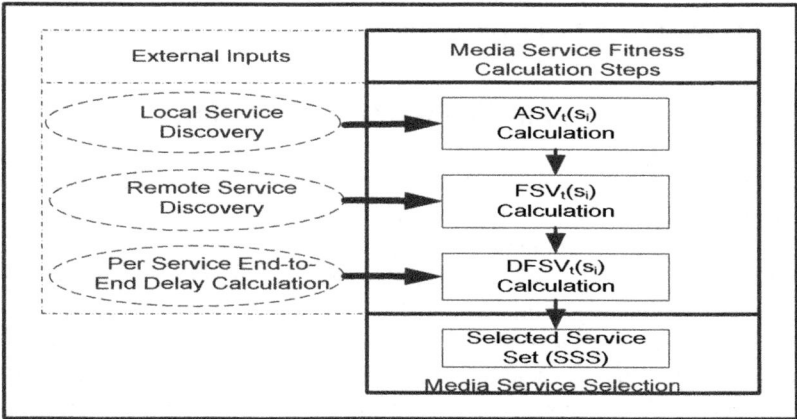

Fig. 3. Service composition system architecture. A number of external entities provide data to the media service composition algorithm. Local service discovery provides all necessary information to calculate and generate tables based on $ASV_t(s_i)$. Remote service discovery provides all data necessary to further calculate and update $ASV_t(s_i)$ ranking tables, producing $FSV_t(s_i)$. Using per service end-to-end delay calculation and $FSV_t(s_i)$ enables the generation of tables based on $DFSV_t(s_i)$. The SSS is calculated based on tables produced after calculation of $DFSV_t(s_i)$.

The $ASV_t(s_i)$ tables are updated by considering existence of potential atomic services of the remote composition resulting in $FSV_t(s_i)$ ranking tables per service type. Again from Fig 3, discovery of remote services is necessary in order to calculate $FSV_t(s_i)$. Detailed information on calculating $ASV_t(s_i)$ and $FSV_t(s_i)$ can be obtained from [2]. This work extends the contribution in [2] to calculate the fitness of local atomic services by considering end-to-end delay. By providing weightings for the services with the lowest end-to-end delays, atomic services that support the most interactive user experience have a greater chance of being selected for composition. As Toshiro et al. have shown in their comparison of the best known group synchronization techniques [15], synchronization, which is necessary in any cluster application like the virtual device increases latency. Considering the delay of a service with respect to its peers results in the generation of updated $FSV_t(s_i)$ ranking tables, namely delay-based final atomic service value ($DFSV_t(s_i)$). This reflects an atomic service's position with respect to other atomic services considering end-to-end delay. As a result of this step, the selected synchronization scheme will be applied to the streams with the lowest end-to-end delays.

4.3   Atomic Service Selection

The final ranking tables comprise atomic services ordered in terms of $DFSV_t(s_i)$. These tables represent the set of all available atomic services for the communication session. The top scoring services in each table are selected for composition. These services make the Selected Service Set (SSS) [2]. Only one service is selected from each of the service tables at any one time. The resultant selected composition reflects the fittest set of services to support the user task based on our service fitness scoring system. The rating for the optimized service composition is a sum of the highest scoring atomic services i.e. sum of the max $DFSV_t(s_i)$ values of each atomic service type.

5   Service Fitness and Composition Algorithm

This section outlines in pseudo code the various steps of the new algorithm introduced as part of this paper. The functionality of this algorithm compliments the algorithms introduced in [2] i.e. calculation of $ASV_t(s_i)$ and $FSV_t(s_i)$. $ASV_t(s_i)$ is calculated by scoring local atomic services in terms of devices resources, user inputted weightings, atomic service similarity with requested service, user preferences and availability. $ASV_t(s_i)$ values are updated firstly by considering existence of remote "partner" services and secondly by determining if commonalities exist between services in terms of the codec's supported. The result is $FSV_t(s_i)$ tables. Pseudo code of algorithms to calculate $ASV_t(s_i)$ and $FSV_t(s_i)$ can be obtained in [2].

Algorithm 1 below compares the end-to-end delay of each of the atomic services within each of the atomic service types. The lowest end-to-end delay values reflect services that will provide the most interactive and real time user experience. Hence, these services are given the highest weighting. $DFSV_t(s_i)$ tables are produced by updating the $FSV_t(s_i)$ tables considering end-to-end delay. For a particular service instance, the delay weighting is multiplied by the respective $FSV_t(s_i)$ value to produce a DFSV$_t$(s$_i$) value. The highest scoring atomic services in each of the final set of fitness tables are then selected for composition. It is the highest quality mapping from the user specification to the actual services available on devices. If for any reason, one of the atomic service types becomes unavailable, the next service is selected for instantiation to support execution of the user task.

Algorithm 1: Comparison of local Atomic Services w.r.t end-to-end delay and Optimal Service Composition
1.   For each Final service value (FSV$_t$(s$_i$)) fitness tables needed to satisfy a task
2.   For each Atomic Service within each table
3.   Compare its end-to-end delay with other atomic services of the same type. //
4.   Determine weighting with respect to peer services.
5.   Update FSV$_t$(s$_i$) based on its value with others. Lowest end-to-end delay gets highest weighting score to determine DFSV$_t$(s$_i$) per atomic service.
6.   Sort all rows in table based on DFSV$_t$(s$_i$) value.
7.   return All Fitness Tables
8.   For each DFSV$_t$(s$_i$) table
9.   SelectTopScoringService()
10. Compose Set of setOfTopScoringServices

6   Simulation Results and Analysis

Simulations are performed on a Windows Vista Ultimate OS with 4.00 GB RAM and Intel® Core™ 2 Quad Q6600 @ 2.4GHz. The simulated environment models ten devices that can potentially provide one-to-many services within the vicinity of the user.

For completeness Table 1 below is included which shows a sample generated ranking table for audio consumption type atomic services. All execution points to determine ASV$_t$(s$_i$) are displayed. The maximum value for each of the parameters is one, hence the highest possible score of ASV$_t$(s$_i$) is also one. Inspecting the values, considering resource capability (R(n)) and similarity of available with requested service Sim(RS$_i$, AS$_i$) [15], the PC speakers and surround sound services score as the strongest candidates. All instances are available (A(s$_i$) column) and hence are given a value of 1. W$_t$(s) [15], is a user inputted weight to signify how important the user views a service type. The user preferences U(s$_i$) are the same (value is 0.7) for both the surround sound and PC speakers so after determination of ASV$_t$(s$_i$), both services have the same fitness score. Formulas for calculation of ASV$_t$(s$_i$) and FSV$_t$(s$_i$) are available in [2].

FSV$_t$(s$_i$) as shown in table 2 is a the product of columns ASV$_t$(s$_i$) and encoding match weighting. Considering encodings supported, the HDTV speakers and surround sound speakers are assigned weightings of 1 as they have common encodings with remote services. The PC speakers do not have a common encoding with remote services, hence it is assigned a weighting of 0.5. At this point the surround sound is considered the best atomic service with a FSV value of 0.7, the HDTV speakers are second strongest with a FSV value of 0.651. These respective values are achieved by multiplying the $ASV_t(s_i)$ and the encoding match weighing to determine the updated service value $FSV_t(s_i)$.

Table 1. Audio Consumption $ASV_t(s_i)$ Fitness table after execution of Algorithm 1

AS Type: Audio Consumer	R(n)	Sim(RS$_i$, AS$_i$)	A(s$_i$)	U(s$_i$)	W$_t$(s)	ASV$_t$(s$_i$)
HDTV Speakers	0.93	1.0	1.0	0.7	1.0	0.65
Surround Sound Network connected speakers	1.0	1.0	1.0	0.7	1.0	0.7
PC Speakers	1.0	1.0	1.0	0.7	1.0	0.7
Smart Phone Speakers	0.73	1.0	1.0	0.3	1.0	0.219
Radio Speakers	0.85	1.0	1.0	0.0	1.0	0.0

The final step taken and novel contribution of this work is the consideration of end-to-end delay of an atomic service in determining service fitness using the calculated $FSV_t(s_i)$ and EED weightings as inputs. Different weightings are obtained for each of the audio consumption atomic services based on end-to-end delay. The delay values for each of the atomic services are compared. The lowest delay value between the services gets the weighting value of one.

Table 2. Audio Service fitness calculation: $FSV_t(s_i)$ and $DFSV_t(s_i)$

AS Type: Audio Consumer	ASV$_t$(s$_i$)	Encoding Match Weighting	FSV$_t$(s$_i$)	EED(s$_i$) Weighting	DFSV$_t$(s$_i$)
HDTV Speakers	0.651	1.0	0.651	0.8	0.5208
Surround Sound connected speakers	0.7	1.0	0.7	0.7	0.49
PC Speakers	0.7	0.5	0.35	0.9	0.315
Smart Phone Speakers	0.219	0.5	0.1095	1.0	0.1095

All other atomic services, ordered in terms of their end-to-end delay, are assigned decremented weightings of 0.1. Hence, the second lowest end-to-end delay is given weighting of 0.9. The third lowest is assigned a weighting of 0.8 etc.

Figure 4a below reflects the ratings of the services as their information is processed through each of the stages of algorithms in this work and in [2]. The services selected for composition are the highest individual scoring atomic services. For explanation purposes, Fig. 4b compares the fitness of the highest scoring SSS with a set comprised of medium scoring atomic services, and two compositions with sets of low scoring atomic services. Comparing the highest SSS with services (s3,s6,s11,s16) and lowest SSS with services (s4,s8,s12,s14) have scores of 2.37 and 0.584 respectively as shown in Fig. 4b.

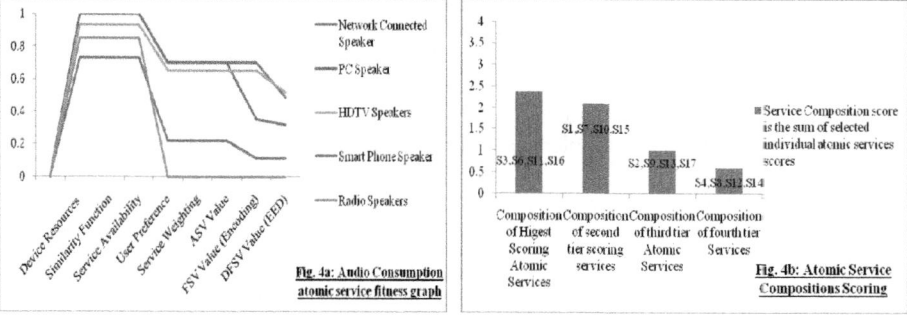

Fig. 4a: Audio Consumption atomic service fitness graph

Fig. 4b: Atomic Service Compositions Scoring

Fig. 4. 4a reflects the change in scores after each of the steps to calculate $ASV_t(s_i)$, $FSV_t(s_i)$ and $DFSV_t(s_i)$. **4b** compares the scores of a set of high fitness services with a set of medium fitness services and a set of low fitness services.

7   Conclusion

This paper has presented an algorithm to calculate service fitness and select services for composition. This novel approach uses the end-to-end delay of local atomic services as input to deciding the local service composition. In addition, it considers device resources, user preferences, remote encodings supported and resource capabilities. The user task is modelled taking into consideration all of the aforementioned factors. A three step suitability process produces ranking tables for required atomic services based on service fitness. The result provides a user with an optimized selection of services, whilst providing an efficient service failure recovery mechanism. The simulation and experimentation show how each of the factors considered; local services, end-to-end delay, remote services, resource capability and user preferences affect the service scoring and how the best possible set of services are selected. In future work, we will investigate synchronization requirements for cluster applications that support multiple correlated media streams, namely audio, video and olfactory data.

Acknowledgments. We would like to recognize the support of Enterprise Ireland through its Applied Research Enhancements program in the financing of this Research. Also Dr. Mark Daly, a lecturer in the AIT School of Engineering for his assistance and advice with system modelling here and in previous work.

References

1. Sousa, J.P., Poladian, V., Garlan, D., Schmerl, B., Shaw, M.: Task-based Adaptation for Ubiquitous Computing. IEEE Transactions on Systems, Man, and Cybernetics – Part C: Applications and Reviews 36, 328–340 (2006)
2. Murray, N., Lee, B., Karanukar, A.K., Qiao, Y., Fallon, E.: Virtual Device: Media Service Fitness, Selection and Composition. In: 4th International ICST Conference on MOBILe Wireless MiddleWARE, Operating Systems, and Applications (2011)
3. Nahrstedt, K., Balke, W.-T.: A Taxonomy for Multimedia Service Composition. In: 12th ACM International Conference on Multimedia (2004)
4. Huang, Z., Wu, W., Nahrstedt, K., Rivas, R., Arefin, A.: SyncCast: Synchronized Dissemination in Multi-site Interactive 3D Tele-immersion. In: Multimedia Systems 2011 (2011)
5. Boronat, F., Lloret, J., García, M.: Multimedia group and inter-stream synchronization techniques: A comparative study. Inf. Syst. 34(1), 108–131 (2009)
6. Nahrstedt, K., Yu, B., Liang, J., Cui, Y.: Hourglass multimedia content and service composition framework for smart room environments. Journal on Pervasive and Mobile Computing 1(1), 43–75 (2005)
7. Basu, P., Wang, K., Little, T.D.C.: Dynamic Task Based Anycasting in Mobile Ad Hoc Networks. Journal Mobile Networks and Applications (2003)
8. Mukhtar, H., Belaïd, D., Bernard, G.: A Graph-Based Approach for Ad hoc Task Composition Considering User Preferences and Device Capabilities. In: GLOBECOM Workshops, Current version (2009)
9. Kalasapur, S., Kumar, M., Shirazi, B.A.: Dynamic Service Composition in Pervasive Computing. IEEE Transactions on Parallel and Distributed Systems 18(7) (2007)
10. Dutra, R.G., Martucci Jr., M.: Dynamic Adaptive Middleware Services for Service Selection in Mobile Ad-Hoc Networks. In: Cai, Y., Magedanz, T., Li, M., Xia, J., Giannelli, C. (eds.) Mobilware 2010. LNICST, vol. 48, pp. 189–202. Springer, Heidelberg (2010)
11. Boronat, F., Montagud, M., Guerri, J.C.: Multimedia Group Synchronization for One-way Cluster-to-Cluster Applications. In: 34th IEEE Conference on Local Computer Networks, LCN (2009)
12. Atrey, P.K., Hossain, M.A., El Saddik, A.: A Method for Computing the Reputation of Multimedia Services through Selection and Composition. Journal Mobile Networks and Applications 13(6) (2008)
13. Hossain, M.A., Atrey, P.K., El Saddik, A.: Gain-based Selection of Ambient Media Services in Pervasive Environments. Springer Journal Mobile Networks and Applications (2008)
14. Karmouch, E., Nayak, A.: A Distributed Constraint Satisfaction Problem for Virtual Device Composition in Mobile Ad Hoc Networks. In: Global Telecommunications Conference, GLOBECOM (2009) Current Version (2010)

15. Perttunen, M., Jurmu, M., Riekki, J.: A QoS Model for Task-Based Service Composition. In: Workshop on Managing Ubiquitous Communications and Services (2007)
16. Mukhtar, H., Belaïd, D., Bernard, G.: User Preferences-Based Automatic Device Selection for Multimedia User Tasks in Pervasive Environments. In: 5th Conference on Networking and Services (2009)
17. Jiang, S., Xue, Y., Schmidt, D.: Minimum Disruption Service Composition and Recovery in Mobile Ad hoc Networks. The International Journal of Computer and Telecommunication Networking 53(10) (2009)
18. Karmouch, E., Nayak, A.: A Distributed Protocol for Virtual Device Composition in Mobile Ad Hoc Networks. In: IEEE International Conference on Communications (2009)
19. Chakraborty, D., Joshi, A., Finin, T., Yesha, Y.: Service Composition for Mobile Environments. Mobile Network and Applications 10(4), 435–451 (2005)
20. Senthivel, K., Kalasapur, S., Kumar, M.: PerSON: A Framework for Service Overlay Network in Pervasive Environments. In: Kuo, T.-W., Sha, E., Guo, M., Yang, L.T., Shao, Z. (eds.) EUC 2007. LNCS, vol. 4808, pp. 671–682. Springer, Heidelberg (2007)
21. Park, H., Park, J.-H., Kim, N.: A Framework for Interoperability of Heterogeneous Devices in Ubiquitous Home. In: 2nd Conference on Advances in Future Internet (2010)
22. Schuster, M., Domene, A., Vaidya, R., Arbanowski, S., Kim, S.M., Lee, J.W., Lim, H.: Virtual device Composition. In: 8th International Symposium on Autonomous Decentralized Systems (2007)
23. Ibrahim, N., Le Mouël, F.: A Survey on Service Composition Middleware in Pervasive Environments. International Journal of Computer Science Issues, IJCSI (2009)
24. Lee, J.W., Kim, S.M., Lim, H., Schuster, M., Domene, A.: A Software Architecture for Virtual Device Composition and Its Applications. In: Ubiquitous Computing Systems (2010)
25. Grigoras, D., Riordan, M.: Service Driven Mobile Ad Hoc Networks Formation and Management. In: 4th Symposium on Parallel and Distributed Computing (2005)
26. Gu, X., Nahrstedt, K.: Distributed Multimedia Service Composition with Statistical QoS Assurances. IEEE Transactions on Multimedia 8(1), 141–151 (2006)
27. Jafarpour, H., Hore, B., Mehrotra, S., Venkatasubramanian, N.: CCD: Efficient Customized Content Dissemination in Distributed Publish/Subscribe. In: ACM/IFIP/USENIX International Conference on Middleware (2009)
28. Qian, Z., Guo, M., Zhang, S., Lu, S.: Service-Oriented Multimedia Deliver. In: Pervasive Space. In: IEEE Wireless Communications & Networking Conference (2009)
29. Xu, C., Muntean, G.-M., Fallon, E., Hanley, A.: Distributed Storage-Assisted Data-driven Overlay Network for P2P VoD Services. IEEE Transactions on Broadcasting (2008)
30. Nahrstedt, K., Balke, W.-T.: Towards Building large Scale Multimedia Systems and Applications: Challenges and Status. In: ACM International Workshop on Multimedia Service Composition (2005)
31. The Spontaneous Virtual Network (SpoVNet), http://www.spovnet.ce
32. Kim, J., Han, S.W., Yi, D.-H., Kim, N., Jay Kuo, C.-C.: Media-Oriented Service Composition with Service Overlay Networks: Challenges, Approaches and Future Trends. Journal of Communications (2010)
33. The Buford, J.F., Yu, H., Lua, E.K.: P2P Networking and Applications. Morgan Kaufmann (2009)
34. Huang, Z., Wu, W., Nahrstedt, K., Arefin, A., Rivas, R.: TSynch: A new Synchronization Framework for Multi-site 3D Tele-immersion. In: 20th International Workshop on Network and Operating Systems Support for Digital Audio and Video, NOSSDAV (2010)

35. Murray, N., Qiao, Y., Lee, B., Karunakar, A.K., Fallon, E.: Design Considerations for Future Multimedia Systems: A Comparison of Approaches Based on SIP and the Advanced Multimedia System. International Journal of Ambient Computing and Intelligence 3(1), 20–32 (2011)

36. Eid, M., Cha, J., El Saddik, A.: Admux: An Adaptive Multiplexer for Haptic-Audio-Visual Data Communication. IEEE Transactions on Instrumentation and Measurement 60(1) (2011)

37. Huang, P., Ishibashi, Y., Fukushima, N., Sugawara, S.: Interactivity Improvement of Group Synchronization Control in Collaborative Haptic Play with Building Blocks. In: 9th Annual workshop on Network and Systems Support for Games (2010)

38. Ghinea, G., Ademoye, O.A.: Perceived Synchronization of Olfactory Multimedia. IEEE Transactions on Systems, MAN, and Cybernetics – Part A: Systems and Humans 40(4), 657–663 (2010)

39. Nunome, T., Tasaka, S.: Inter-Destination Synchronization Quality in a Multicast Mobile Ad Hoc Network. In: 16th IEEE International Symposium on Personal (current version 2006)

40. webinos (Secure Web Operating System Application Delivery Environment), FP7-ICT-2009-5 – Objective 1.2, http://webinos.org/

41. Pisinger, D.: An Exact algorithm for Large Multiple Knapsack Problems. European in Journal of Operational Research 114 (1999)

Location Based Abstraction of User Generated Mobile Videos

Onni Ojutkangas, Johannes Peltola, and Sari Järvinen

VTT Technical Research Centre of Finland
Kaitoväylä 1, 90570 Oulu, Finland
{onni.ojutkangas,johannes.peltola,sari.jarvinen}@vtt.fi

Abstract. Demand for efficient ways to represent vast amount of video data has grown rapidly in recent years. The advances in positioning services have led to new possibilities of combining location information to video content. In this paper we present an automatic video editing system for geotagged mobile videos. In our solution the system creates automatically a video summary from a set of unedited video clips. Geotags are used to group video clips with the same context properties. The groups are used to create a video summary where the videos from the same group are represented as scenes. The novelty in our solution lies in the combining of geotags with low level content analysis tools in video abstraction. Evaluations of the system prove the concept useful as it improves coherence and enjoyability of the automatic video summaries.

Keywords: video summarization, context awareness, video content analysis, temporal segmentation.

1   Introduction

The significance of digital video applications and their market potential has grown remarkably in past years. Drivers for this change can be found from increased processing power of computers and portable devices, larger capacity of storages, faster networks, and digitalization of video. Consumers have moved from using old tape camcorders to digital video recording with devices such as mobile phones and digital cameras. Video recording with mobile phones is fundamentally different than with traditional camcorders. Traditionally videos were recorded at special occasions such as birthdays, festivals and when travelling and the use of the recorder was planned beforehand. Also more effort was given to editing the content. In contrast, as people carry mobile phones with them video recording has become more spontaneous and video is recorded at situations where it previously was not.

The growing amount of personal video content has led to problems with management and consumption of videos. Tools and methods for video editing available to consumers are modeled after those of professional video production, even if consumers usually possess neither time, money, nor expertise that professional production methods require [19]. In order to overcome this obstacle for personal

L. Atzori, J. Delgado, and D. Giusto (Eds.): MOBIMEDIA 2011, LNICST 79, pp. 295–306, 2012.
© Institute for Computer Sciences, Social Informatics and Telecommunications Engineering 2012

video management a number of solutions for automatic video editing have been presented. The published work includes systems used for various video domains: sport [3], news [4], documentaries, movies [5], lecture recordings and home videos [1],[2][6],[7]. However, automatic video editing of mobile videos is rarely discussed. In [8] the authors present a video editor for mobile phones based on user studies. They identify the following user goals for video editing in mobile context: select the clips to be edited from the raw source material, combine several separate video clips into one video, cut a clip, enhance the video with text, images, music, and special effects, store the completed video in the device and share the created videos with family or with peer group.

An effective way to make the viewing experience more entertaining is to present the video in as compact form as possible. Hua et al. [2] presents a system, which automatically selects suitable or desirable highlight segments from a set of unedited home videos and aligns them with a given piece of incidental music.

Truong et al. [17] urges more work to be done on maintaining the context and coherence of generated video summaries. Usually shots are joined together simply by their temporal order. However when a video summary is done from a large video collection of material from multiple users from multiple locations, this approach is not feasible.

The targeted domain in our work is personal mobile videos. Techniques proposed for video summarization often use domain specific features, for instance applause or cheering of audience in sport videos. Lienhart [3] uses an empirically motivated approach to cluster time-stamped shots of home video. The approach was encouraged by the fact that the moment of shooting plays a significant role in the home video domain. Lienhart identifies the following unique features for home videos that also apply to personal mobile videos:

- home videos don't have artificial story, plot or structure and
- home video consists of unedited, raw footage.

Our own observations from previous work on mobile videos [18][20][21] revealed that mobile videos include also unique features. Mobile videos are recorded freehanded with small and light devices and this sometimes leads to jerkiness or unstableness in the video. Another observation is that instead of recording several shots average user tends to use panning to cover the surroundings of scenery. Videos recorded with a mobile phone are also often relatively short as the motivation for capturing a mobile video usually is to record a surprising and interesting event or happening for reminiscing or sharing. In addition to this the capturing of mobile videos is not usually planned beforehand unlike use of camcorders. One feature of current video recorders in mobile phones is that they can shoot high quality video in good lighting conditions but they suffer from poor quality in the gloom.

A mobile phone as video recorder provides also means for sensing of context and associating this information to video content. Dey and Abowd [13] [14] regard identity and activity as primary types of context. These primary context types can be used as indices to access secondary context metadata such as a list of friends, user preferences or other people at the same location. This kind of metadata could be used for personalizing the video abstraction.

Geotagging is becoming more popular as integrated GPS modules are found more and more in consumer products such as Nokia N-series multimedia phones, Apple iPhone and Nikon Coolpix P6000 digital camera. Popular multimedia services such as Flickr offer their users various ways to associate location information into their multimedia content. On the other hand map services like Google Maps or Google Earth can be used to visualize the multimedia content using a map interface.

Video clips from same context

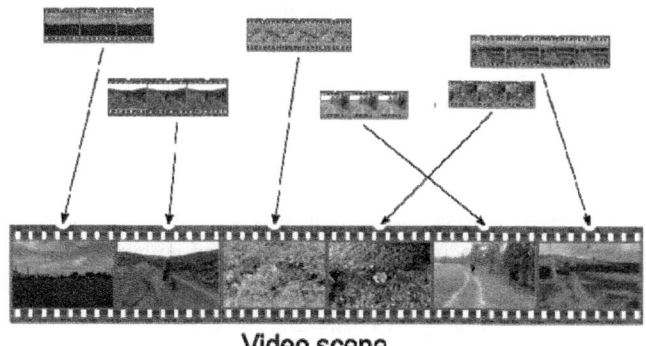

Video scene

Fig. 1. Concept of a video scene

A novel approach for abstraction of video content is introduced in this work. The goal was to build a system to automatically edit a coherent video summary from a set of unedited user generated mobile video clips created by a group of users. Our aim is to enhance community-based communication by creating an easy-to-use service that allows composing different video summaries based on commonly created video content. Creation location (GPS coordinate) and time information associated with the video clips are used to group shots from same context together to collections of video clips for creation of scenes (Fig. 1). This approach follows the definition [16] of a video scene which depicts action in one location and time. The video clips are segmented and shots to be included in the summary are selected based on results from audio and video analysis taking into consideration parameters such as segmentation threshold, motion intensity weight, motion variance weight and sound weight. A summary of videos is presented as a sequence of scenes. Fading transition effect is used between video scenes to express passage of time and changing of context to the viewer.

The system functionality has been evaluated using a set of video clips acquired from real users of the content creation platform presented in [18]. The video content has been used to generate various video summaries for a small scale user evaluation. The purpose for the evaluation was to get a first impression of the applicability of the chosen approach in enhancing the quality and the amount of information of the video summaries.

The paper is organized as follows; in section 2, we describe the developed video summarization system. In section 3, we present results from an evaluation performed with a small user group. Finally, in sections 4 and 5, we provide concluding remarks and our thoughts of future work.

2   Video Summarization System

Basic steps of video summarization in this system include video segmentation, subshot selection, scene creation and video rendering. Video segmentation is done by analyzing the coherence of video content in frame level. In subshot selection phase a score that reflects the level of interest in a particular subshot is calculated and the most interesting subshots are selected. Video scenes are created by clustering selected subshots by their creation location coordinates and timestamps. Final video summary is rendered by connecting video scenes together and using fading transition effect between them. These steps are described in more detail in the following chapters.

2.1   Feature Extraction

Segmentation of video cannot be done without analyzing the content of video. Video content is analyzed by calculating feature (Table 1) values from audio and video stream. Image brightness, image contrast and similarity features are all calculated from image histograms. Similarity of consecutive frames is measured with match value

$$match(h_1(i), h_2(i)) = \frac{\sum_{j=0}^{L-1} \min\{h_1(j), h_2(j)\}}{\sum_{j=0}^{L-1} h_1(j)}$$

where h_1 and h_2 are compared histograms with L bins. These features are lightweight and give general information of visual content. Camera and object motion are calculated from optical flow of two consecutive frames. Optical flow is calculated with Lucas Kanade method utilising OpenCV library [8]. Root-mean-squared (rms) value of sound signal is calculated over a sequence of n samples, which is set to match the video frame rate:

$$n = \left\lfloor \frac{number\ of\ audio\ samples}{number\ of\ video\ frames} \right\rfloor = \left\lfloor \frac{audio\ sample\ rate}{video\ sample\ rate} \right\rfloor$$

Table 1. Extracted features

Feature	Measure
Image brightness	Mean of image Y channel intensity values
Image contrast	Variance of image Y channel intensity values
Similarity of consecutive frames	Match values of Y, U, and V components
Camera motion	Mean of optical flow vector x and y components
Object motion	Variance of optical flow vector magnitudes
Sound power	Sound RMS power

2.2   Video Segmentation

In video segmentation phase the whole video sequence is segmented to base units, subshots. Low level video parsing can be seen as the analysis of content coherence of a video stream. Subshot boundaries are detected by computing the values of features for each frame of the clip and applying a coherence discontinuity detection mechanism. Coherence of a subshot is analyzed by comparing feature values of two consecutive frames and also comparing average of values before and after the frames. This method is presented in Fig 2.

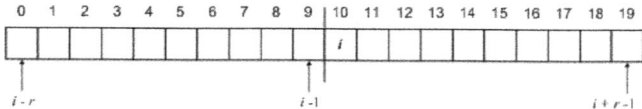

Fig. 2. Segmentation decision is made comparing feature vectors of two consecutive frames, i and i-1, and average vector of r previous frames and r next frames

Similarity of feature vectors is measured with normalized Euclidean distance

$$D(\bar{x}, \bar{y}) = \sqrt{\sum_{i=1}^{p} \frac{(x_i - y_i)^2}{\sigma_i^2}},$$

where x_i and y_i are elements of feature vectors $\bar{x}$ and $\bar{y}$.

Average feature values represented in Table 2 are calculated for each subshot. Personalization of the video summary is accomplished by weight factors a_j given for each feature.

Table 2. Features of a subshot and expected effect of weight factor to the final video summary

Feature of subshot	Expected effect on final summary
Brightness	Giving greater weight to this feature subshots recorded in poor lighting conditions are not likely to be selected to the video summary.
Contrast	Subshots with narrow contrast value are not likely to be included to the video summary if this feature is given weight.
Motion intensity	This feature should be given weight if a video summary with camera movement is desired.
Motion variance	If this feature is given weight scenes with moving objects are expected to be found in the final video summary.
Sound power	By giving weight to this feature subshots including sounds are more likely to be included in the video summary.

2.3   Subshot Selection

In order to create a video summary with the best possible user experience the system tries to preserve the most interesting subshots. The level of interest is modeled with a linear combination

$$\sum_{i=1}^{m} a_i z_i \, ,$$

where z_i is the subshot feature value and a_i the feature weight.

Selection of the subshots can be seen as a 0-1 knapsack problem: maximizing the total perspective score while not exceeding the upper bound of the summary length. In this work a summarization ratio is known a priori information. Summary length can be easily calculated by multiplying total duration of videos with the summarization ratio. The knapsack problem can be formulated as an optimization problem:

$$\text{maximize} \quad \sum_{j=1}^{n} p_j x_j$$

$$\text{subject to} \quad \sum_{j=1}^{n} w_j x_j \le c, \quad x_j = 0 \text{ or } 1, \quad j = 1, \ldots, n.$$

Here p_j is a profit, w_j is a weight of an item j and c is capacity of the knapsack. In subshot selection phase this problem needs to be solved.

In this work greedy algorithm is used to solve the optimization problem. Algorithm arranges items to descending order according to their p_i/w_i ratio. Then items are added to the sack starting from the first item and stopped when no items fit to the sack anymore. In this case the duration of a subshot is the weight value w_j. The profit value p_j represents the level of interest value of a subshot.

Duration of a subshot cannot be used as subshot feature in this case, because of the greedy algorithm used for solving the subshot selection. This can be simply discovered by assuming duration feature to be z_1, which means that p_i/w_i ratio would be

$$\frac{a_1 z_1 + a_2 z_2 + \cdots + a_m z_m}{z_1} = a_1 + \frac{a_2 z_2 + \cdots + a_m z_m}{z_1} \, ,$$

where duration weight value a_1 would have no effect on ordering of subshots.

2.4   Subshot Clustering

After the segment selection the selected segments are organized in two-level hierarchical manner. First the subshots are divided into groups according to their creation location and time information. This approach is following the definition of a scene which depicts action in one location and time.

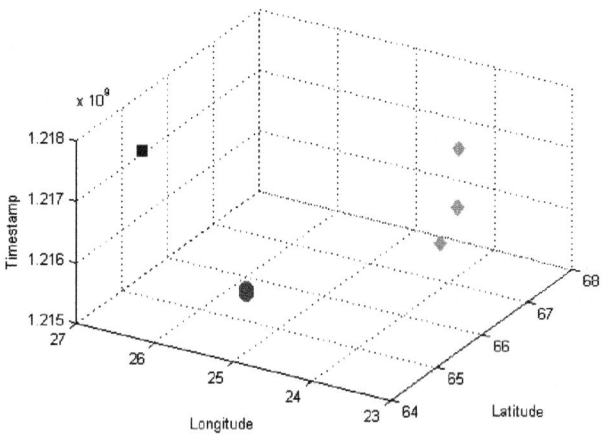

Fig. 3. Result of the clustering algorithm using video creation timestamp and location (latitude and longitude). Subshots from 6 video clips were assigned to three clusters visualized with squares, circles and diamonds.

Subshot grouping is accomplished with an iterative K-means clustering algorithm which clusters the subshots in three-dimensional vector space using the creation time and location information (latitude and longitude) of the clip (Fig. 3). The algorithm finds iteratively natural mean points for the clusters and classifies data points to a cluster with nearest mean. Data points are assumed to be defined in a vector space where the chosen metric can be applied. In this work normalized Euclidean distance is used as metric for distance of two vectors.

K-means clustering algorithm assumes the number of clusters K as a priori information. The algorithm tries to minimize a squared error

$$V = \sum_{i=1}^{k} \sum_{x_j \in S_i} (\overline{x}_j - \overline{\mu}_i)^2$$

where $\overline{x}_j$ is data point j and $\overline{\mu}_i$ is mean or center of cluster S_i.

In this work a variation of this algorithm, K-medoids, is used. In the algorithm the cluster means are forced to be one of the data points, a medoid.

Performance of the algorithm is heavily influenced by the initialization of cluster means. It is impossible to know beforehand where the clusters are located, after all that is what the algorithm is supposed to solve in the first place. This issue is handled by choosing initial means randomly and running the algorithm several times to find the best clustering.

Another challenge is to choose the right number of clusters K. Approach taken in this work is to iteratively run the algorithm ten times and then increment the value of K. This is repeated until the squared error V goes below a predefined threshold.

2.5   Video Rendering

In video rendering phase the objective is to assemble video scenes to a video sequence and render it to a final video summary. The scenes are ordered chronologically by comparing creation times of the median subshots of clusters. At the second level subshots are ordered inside each cluster by their temporal order. This is particularly important if subshots from the same original video clip are represented in a video scene.

Subshots can be assembled together with various types of transition effects, such as fades, wipes and dissolves. Proper use of shot transition effects can greatly improve the overall appearance of video summary. However some transition effects can have a unwanted effect of drawing attention to the actual effect rather than to the actual transition and therefore the use of effects should be carefully thought over in design of such a system for a specific application. In this work fading transition effect is chosen to be used between video scenes because it generally represents passage of a time or change of context to the user. Simple cuts are used between shots inside a video scene.

Problematic integration of audio modality must be considered when creating video summaries. The problem is that the subshot segmentation is hard to make in a way that would take into account both audio and video modality coherences at the same time. This sometimes leads to awkward cuts in either audio or video stream. Simply combining original audio stream of the subshots to video summary may not be the most appealing solution. In this work a completely unrelated audio stream, for example a music clip, can be added to the summary. This way the video summary should feel more consistent for the viewer.

3   Experiments and Results

In order to have a first impression on the feasibility of the automatic video editing system we organized a small scale user evaluation. In the user evaluations effect of system parameters on experienced *amount of information* and *enjoyability* was measured. *Amount of information* is described as how well the generated summary is considered to represent the whole original video material and *enjoyability* as how pleasing the summary is to watch.

The video collection used in the user evaluations included 48 video clips. Videos were gathered from mobile multimedia content creation platform [18] test conducted in summer 2008. In the tests 7 users got Nokia N95 multimedia phones with context metadata acquiring functionality installed in them and an assignment to shoot and share videos with the content sharing platform. Total video material used in these user evaluations sums up to 11 minutes and 30 seconds.

Five video summaries were created for the user evaluations using the whole video collection as starting point. First summary was generated with random segment selection and assembly. This summary is used as a reference when rating other summaries. Summaries were created with summary ratio of 0.1 so each video is about one minute long summary of total 11 minutes and 30 seconds of video material.

System parameters chosen to be tested were segmentation threshold, motion intensity weight, motion variance weight and sound weight. Parameters used in generation of each video summary are presented in Table 3.

Table 3. Summary parameter values. Parameter t_{segm} is the segmentation threshold, a_{mi} is the motion intensity weight, a_{mv} is the motion variance weight, and a_s is the sound weight.

Summary	t_{segm}	a_{mi}	a_{mv}	a_s
2	2.2	-0.58	-0.11	0.0
3	1.6	-0.58	-0.11	0.0
4	2.2	-0.58	-0.11	30.0
5	2.2	5.8	1.1	0.0

With these four parameters the pace and camera movement of the video summary are changed. Second video summary includes long duration shots with static camera, and fifth summary includes lot of camera movement and shorter shots. Third summary is similar to the second but with shorter shots and thus faster pace. In video summary 4 some different shots were chosen than in the second video summary. Notable difference is that many shots were selected from a football match where sound power was higher than in general shots.

We recruited 8 students of which 7 were men and one woman for these evaluations. Ages of the evaluators were between 20 and 24 years. We asked some details of their video creation and consumption habits in order to have an idea on their expertise in video recording and editing. Six of them said they recorded videos with their mobile phone. Six evaluators also said they recorded videos with digital cameras or digital camcorders. Only three of them were editing their videos afterwards. Software that these evaluators used was Apple iMovie [23], Pinnacle Studio [25], Kino [26] and Windows Movie Maker [27].

The evaluators were first given a short introduction to the subject and the video collection. Then they were asked to view the video summaries from 1 to 5 in consecutive order and rate the experienced level of *amount of information* and *enjoyability* with a scale from 0 to 9 where zero is the lowest level. Values of first video summary were fixed to 4. Results of the evaluations are represented in Table 4.

Table 4. Results from the evaluation of *amount of information / enjoyability*

Summary	0	1	2	3	4	5	6	7	8	9	mean
1 (ref.)					8/8						4.0/4.0
2				2/1	1/0	0/3	2/2	2/2	1/0		5.3/5.5
3					0/1	2/1	1/1	1/2	4/2	0/1	6.9/6.6
4			2/0	1/2	1/2	1/2	2/1		1/1		4.5/4.8
5			1/0	1/1	0/2	2/3	1/1	1/0	2/1		5.5/5.0

All created summaries were generally considered to be more informative and enjoyable than the summary with random segment selection and assembly. Results show that video summary three was experienced to be the most informative and enjoyable. In summary 4 the *amount of information* and *enjoyability* probably suffered from the multitude of football scenery. Summaries with more camera movement were generally considered less enjoyable than summaries that included shots with static camera. However the experienced *amount of information* did not seem to be lower in summary with more camera movement.

4   Conclusion

Video abstraction tools can be effectively used to visualize large amount of video content, such as user generated mobile videos. The advances in positioning services have led to new possibilities of combining location information to video content and use it for content management purposes.

In this paper we introduced a novel approach to utilize location and time metadata in video abstraction together with more traditional content analysis tools. Video clips are segmented to subshots by analyzing video content in frame level. A level of interest score is calculated for each subshot and the most interesting subshots are selected for the summary. The location and time related metadata is used to cluster subshots from the same context together to form video scenes. These scenes are rendered to a video summary by using fading transition effect between scenes and hard cuts between shots inside a same scene.

This approach was found to be very useful as the coherence and *enjoyability* of generated video summary was greatly enhanced based on our small scale user evaluation. In these evaluations the generated video summaries were considered to be more enjoyable and informative than summaries generated with random subshot selection and assembly. Video summaries that included static camera were found to be the most enjoyable.

5   Future Work

The performed user evaluations showed that the idea of automatic video editing system based on geotagging and timestamps of the original video clips is a feasible idea. From the user perspective it is essential to define the correct ratio of automation and manual work in creation of the video summaries. We need to study further the effect of system parameters in the final summary taking into account the characteristics of mobile video clips in order to be able to provide best possible service and user interface for the end user to create his or her personal and community video summaries. The result of this work will need to be examined with a larger scale user test to validate the functionality and usability of overall automatic video editing system and user interface.

There is also a lot to do on algorithm development to ensure the scalability of our system. Scalability requirements can be met by optimizing content analysis and encoding algorithms as well as distributing processing e.g. in a cloud environment. The results from various audiovisual content analysis methods can be integrated to the automatic video editing system in order to add more features to the content clustering process. In order to provide the best possible summaries for specific use cases the logic for summary generation should be designed separately for each application domain. Adding an audio track to the background will surely increase the *enjoyability* of the video summary.

Acknowledgements. This work was part of ITEA2 project ExpeShare – Experience Sharing in Mobile Peer Communities (2007-2009) partially funded by TEKES the National Technology Agency of Finland.

References

1. Hua, X.-S., Lu, L., Zhang, H.-J.: Optimization-Based Automated Home Video Editing System. IEEE Transactions on Circuits and Systems for Video Technology 14(5) (2004)
2. Mei, T., Hua, X.-S., Zhu, C.-Z., Zhou, H.-Q., Li, S.: Home Video Visual Quality Assesment With Spatiotemporal Factors. IEEE Transactions on Circuits and Systems for Video Technology 17(6) (2007)
3. Ekin, A., Tekalp, A.M., Mehrotra, R.: Automatic soccer video analysis and summarization. IEEE Transactions on Image Processing 12(7) (2003)
4. Huang, C.-L., Hsieh, C.-H., Wu, C.-H.: Audio-video summarization of TV news using speech recognition and shot change detection. In: 9th European Conference on Speech Communication and Technology, International Speech and Communication Association (2005)
5. Evangelopoulos, G., Rapantzikos, K., Potamianos, A., Maragos, P., Zlatints, A., Avrithis, Y.: Movie summarization based on audiovisual saliency detection. In: 15th IEEE International Conference on Image Processing (2008)
6. Lienhart, R.: Abstracting home video automatically. In: Proceedings of the 7th ACM International Conference on Multimedia, Part 2 (1999)
7. Zhao, M., Bu, J., Chen, C.: Audio and video combined for home video abstraction. In: Proceedings of IEEE International Conference on Acoustics, Speech, and Signal Processing, ICASSP 2003 (2003)
8. Jokela, T., Karukka, M., Mäkelä, K.: Mobile Video Editor: Design and Evaluation. In: Jacko, J.A. (ed.) HCI 2007. LNCS, vol. 4551, pp. 344–353. Springer, Heidelberg (2007)
9. OpenCV documentation of optical flow algorithms OpenCV Wiki, http://opencv.willowgarage.com/wiki/CvReference#OpticalFlow (accessed April 9, 2009)
10. Yu, B., Ma, W.-Y., Nahrstedt, K., Zhang, H.-J.: Video summarization based on user log enhanced link analysis. In: Proceedings of the Eleventh ACM International Conference on Multimedia (2003)
11. Adami, N., Benini, S., Leonardi, R.: An overview of video shot clustering and summarization techniques for mobile applications. In: Proceedings of the 2rd International Conference on Mobile Multimedia Communications (2006)
12. Huet, B., Merialdo, B.: Automatic Video Summarization. In: Interactive Video, Signals and Communication Technology. Springer, Berlin (2006)

13. Dey, A.K.: Understanding and Using Context. Personal and Ubiquitous Computing 5(1) (2001)
14. Abowd, G.D., Dey, A.K., Brown, P.J., Davies, N., Smith, M., Steggles, P.: Towards a Better Understanding of Context and Context-Awareness. In: Gellersen, H.-W. (ed.) HUC 1999. LNCS, vol. 1707, pp. 304–307. Springer, Heidelberg (1999)
15. Davis, M., King, S., Good, N., Sarvas, S.: From context to content: leveraging context to infer media metadata. In: Proceedings of the 12th Annual ACM International Conference on Multimedia (2004)
16. Goodman, R.M., McGrath, P.: Editing Digital Video: The Complete Creative and Technical Guide, 1st edn. McGraw-Hill, Inc. (2002)
17. Truong, B.T., Venkatesh, S.: Video abstraction: A systematic review and classification. ACM Transaction on Multimedia Computing, Communications, and Applications 3(1) (2007)
18. Järvinen, S., Peltola, J., Plomp, J., Ojutkangas, O., Heino, I., Lahti, J., Heinilä, J.: Deploying mobile multimedia services for everyday experience sharing. In: Proceedings of IEEE International Conference on Multimedia and Expo., ICME 2009 (2009)
19. Davis, M.: Editing out video editing. IEEE Multimedia 10(2) (2003)
20. Pietarila, P., Westermann, U., Jarvinen, S., Korva, J., Lahti, J., Lothman, H.: CANDELA – storage, analysis, and retrieval of video content in distributed systems. In: Proceedings of the IEEE International Conference on Multimedia and Expo. (ICME 2005), pp. 1557–1560 (2005)
21. Järvinen, S., Peltola, J., Lahti, J., Sachinopoulou, A.: Multimedia service creation platform for mobile experience sharing. In: Proceedings of the 8th International Conference on Mobile and Ubiquitous Multimedia, MUM (2009)

Modeling of Network Connectivity in Multi-Homed Hybrid Ad Hoc Networks

Michele Nitti and Luigi Atzori, *Senior Member, IEEE*

Department of Electrical and Electronic Engineering,
University of Cagliari, 09123 Cagliari, Italy
{michele.nitti,l.atzori}@diee.unica.it

Abstract. A Hybrid Ad Hoc Network consists of self-organized and self-configured mobile nodes, which make use of a fixed gateway to connect to the Internet. When there are two or more gateways to the fixed network, this is referred to with MultiHomed Hybrid Ad Hoc Network. In this scenario, different networks are formed, each one associated with a different gateway. A node can maintain its connectivity to the Internet when moving from a network to another by performing handover procedures and changing its gateway to the Internet. This scenario is quite interesting for its capacity of increasing the geographical extension of a single mobile network. The major contribution of this work is to provide a preliminary modeling of the node connectivity in this framework. We consider a typical architecture with gateways organized in a honey cell structure, where nodes move according to the RDMM (Random Direction Mobility Model), and present a three-state Markov model that describes the moving node behaviour: mobility without route changes, route change, and handover. Notwithstanding the simplicity of the underlying assumptions, the proposed model represents a valid basis for the analysis of the connectivity performance in this scenario, whose accuracy has been proved by means of extensive simulations.

Keywords: MANETs, hybrid ad hoc network, multi homed, Markov model, mobility modeling.

1 Introduction

Mobile Ad Hoc Networks (MANETs) are networks without infrastructures where mobile nodes are self-organized and self-configured making use of ad hoc routing protocols. These characteristics make these technologies good solutions for nodes that need to communicate with a host in a fixed infrastructure, but are away from it. Communication can be made through special nodes, called gateways, which are equipped with fixed network and MANET interfaces. When a MANET is connected to two or more gateways it is referred to with Multi-Homed Hybrid Ad Hoc Network. In this scenario, different MANETs are formed,

L. Atzori, J. Delgado, and D. Giusto (Eds.): MOBIMEDIA 2011, LNICST 79, pp. 307–320, 2012.
© Institute for Computer Sciences, Social Informatics and Telecommunications Engineering 2012

each one associated with a different gateway. A node can maintain its connectivity to the Internet when moving from a network to another by performing handover procedures and changing its gateway to the Internet. This scenario is quite interesting for its capacity of increasing the geographical extension of a single mobile network, but at the same time it makes quite complex the management of node connectivity, especially if the active connections are to be kept alive during handover.

The contribution of this paper is a *preliminary modeling of the node connectivity in the scenario of Multi-Homed Hybrid Ad Hoc Networks*, considering both the events of nodes changing the route to the gateway and nodes performing handoff from a mobile network to another. Specifically, the objective is to highlight the effects of mobility on connection retainability. We consider a typical architecture with gateways organized in a honey cell structure, where nodes move according to the RDMM (Random Direction Mobility Model), and present a three-state Markov model that describes the moving node behaviour: mobility without route changes, route change, and handover. To the best of our knowledge the literature is missing this study, whereas several works deal with the modeling of single link lifetime.

Of particular relevance to our work are the studies in [1] and [2], where Samar and Wicker created a model to characterize the statistics for link dynamics in MANETs assuming the nodes maintain constant speed and direction. In [3], Wu, Sadjadpour and Garcia-Luna-Aceves improve this model with a two-state Markov chain, where nodes move according to the RDMM described in [4] and [5]. They also demonstrate how Samar and Wicker's work is a particular case that gives good approximation only when the ratio between the radio range and the node's speed is small. Preliminary works that extend the link lifetime analysis to model route retainability are [6] and [7]. They both rely on the Random WayPoint model (RWP) as mobility model to evaluate network connectivity, which shows some unrealistic movement behaviors. When studying our scenario of Multi-Homed Hybrid Ad Hoc Networks, we mainly exploit the results in [3] to evaluate the performance during route changes and handover.

The paper is organized as follows. Section 2 describes the considered scenario and the mobility model. In Section 3 we describe the proposed three-state Markov model, while Section 4 presents the simulation results. Conclusions are drawn in Section 5.

2   System Description

2.1   Routing and Gateway Discovery

The performance in a MANET is highly influenced by the type of the routing protocol implemented, which can roughly belong to either the proactive or reactive categories. In proactive routing protocols, every node keeps routing information about its neighbors so it can respond to a topology change as soon as it detects a link fault, but this leads to significant and sometimes prohibitive signalling overhead; to overcome this problem a reactive routing protocol can be used to discover routes only when they are needed; the price to paid is a longer time to set the route when the active one is failed.

When considering a Multi-Homed Hybrid Ad Hoc Network the type of the handoff trigger also influences the network performance [8]. In a proactive approach, nodes periodically receive advertisements from the reachable gateways and can choose every moment the most convenient one, which is usually done on the basis of the gateway distance in terms of the number of hops measure. Differently, in a reactive approach, nodes receive advertisements only when they require them so they need more time to handoff.

Routing and handoff strategies can be selected independently each other so that we may have four possible combinations. However, a good compromise in terms of reactivity and time of service interruption is to make use of a reactive routing protocol and a proactive gateway discovery. This is the scenario we consider in this work so that the computation of a new route is performed when the transmitting node identifies a route failure, whereas gateway handover is performed when the end-node receives advertisements from a gateway that is closer then the current one.

2.2   Network Architecture

We consider a network architecture like the one in Figure 1, with gateways organized in a honey cell structure of side L. We consider the number of nodes following a two-dimensional Poisson Process with intensity σ so that for a region D of area A the probability to have k nodes within is the following [1]:

$$Pr(k\ nodes\ in\ D) = \frac{(\sigma A)^k e^{-\sigma A}}{k!} \tag{1}$$

where σA represents the expected number of nodes in D.

We consider a generic node having an ongoing communication to a host in the Internet. A node can establish bidirectional links with every node if it is R meters far from it. Indeed, we consider every node to have the same transmission power, whereas we are not considering shadowing and multipath fading that change transmission range from a symmetrical shape to an asymmetrical one.

Mobility in the network is modelled by the random direction mobility model (RDMM); this model assumes that nodes movement is divided in temporal windows, called epochs, whose length is exponentially distributed with mean λ^{-1}, so that the cumulative distribution function (CDF) is the following [4]:

$$F(x) = P\{Epoch\ lenghts\ \leq\ x\} = 1 - e^{-\lambda x} \tag{2}$$

During each epoch a node has constant speed and direction, but these parameters change from one epoch to another; direction and speed are uniformly distributed respectively between 0 and 2π and between a minimum speed v_{min} and a maximum speed v_{max}. Since epoch times and nodes directions and velocities are mutually independent, we can consider their movement independent and identically distributed (i.i.d.) so we have a uniform distribution of node locations at

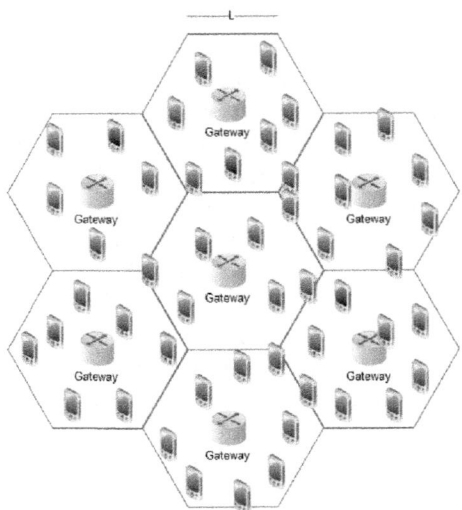

Fig. 1. Reference architecture for a Multi-Home Hybrid Ad Hoc Network

any point of time and the number of nodes distribution at the start and in every moment is still the same as described by (1).

2.3   Link Lifetime

To better understand the model we propose, we want to briefly summarize the results obtained from Wu, Sadjadpour and Garcia-Luna-Aceves in [3] for the link lifetime (LLT) T_L, which represents the duration of the link that can be used for data transfer.

They consider a link to be up if two nodes (e.g. a and b) are within range of each other during a communication session, so that:

$$T_L = min(T_a, T_b) \tag{3}$$

T_a and T_b are defined as Single-Node Link Lifetime (S-LLT) which measures the duration of time for a node to stay inside the communication circle of another node; since the nodes are random located, they have the same distribution and it is possible to calculate the complementary cumulative distribution function for T_L:

$$F_L(t) = F_S^2(t) \tag{4}$$

where $F_S(t)$ is the S-LLT complementary cumulative distribution function. To calculate it, the characteristic function $U_{T_S}(\theta)$ is computed as:

$$U_{T_S}(\theta) = \frac{U_1(\theta)}{1 - U_0(\theta)P_S} \tag{5}$$

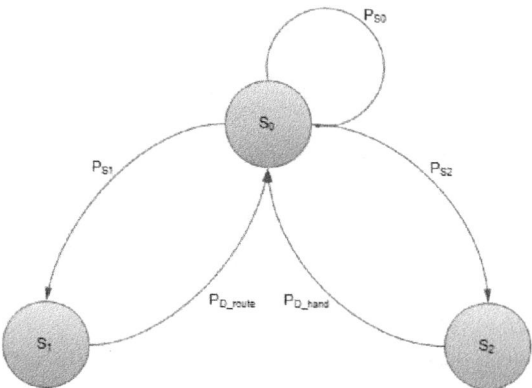

Fig. 2. The proposed three-state Markov chain

where P_S is the residence probability (i.e., the probability a node will be inside the communication circle at the end of a time epoch), $U_1(\theta)$ represents the characteristic functions of $p_{S_1}(t)$ (i.e., the probability of a node moving out of the communication circle at time t), and $U_0(\theta)$ is the characteristic function of $p_{S_0}(t)$ (i.e., the probability of a node being inside the communication circle at time t). We refer to [3] for a complete analysis of the model.

3 Proposed Model

To model the node behaviour in a Multi-Homed Hybrid Ad Hoc Network we introduce the time-continuous three-state Markov chain shown in Figure 2. S_0 represents the desired state of a node with a working stable route to the Internet gateway. It remains in this state for K epochs of the RDMM until either one of the links that connects it to the destination breaks or it finds a closer gateway and decides to handoff. S_1 describes a node looking for a new route to reach the same gateway after a link failure. S_2 represents a node trying to register to a new gateway because a shorter route to the Internet has been found. The transition probabilities from S_0 to the other states ($P_{S_0}(t)$, $P_{S_1}(t)$ and $P_{S_2}(t)$) are influenced by the two possible events that can occur in the network: a node losing its route to the Internet and a node changing its gateway for a closer one. We describe these events by computing the route and the handoff lifetimes.

In subsection A we obtain an estimation of the average number of hops to the gateway. It is then used in subsection B to present the computation of the proposed Markov model transition probabilities.

3.1 Number of Hops to the Gateway

The number of hops to the gateway is a variable that depends on the distance from the destination and on the distribution of the nodes. Herein, we are interested in estimating the expected number of hops N for a generic node in the considered scenario. We can easily observe that N can be expressed as

$$N = \frac{mean\ distance\ from\ the\ gateway}{mean\ distance\ covered\ by\ a\ single\ hop} \tag{6}$$

To calculate the mean distance d_g between a node and the gateway we approximate the hexagonal cell with a circle with the same area. Then the radius L_{eq} of the approximating circle is:

$$L_{eq} = \sqrt{\frac{3\sqrt{3}}{2\pi}} L \approx 0.91L \tag{7}$$

and since the mobile nodes are spread over the area of the cell uniformly, we can calculate d_g, which is equal to $\frac{2}{3}L_{eq}$.

To estimate the distance covered by a single hop in the gateway direction, d_c, we assume that the shortest route is always selected among the possible ones, as it is done by most of the available ad hoc routing algorithms. In this scenario, on average, each hop is the one that allows for the longest run in the direction of the gateway. The next hop node will be located in the semicircle of the node coverage area in the direction of the gateway. To determine d_c, we divide this semicircle in circular segments with parallel bases, so that the difference of two consecutive circular segments has a constant area A. This area is chosen so that the highest probability is reached for only one node within. With reference to Figure 3, A_1, A_2 and A_3 are the differences of the considered circular segments, whose area is equal to A.

Each area has a barycentre indicated with b_i as shown by the red dot in Figure 3. So since we are interested in the longest run, we weight each barycentre with the probability p_i to have a node in the area A_i and none in the areas A_j with $j = 1 \cdots i - 1$:

$$p_i = Pr(1\ node\ in\ A_i,\ 0\ nodes\ in\ A_1, \ldots, A_{i-1}) \tag{8}$$

so that d_c can be finally computed as:

$$d_c = \sum_{i=1}^{M} b_i p_i \bigg/ \sum_{i=1}^{M} p_i \tag{9}$$

However, even if this path is the shortest one, it is unlikely that all the hops are aligned in the gateway direction. In most cases, the next hop is in a different direction, creating an angle α with respect to the direction between source node and gateway (Figure 3).

The receiver node has coordinates d_c and x_c, calculated considering the barycentre of half equivalent circular segment with barycentre d_c (the grey area in Figure 3).

The angle α can then be calculated as:

$$\alpha = \frac{\pi}{2} - tan^{-1}\left(\frac{d_c}{x_c}\right) \tag{10}$$

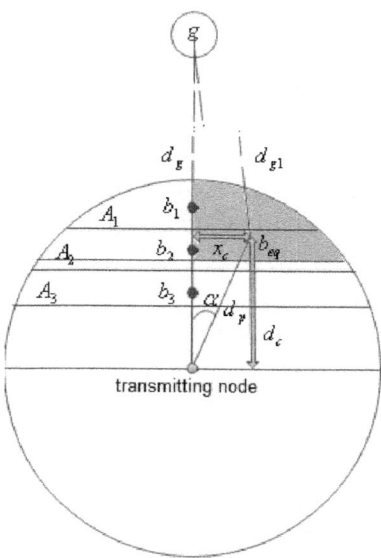

Fig. 3. Distance progress in the gateway direction

The distance covered by a single hop d_p in Figure 3 is:

$$d_p = \frac{d_c}{\cos\alpha} \tag{11}$$

The distance to the gateway is:

$$d_{g_i} = \sqrt{d_{g_{i-1}}^2 + d_p^2 - 2d_{g_{i-1}}d_p\cos\alpha} \tag{12}$$

where i indexes the hops to the gateway, with $i = 1\cdots N$ and d_{g_0} equals to d_g. This distance is not the same in every hop, since it depends on the distance to the gateway in the previous hop. We can finally say that the average distance covered by a single hop can be calculated as:

$$\frac{(d_g - d_{g_1}) + \cdots + (d_{g_{N-1}} - d_{g_N})}{N} = \frac{d_g - d_{g_N}}{N} \tag{13}$$

Unluckily, to compute exactly this formula we need a priori knowledge of the number of hops N; a good approximation can be obtained if we consider the minimum number of hops to reach the gateway from the generic node at distance d_g:

$$N_{min} = \left\lfloor \frac{d_g}{R} \right\rfloor \tag{14}$$

Finally the number of hops is:

$$N = \frac{d_g}{\dfrac{d_g - d_{g_{N_{min}}}}{N_{min}}} \tag{15}$$

3.2 Route and Handoff Lifetime

We are interested in computing the probability for a route to stay alive. To this we define T_R as the route lifetime, i.e., a random variable representing the time the route remains up before one link within breaks. Since the nodes movement are i.i.d. in the considered RDMM, also the links between each pair of nodes that set the route will be i.i.d.; we can evaluate the route lifetime complementary cumulative distribution function (CCDF) $F_R(t) = P(T_R \geq t)$ starting from the single-node link lifetime distribution $F_S(t)$ described in Section 2.C:

$$F_R(t) = F_L(t)^{N-1} F_S(t) \tag{16}$$

where N specifies the expected number of links between the source and the gateway; only $N - 1$ links can be described with the LLT distribution since they have two mobile nodes involved in the communication, while in the last link (mobile node - gateway) there is only one mobile device, since the gateway position is fixed, so we use the S-LLT.

We also define T_H as the handoff lifetime, i.e. the time a node will stay near its gateway without handoff to another one; we can calculate the complementary distribution function $F_H(t) = P(T_H \geq t)$ as:

$$F_H(t) = F_G(t) \tag{17}$$

where we define $F_G(t)$ as the single-node gateway lifetime. It can be evaluated starting from the single-node link lifetime, considering a communication circle with a radius of L_{eq} instead of R. This model represents a single node moving away from the gateway, which does not change its position, and becoming closer to another gateway.

It is now possible to calculate at any time t_1 the transition probabilities:

$$P_{s_0}(t_1) = F_R(t_1) F_H(t_1)$$

$$P_{s_1}(t_1) = F_H(t_1)(1 - F_R(t_1)) \tag{18}$$

$$P_{s_2}(t_1) = (1 - F_H(t_1))$$

where the probability $P_{s_0}(t_1)$ to stay in the state S_0 after time t_1 is obtained by simply considering routes and gateway unchanged; the probability $P_{s_2}(t_1)$ to handover before time t_1 is calculated taking into account only the handoff lifetime as a result of the particular handoff trigger chosen, since the handoff is only based on the nearest gateway regardless whether the route is active or not. $P_{s_1}(t_1)$ represents the probability of a route lifetime lesser than t_1 considering the node is still register to the same gateway and it is calculated as the complement of the other two probabilities and shows how it depends from $F_H(t_1)$ since if a node decides to handoff it will also look for a new route.

These probabilities are then what we needed to design a variety of applications; the knowledge of the probability to change state can be used to adapt the bit rate in order to reduce packet loss or to meet deadlines.

Table 1. Time Components

Signalling Message	Transmission Time per hop	Processing and Queuing Time per node
Router Solicitation	T_{r_sol}	T_{pq_sol}
Router Advertisement	T_{r_adv}	T_{pq_adv}
Binding Update	T_{r_upd}	T_{pq_upd}
Binding Acknowledge	T_{r_ack}	T_{pq_ack}
Link Error	T_{r_err}	T_{pq_err}
Route Request	T_{r_req}	T_{pq_req}
Route Reply	T_{r_rep}	T_{pq_rep}

Table 2. Parameters Definition

Propagation Time / hop (Wireless Network)	T_{p_wir}
Propagation Time / hop (Infrastructure Network)	T_{p_inf}
Link Recovery Reactive Protocols	T_{lr}
Number of wireless hop	N_{wir}
Number of wired hop (Infrastructure network)	N_{inf}

3.3 Route and Handoff Delay

Many factors influence the time a node needs to change its route or its affiliation from one agent to another.

In this section we show the parameters that influence these delays, leaving to another time the exact calculation of the transition probabilities, since these will become easier to derive once defined more precisely the usage scenario.

In Table 1 we show the different time components associated with the processing and transmission of handover and route change signalling messages, while Table 2 presents some parameters definition.

The handover delay for the proactive approach can be expressed as follow:

$$D_{hand} = N_{wir}(T_{r\_sol} + T_{pq\_sol} + T_{r\_adv} + T_{pq\_adv}) +$$

$$N_{inf}(T_{r\_upd} + T_{pq\_upd} + T_{r\_ack} + T_{pq\_ack}) + \qquad (19)$$

$$2N_{wir}T_{p\_wir} + 2N_{inf}T_{p\_inf}$$

while the delay for a route change can be derived from:

$$D_{route} = T_{lr} + \frac{N_{wir}}{2}(T_{r\_err} + T_{pq\_err}) +$$

$$\frac{N_{wir}}{2}(T_{r\_req} + T_{pq\_req}) + \frac{N_{wir}}{2}(T_{r\_rep} + T_{pq\_rep}) + \qquad (20)$$

$$N_{wir}T_{p\_wir}$$

Table 3. Simulation parameters

Parameter	Value
Number of gateways	7
L	1000 m
σ	$\frac{1}{\pi L} nodes/m^2$
Average number of nodes	5789
λ	4

The signalling message transmission delay T_r depends on the transmitted packet length and the transmission speed, while the process and queuing delay t_{pq} is a random variable characterized by the traffic load in the network and the queue length at each node.

The number of wireless hops N_{wir} is a random variable depending on the particular mobile nodes distribution, and in our scenario, with a Poisson distribution, its mean is represented by (15) calculated in Section 4.B.

The propagation time T_p, both in the wireless and in the wired network, depends on the hop distance, while T_{lr} represents the time a node needs to recognize a link fault and depends on the particular routing protocol implemented.

4   Simulation Results

To evaluate the reliability of the proposed model, we have performed simulations with different scenarios using the Matlab environment. The simulation parameters are shown in Table 3: the gateways are arranged as shown in Figure 1, where nodes are placed randomly. The number of nodes is not fixed but it is decided by (1), where node density σ has been chosen so that in every moment there is at least one route from each node to a gateway. The nodes move according to the RDMM with epoch lifetime controlled by parameter λ.

We have considered different scenarios changing the transmission radius and the speed of the nodes. Two different profiles have been chosen for the transmission radius: $100m$ and $200m$. The minimum velocity has been chosen equals to $0m/sec$, while we have set the maximum velocity to: $1m/sec$, $10m/sec$ and $20m/sec$, i.e., from walker to car speed. For each scenario, five simulations were run, with statistics recorded for $3600sec$ for the handoff lifetime and for $180sec$ for the route lifetime.

We assume to have a perfect MAC and routing layer and that the hidden and exposed terminal problems do not affect the communications, so that the simulations only show the behaviour of node mobility. This is done because we want

Table 4. Number of hops

R (m)	N		percentage error
	Simulated	Theoretical	
100	9.07	9.02	0.55 %
150	5.39	5.23	3 %
200	4	3.8	5 %

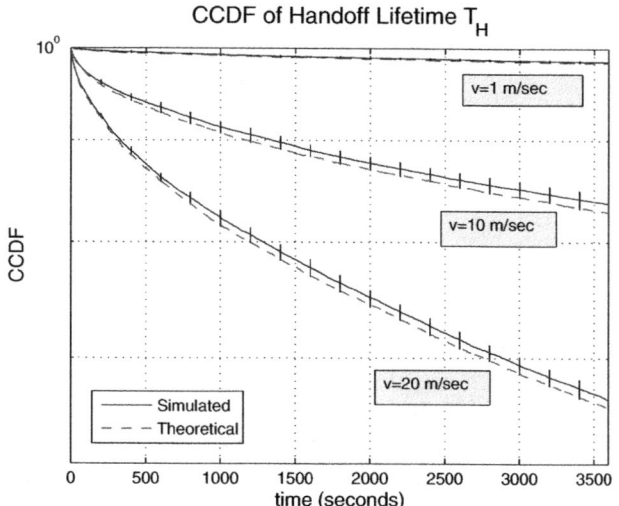

Fig. 4. Handoff lifetime T_H. For the simulation results the curves show the 99% confidence interval.

to analyze how mobility affects connectivity regardless of how the connections were established.

Table 4 describes the number of hops calculated with (15) and the simulated one, for three different values of the radio transmission range R. It is worth to note that even if our formula is only a first approximation, the results are quite accurate and with an error always lower than 5%. The differences are due to the fact that our formula consider only N_{min} hops to compute the mean distance covered by a single hop. This implies that the mean distance covered by a single hop is slightly higher than the real one (simulated) and consequently we have a slightly lower number of hops with our formula.

Figures 4 and 5 show the validity of our model for different combinations of the transmission radius and maximum velocity. This can be seen from Figure 4 where the 99% confidence interval is shown along with the simulation curves (blue lines). In Figure 5 the confidence interval is not shown for readability issues.

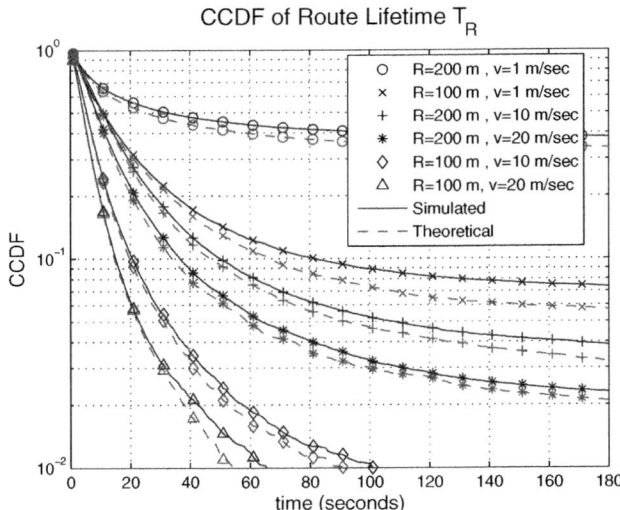

Fig. 5. Route lifetime T_R

These clearly show how the dynamics of a network are highly influenced by the route lifetime. For a given time, in fact, the probability to lose a route is much higher than the probability to handoff.

The route lifetime is characterized by a dependency from the R/v ratio. As expected, decreasing this ratio brings to a quick decrease in the $F_R(t)$ probability, which means frequent changes in the network topology. This is because this ratio represents the maximum time interval a node needs to pass through the trasmission radius of another node; decreasing this ratio means reducing the available time for two nodes to communicate. Moreover, for every link that makes up the route, two mobile nodes are involved in the communication so this further reduces the route lifetime.

Differently, the handoff lifetime does not depend on the transmission radius since the decision for a node to handoff is only affected by the distance from the gateways. Moreover, during handover, only a node is moving since the gateways don't change their position. Accordingly, since we consider a fixed cell size L, the only parameter that influences the handoff lifetime is the velocity of the node.

4.1   Use Case

We consider the scenario of a mobile node connected to a streaming server on the Internet, running a video application with a low frame rate: the mobile node can freely move in an architecture like the one in Figure 1, using a reactive protocol, such as AODV, to discovery the route while connection with the gateways follows a proactive approach.

Video streaming applications are one of the most challenging among the multimedia services, first of all due to time constraints. Being able to know when a route is going to break or a handover to occur can be really useful to help an application to modify its bit rate and achieve deadlines.

We want to show the possible benefits of an application that makes use of our model; we don't examine problems of collisions and interference for both the applications, so that the only difference is the knowledge of the network connectivity due to the mobility.

We can assume to know the maximum bit rate the network can support for the considered application. In a normal video streaming application, after the communication is established, the source starts to send packets with a constant bit rate, using the maximum link capacity. When the connection is interrupted, a certain number of packets will be lost as the source does not notice immediately the fault and will continue to send data until it receives an error notification. These packets can be retransmitted when the connection is established again or otherwise they are considered lost packet and the result is a worse quality video.

In a mobile network, like the one we are considering, this approach is hazardous because the connections are likely to be interrupted; as we have demonstrated, the network topology becomes more dynamic and the routes are more instable as the velocity increase.

If the application knows that a connection is going to break with a certain probability at the time t, it can adapt its transmission in different ways. One possibility should be to reduce the frames coding rate when the probability for the streaming video to stop is above a certain threshold. In this way, we lower the quality video, but we accomplish a better recovery capability. Another possibility should be to send important data, for example I-frames, when the probability to stay connected is high to be sure to respect a deadline.

5 Conclusions

In this paper, we have proposed a three-state Markov model to study the dynamics of a Multi-Homed Hybrid Ad Hoc Network. It provides the probability for a node to remain in the stable working connection to the Internet for the next desired interval. This can be useful when implementing application rate-control algorithms, which would modify the source and channel rates according to the node state.

Simulations have shown the model represents a valid basis for the analysis of the connectivity performance.

Even if our formula for the number of hops is only a first approximation, the results are quite accurate and with an error always lower than 5%. The estimated transition probabilities fall within the 99% confidence interval with less than 1% errors.

References

1. Samar, P., Wicker, S.B.: Link dynamics and protocol design in a multihop mobile environment. IEEE Transactions on Mobile Computing 5(9), 1156–1172 (2006)
2. Samar, P., Wicker, S.B.: On the behavior of communication links of a node in a multi-hop mobile environment. In: MobiHoc 2004: Proceedings of the 5th ACM International Symposium on Mobile ad Hoc Networking and Computing, pp. 145–156. ACM, New York (2004)
3. Wu, X., Sadjadpour, H., Garcia-Luna-Aceves, J.: Link dynamics in manets restricted node mobility: modeling and applications. IEEE Transactions on Wireless Communications 8(9), 4508–4517 (2009)
4. Jiang, S., He, D., Rao, J.: A prediction-based link availability estimation for mobile ad hoc networks. In: INFOCOM 2001: Proceedings of IEEE Twentieth Annual Joint Conference of the IEEE Computer and Communications Societies, vol. 3, pp. 1745–1752 (2001)
5. Liang, B., Haas, Z.: Predictive distance-based mobility management for pcs networks. In: INFOCOM 1999: Proceedings of IEEE Eighteenth Annual Joint Conference of the IEEE Computer and Communications Societies, March 21-25, vol. 3, pp. 1377–1384 (1999)
6. Xiang, H., Liu, J., Kuang, J.: Minimum node degree and connectivity of two-dimensional manets under random waypoint mobility model. In: IEEE 10th International Conference on omputer and Information Technology, CIT 2010, June 29-July 1, pp. 2800–2805 (2010)
7. Luo, H., Laurenson, D.: Link-duration-oriented route lifetime computation for aodv in manet. In: International Conference on Wireless Communications and Signal Processing, WCSP 2010, pp. 1–4 (October 2010)
8. Vogt, C., Zitterbart, M.: Efficient and scalable, end-to-end mobility support for reactive and proactive handoffs in ipv6. IEEE Communications Magazine 44(6), 74–82 (2006)

Author Index